D0983626

LINNAEUS: NATURE AND NATION

Lisbet Koerner

Linnaeus:
Nature and Nation

HARVARD UNIVERSITY PRESS

Cambridge, Massachusetts, and London, England   1999

Library of Congress Cataloging-in-Publication Data

Koerner, Lisbet.
    Linnaeus : nature and nation / Lisbet Koerner.
        p.   cm.
    Includes bibliograpical references (p.   ) and index.
    ISBN 0-674-09745-9 (alk. paper)
        1. Linné, Carl von, 1707–1778.   2. Naturalists—Sweden Biography.
    3. Botanists—Sweden Biography.   4. Nature.   5. Economics.
    I. Title.
    QH44.K58   1999
    580'.92—dc21
        [B]   99-34570
                                                                  CIP

To my dearest husband, Joseph Leo Koerner

Contents

LINNAEUS: NATURE AND NATION

"To Apply Nature to Economics and Vice Versa"

Late in the eighteenth century, classical economists formulated a concept of global modernity. They argued that an efficient economy
demanded a liberal political order, with well-defined limits to state power
and to national sovereignty. Around the same time, Romantic antimodernists rejected the material benefits of mass production. They wanted instead
to safeguard intangible values such as moral virtue and local independence.
Both types of thinkers assumed a trade-off between global economic efficiency and the values of local elites. Classical economists opted for an
efficient economy; Romantic antimodernists, for local elites.

Earlier secular philosophers of statehood had not considered such a trade-
off necessary. Cameralists—which is to say, at their origin, seventeenth-
century German fiscal theoreticians employed as princely financial advis-
ers—had hoped to link the political order to a new and rationalized economic order. They aimed to improve manufacturing and agriculture by
means of protectionist legal measures and technological innovations. More
broadly, they worked to preserve the political power and social prestige of a
rent-seeking state elite, which would now supervise (and live off) a refashioned and largely self-sufficient domestic economy.

Whereas classical economists advocated one single, ungoverned, yet self-
regulating global modernity, and whereas Romantic antimodernists hoped
for an infinitude of custom-governed, local, traditional communities, cameralists strove for rationalistically governed autarkies. Their now largely forgotten model of society I term here the cameralist concept of a local modernity.

This book investigates that concept as it is fleshed out in the life and work
of one of the eighteenth century's most famous naturalists, the Swedish botanist Carl Linnaeus (1707–1778). *Linnaeus: Nature and Nation* is thus both a

biography and a case study of the relation between natural knowledge and political economy in the early Enlightenment. It asks how one thinker posed and answered the question: what is the relation between nature and the economy?

In 1746, writing as an Uppsala University professor to the Swedish Academy of Science, Linnaeus summarized his answer to this question: "Nature has arranged itself in such a way that each country produces something especially useful; the task of economics is to collect [plants] from other places and cultivate [at home] such things that don't want to grow [here] but can grow [here]."[1]

In his letter, Linnaeus pondered the fact that natural resources are unevenly scattered over the globe. "Each country produces something especially useful." Later in the century, Adam Smith argued that because countries differ in how they are endowed, their economies would become more productive if each specialized in what it did best, and then traded goods internationally. Linnaeus, however, thought that trade could never overcome the problems of manifold nature and crafts.

Linnaeus thus rejected Adam Smith's laissez-faire theory. Of course he never came across the fully formulated doctrine. But he was cognizant of earlier and quite elaborate pleas for a freer economy, common in the Swedish pamphlet literature during the Era of Freedom between 1718 and 1772. He would have disagreed entirely with Adam Smith's later thesis that voluntary trade benefits both parties. Like other cameralists, he had a zero-sum view of the economy. He regarded trade as parasitic. His benchmark of success was for a country to keep its trade balance positive, and this benchmark he quantified in "barrels of gold."

Nor did Linnaeus subscribe to the *other* political answer of his era to the inescapable reality of natural and technological diversity. Hence he scorned mercantile imperialism and the military conquest of tropical regions to serve as tributaries. We know in retrospect that by 1750 Europe was on the verge of acquiring global economic hegemony. But Linnaeus, like his contemporaries, regarded non-European powers—the Ottoman, Mughal, Chinese, and Japanese empires—as matching Europe's military skills and productive powers.

In his era, Europeans fretted over the drain of bullion to Asia. No government had the probity or public trust necessary to guarantee a conventional money base. Credit institutions were fledglings. Gold and silver had always been perceived as having intrinsic value (as well as exchange value and use value). For all these reasons, the bullion drain implied deflation, and economists theorized that states' main macro-economic goal must be a perpetually positive trade balance, as measured in precious metals.[2]

A century before Linnaeus, to justify the Asia trade English seventeenth-century mercantilists had argued that reselling Oriental goods abroad netted England more specie than its original outlay in Asia.[3] German economists and their Scandinavian followers noted that England's gain was the Continent's loss. After the end of the Thirty Years War in 1648, Germany's polities struggled to reconstruct their war-shattered economies. They also hoped to compensate for the decline of the cross-Alpine German-Italian trade, which had flourished in earlier centuries. Thus they developed cameralism, a continental sister doctrine to mercantilism.[4]

Cameralists and mercantilists built their doctrine on the traditional goal of the European courts: to appropriate the wealth of their subjects. Indeed, the word "cameralist" was originally a term for the officeholders who could enter the princely *camera*, or treasure chamber. Cameralists and mercantilists alike measured national wealth in rulers' gold hoards. (The more sophisticated among them also discussed currency circulation. Others—ranging from Cardinal Richelieu to King Gustav III of Sweden—attempted to solve their budget deficits more directly, with the help of gold-producing alchemists.)[5]

Cameralists and mercantilists also theorized the advantages that accrued to the ruler from an economic symbiosis with his subjects. They did so in the secular language of self-interest, and in the practical format of handbooks.[6] (Preaching to coarse and drunken German princes, they worked with simple analogies. Your subjects, they wrote, are like sheep. If you kill them, you will have no wool.)

Both sets of early economists thus aimed for prosperous, populous, and unified national economies. They hoped to draw together the state's various domains into one administrative unit, and to regulate that state's material links with other polities. To that end, they employed many tools: producer monopolies and state manufactures; import barriers; navigation acts; pronatalist legislation; export bans on gold; funding of science and technology; and improvements to infrastructure.

Historians of cameralism traditionally study cameralists' attempts to integrate regions within a state. Noting that in the early modern era the economic, political, and cultural maps of continental Europe overlapped only haphazardly, they regard cameralists as nation builders.[7] In this book, I attend instead to cameralists' efforts to isolate the individual state commercially. In particular, I am interested in the policies they developed when older legal measures failed.

In the early 1700s, smaller European countries licensed interloping East India companies. Smugglers retailed their stocks of sugar, tea, and spices from country to country. As a matter of economic theory, however, state elites still assumed that smuggled imports and tax revenues were inversely correlated. And on a material level, they collected the monopoly rents generated in the trade environment that they regulated. Therefore, these rent-seekers eagerly enforced legal trade restrictions and fiercely fought against smuggled goods.

Yet not even the cruelest punishments worked. In late seventeenth-century and eighteenth-century France, for example, about sixteen thousand people were broken on the wheel, hung, or quartered for selling printed cottons.[8] Still French women continued to wear calicoes. When the rent-seeking state elites of early modern Europe battled the comparative advantage that fuels international trade, they also needed—and they understood this—a positive economic policy, a direction distinct from restrictions and penalties.

Most typically, early modern state elites crafted such positive policies into the form of a search for colonies. As producer monopolies, tropical tributaries could supply the spices, textiles, and stimulants that the motherland consumers now bought tax-free on the black market. As captive markets, they could consume the products that the motherland's state monopolies and state licensees made, but could not sell.[9]

The colonial option was not open to all rent-seeking state elites, however. As one of Linnaeus' Swedish students enviously observed from London in 1761, "here there is such an unnatural bragging over the successes of their weapons, so one hears nothing but that the English can conquer the whole world if they want to."[10] By 1700, the smaller, non-Atlantic European countries could hardly acquire empires overseas, given that the Dutch, French, Spanish, and English ruled over the high seas.

In Linnaeus' homeland, Sweden, state elites had pursued another expansionist design—a European land empire—throughout the seventeenth century. Indeed, they continued to do so for seventy years after the Peace of Westphalia (1648), when Sweden conceded its losses in the Thirty Years War, and through famines (1710s), bubonic plague (1710–11), and Russian invasions (1719 and 1720). It was only in 1721, when Sweden's population had fallen by one third, the king had been murdered, and the country's warlords checked by a newly confident parliament, that the state elites—though still skirmishing with Russia—turned toward peaceful reconstruction.

In 1773, a minor nobleman summarized this change of heart, as he delivered a funeral oration over one of Linnaeus' colleagues at Uppsala University. Looking back over half a century, he remembered how, in his youth,

> people were beginning to tire of the warfaring life. They no longer succeeded in conquering new provinces and coming home with rich war booty; on the contrary, they had experienced the clearly destructive consequences of War. There was no other alternative except to try to do the best with what they already owned and had. Prosperity [*välmåga*] was only to be reached by work and perseverance, and for that they needed a new kind of thinking and a growing population. People found that a pure Warrior People needed a different kind of education to be able to feed itself.[11]

In 1738, when Linnaeus returned home from his journeys abroad and added to his classificatory work the domestic career of cabinet-sponsored experimenter, Sweden was a post-imperial state. He knew that as an economic policymaker he had to make do with the heartlands that Sweden had managed to retain. His country had no military control over non-European, or even non-Baltic, natural or manmade wealth. And it had little hope of acquiring such imperial powers. Sweden could expand neither into a western ocean empire, nor into an eastern land empire. At best, his "dear fatherland" could add to its realm only a few neighboring provinces that would be ecologically and economically similar to the mainland.

At the same time, from Linnaeus' cameralist point of view, trade regulations, and especially restrictions on the Asia trade, had grown more imperative. European consumers avidly bought Asian goods such as silks, porcelain, and tea. But excepting a few mechanical trinkets, and of course guns, Europe had few goods that Asian peoples were willing to buy in turn. (The English later hit upon the commodity of opium.) In the meanwhile, the Eurasian trade was financed by a flow of New World silver to Asia. Linnaeus regarded this exchange as immensely damaging to Europe. As he put it in a *vita* (a sheet of notes for his funeral orators), he "considered no thing more important than to close that gate, through which all silver of Europe disappears."[12]

Linnaeus voiced views common to continental civil servants in his generation. Given their reverence for metal-based forms of currency, they were eager, yet unable, to restrain the Asia trade. Indeed, from a cameralist perspective, their countries' efforts to capture the benefits of mercantile trade worsened the European situation as a whole. Europe's competing East India companies insured that consumers everywhere could buy smuggled Asian goods. And against that material availability, civil servants' moral injunctions and legal bans were in vain.

For reasons both of cameralist theory and geopolitical realities, then, Linnaeus advocated an economic strategy for his homeland based neither on international trade nor on colonial conquest, but on import substitution. Manufacturing had stood at the heart of cameralist thinking from its seventeenth-century beginnings in Germany. In that sense, Linnaeus only followed some eighty years of German reform efforts. The legal framework he expected—and enjoyed—for his own reform plan was also conventionally cameralist. It made use of such time-honored devices as domestic producer monopolies, tariff and non-tariff trade barriers, and government subsidies.

But the technologies Linnaeus used, and the science that informed them, were historically new. Indeed, they involved a particular (even peculiar) science, which was predicated on Linnaeus' hypothesis that colonial plants could be "tamed" to grow in the Baltic realms. Benefiting from Linnaeus' prestige as a floral classifier, between c. 1740 and c. 1760 this transmutational botany was supported by the Swedish court, parliament, universities, and scientific academies and societies.

The Swedish state elites' cameralist plans for domestic manufactures and cash crops thus drew on their most famous naturalist's theories of floral acclimatization. *Linnaeus: Nature and Nation* tells the story of the "task of economics" Linnaeus put before the Swedish Scientific Academy, and of how, in the course of working it out, its author came to project Lapland cinnamon groves, Baltic tea plantations, and Finnish rice paddies.

Chapter 1, "A Geography of Nature," argues that Linnaeus understood himself as a Lutheran and as a civil servant, that this local cultural reference frame was imaginatively misunderstood by his Romantic followers on the Continent, and that this misreading lay at the heart of his fame.

Chapter 2, "A Clapper into a Bell," spans 1707–1731 and part of the 1740s. It explores how the young Linnaeus began charting his natural science (most famously, his sexual system of plant classification). Looking at how Linnaeus introduced binomials for flora and fauna between 1748 and 1752, it argues that he first used these brief, stable, and arbitrary species names as a stopgap measure, to make his students into a more efficient support staff in his project of national self-sufficiency.

Chapter 3, "The Lapp Is Our Teacher," covers Linnaeus' first voyage of discovery, the 1732 Lapland journey, and his three years in Holland, from 1735 to 1738. Linnaeus mythologized his sub-Arctic travels into a formative encounter with an Edenic "wild nation" (Sami reindeer herders) and a cross-cultural encounter between "high" and "folk" science. He then used this performative narrative to enter Dutch learned circles. Yet the Lapland journey was also part of the colonization of "our West Indies," as Scandinavians termed their Arctic frontier. And this colonial venture in turn was predicated on erasing indigenous culture, as the "wild" Sami and their herds were chained to the engines of cameralist industry.

Chapter 4, "God's Endless Larder," begins in 1738, when Linnaeus returned to Sweden from Holland. It addresses his overall, framing theory of nature, and the broad program for science that he drew up as a consequence of that understanding. Linnaeus conceptualized nature as a prelapsarian paradise and as a single self-regulating mechanism, with each nation containing all the natural products necessary for a complete and complex economy.

Chapter 5, "A New World—Pepper, Ginger, Cardamon," probes into how Linnaeus meshed his theology and his economics to form an order at once moral and material, and how that conceptual order framed his scientific practice and underwrote his engagement in politics. It also explores his work on the scaffolds and sites of science: co-founding the Swedish Academy of Science in 1739, reforming Swedish universities, and popularizing his natural knowledge in almanacs, pamphlets, newspapers, lectures, and sermons.

Chapter 6, "Should Coconuts Chance to Come into My Hands," opens in 1741, when Linnaeus was installed as a professor at Uppsala University. It investigates how in practice he sought to ensure his country's self-sufficiency. He searched the Baltic region for domestic flora and folk knowledge. He sent students on long-distance collecting voyages. And he sought to acclimatize foreign flora to Baltic climates—all to replicate within the homeland what was harvested and husbanded under the more favorable conditions of abroad, and thus re-create within his national borders a trans-oceanic empire.

Chapter 7, "The Lord of All of Sweden's Clams," A Local Life, spans the 1740s to the 1760s. It discusses the research generated in Linnaeus' newly created scientific institutions by his first generation of Scandinavian students. It also shows how the mature Linnaeus presented himself at court, in parliament, in the Academy of Science, and at the university as an economist as much as a naturalist. Or rather, he posed as the inventor of a new science blending knowledge of the human and the divine order. Yet the material success of his economics was paltry.

Chapter 8, "His Farmers Dressed in Mourning," outlines the decline of his economics, which began with his illness in 1762 and grew precipitous after his death in 1778. It goes on to discuss the Romantic reception of Linnaeus, and how, around 1900, Swedish conservatives recast him as a national icon who recaptured in his science the military victories won by Sweden's famous warrior kings. In turn, this glorified—and reactionary—portrayal of Linnaeus helps explains why, after the advent in 1932 of a Social Democratic government, he dwindled into obscurity in his "most beloved fatherland."

Finally, the Conclusion, "Without Science Our Herrings Would Still Be Caught by Foreigners," reflects on the wider implications of this book's research findings, outside the limits of Linnaean historiography. It especially discusses how the problem of mediating between local and global economic spheres has been conceptualized since the eighteenth century.

The themes of *Linnaeus: Nature and Nation* go against the grain of Linnaean scholarship new and old. The earlier literature typically concerns itself with evaluating and dating Linnaeus' science. In 1875, the German plant physiologist Julius Sachs, in the world's first source-critical history of botany, condemned Linnaeus as a medieval scholastic. In 1903, T. M. Fries presented Linnaeus as an innovator, foreshadowing developmental nineteenth-century biology. In 1926, Henri Daudin argued that Linnaeus belonged to the Enlightenment with which he coincides chronologically. In 1966, Michel Foucault suggested that Linnaeus is the foremost representative of a world of thought that is "neither an epilogue to the Renaissance nor a prologue to the nineteenth century." In 1990, Scott Atran proposed that Linnaeus' flora "emerges as an elaboration of [the] universal cognitive schema" structuring folk-biological taxonomies.[13]

Linnaeus' science may have been belated, of its own time, forward-looking, or outside history proper. Yet both Julius Sachs and Michel Foucault bracketed Linnaeus as a "a great life from the baroque."[14] In turn, Foucault abbreviated the work of this "classical age" as a "pure tabulation of things." Indeed, Foucault, who studied under Sten Lindroth, a historian of science at Uppsala University, followed his teacher in describing Linnaeus as a collector and classifier, unable to investigate the functions or histories of organisms and therefore cut off from modern biology. Foucault parted from Lindroth only by suggesting, hyperbolically, that the entire era was Linnaean in spirit.

This bracketing of Linnaeus as an antiquated natural historian marked most post-World War II scholarship. Already by the 1930s the Linnaeus cult in Sweden was tapering off, and his considerable scholarly achievements were largely forgotten. The sugary image of the "flower king" in older Scandinavian historiography, the conservative cultural impetus behind that image, and the principle of generational contrarianism led modern Scandinavian scholars to downplay Linnaeus' importance. Sten Lindroth famously complained that pre-war Scandinavian scholars made Linnaeus into a "saint, whose soul hovers even today over the scientist's worktable and the schoolboy's vasculum." He went on to mock Linnaeus' quirky ways, castigating him as "a pure scholastic . . . old-fashioned and lost in his century . . . demonic in his desire to order all things."[15] (These charges, made in 1953, echoed Sachs's accusations of 1875.)[16]

Early post-war historians of biology typically present Linnaeus as a last Aristotelian. His redemption lies outside of science proper, in a "childlike sense of identification with the world of nature," an "almost instinctive way of naming things," and a "fascinating style" heralding "the romantic school of literature in the Scandinavian countries."[17] As is indicated by their book titles—*Growth of Biology, Forerunners to Darwin, Overtures to Biology,* and *Origins of Modern Biology*—these historians have interested themselves in natural history in order to find the origins of Darwinism.[18] Unable to find any such origins in Linnaeus' work, they have instead, by that slip of the imagination which engenders progress as male, cast Linnaeus as the eternally feminine, or as a Romantic poet in dialogue with living nature.

To more recent historians, Linnaeus personifies not his era's epistemic inability to practice modern biology, but its deplorable lack of gender equity, racial equality, and ecological sensibility. In this portrayal Linnaeus engages in an "explicit attempt to 'naturalize' the myth of European superiority," builds an "imperial model of ecology, dovetail[ing] neatly with the needs of the new factory society," and crafts floral systems "making [gender] inequalities seem natural."[19]

Narrating how pluralist, organic, and gynocentric world views are overtaken by unitary, industrializing, phallocentric systematics, these scholars write Linnaeus into a *Verfallsgeschichte* (history of decline). They thus share with traditional Linnaean scholars the founding premise of a central moral narrative and a teleological philosophy of history (although philosophy may not be the right word for the trade practices and mental habits diffused throughout their texts).

In this book, I do not analyze how modern Linnaeus was. (It is, I suspect, a meaningless question.) Nor do I seek to pass moral judgment on him. Rather, I look at how he *himself* understood his science, through his ideas about economics and nation. As I describe how Linnaeus compounded public chores, everyday work practices, and overarching natural doctrines, and how he understood his science to be a legitimation and a technique of state governance, I analyze an early attempt to govern a state economy according to principles of science—an idea that has become co-extensive with our concept of modernity.

In examining Linnaeus' own reasons for studying natural history, I also write about what might be called "a future of the past." I write about the trajectories of progress not as we now see them, with the benefit of hindsight, but as they were projected by the historical actors themselves. The methodology that supports this argument is, I hope, one that resists at every turn a teleological history. It is borrowed most immediately from cultural anthropology, although its longer genealogy is of course to be found in the German hermeneutical tradition.

Linnaeus: Nature and Nation reads as a set of stories about Linnaeus' life and thought. Here, and again in the conclusion, I hint at an analytic framework that helps us to understand him better. But in the chapters sandwiched between these frames, I focus microscopically on Linnaeus and his world. My aim is to introduce him in his manifold and artfully constructed personae—as a son and student, traveler, physician, botanist, economist, theologian, teacher, husband, and father. This has led me to Linnaeus' vernacular, local, and popular science, for there he inscribed his most typical thoughts. I also draw on Enlightenment high theory (political economy, natural philosophy), and on the historical records of the everyday mess and clutter, the material cultures and social customs, that structured quotidian life and thought in eighteenth-century Balticum.

Most immediately, this book belongs to the historiographical school of

Linnaeus research that was initiated in 1908 by Oscar Levertin, author of what remains the most brilliant biography of Linnaeus. An expert on eighteenth-century literature, Levertin was a member of a circle of *fin-de-siècle* literary scholars that included Henrik Schück, Karl Warburg, and Martin Lamm. These scholars, most of them members of Stockholm's small Jewish community, distanced themselves from the reactionary sloganeering common in the cult of Linnaeus. They also discerned proto-Romantic aspects in the piety and literature of the Enlightenment itself. And they turned their studies towards Linnaeus' poetic talents, religious theories, and "folk" beliefs. In turn, their works have influenced wonderful work by later Linnaean scholars, such as Elis Malmeström, Karl Robert Wikman, Ragnar Granit, Tore Frängsmyr, Gunnar Broberg, and Wolf Lepenies.

I admire Levertin's and his followers' sympathetic attempts to understand Linnaeus' mentality. Scholars in this tradition have typically specialized in Linnaeus' religion and anthropology, whereas the explanatory thrust of this book is directed toward another subject matter. We could call it in short the study of Linnaeus' "economics." Less anachronistically, we might call it his notion of nature and nation. (In Latin, Linnaeus used the terms *natura* and *patria*. In Swedish, he wrote *natur*, and variously *nation, rike*, or *fädernesland*). If my book has a central theme, it is how Linnaeus and his students understood the dynamic of history to be the interplay of *natura* and *patria*, and how they (Enlightenment improvers to a man) cast *themselves* as agents of historical change.

Thus I have attempted to extricate myself from Linnaean scholars' longstanding consensus on how to frame problems by addressing a new topic and quarrying new primary material. I have also turned to the older Scandinavian literature on Linnaeus, which was spurred by the 1907 bicentennial of his birth. In those days, scholars worked to reconcile disturbing aspects of the man and his science with their founding hagiographic premises. (On the fringes, they collected supposed "folk myths" about Linnaeus, asked how his nationality and race influenced his science, and even queried whether or not he was a Jew.)[20]

The sentiments of these Romantic nationalist (*nationalromantiska*) scholars have obviously been superseded. But their works remain eminently serviceable. Indeed, T. M. Fries's biography of Linnaeus of 1903 and the Swedish Academy of Science's joint volume on Linnaeus' medical work of 1907 are still standard. Invaluable, too, is the detailed research published in *Svenska*

Linnésällskapets Årsskrift since 1918. The Romantic nationalist school also inspired the publication of Linnaeus' most important autograph manuscripts, as well as many of his lesser scientific writings.

For this book I have often relied on the outstanding scholarly editions prepared by Telemak Fredbärj, J. M. Hulth, Elis Malmeström, T. M. Fries, Arvid H. Uggla, and Ewald Ährling. They include Linnaeus' early botanical works, his students' notes from his Uppsala lectures and botanical excursions, his dietary and etiological writing, his pamphlets, sermons, and broadsides on natural theology, his economic botany, his travel diaries, his autobiographical writings, and many of his letters, including all extant correspondence with Swedish government agencies, the Swedish Academy of Science, and the Uppsala Science Society.

Yet Linnaeus' "economics" is a subject largely without a historiography. Early in this century, excellent monographs were produced on economic theory in Sweden during the era of parliamentary rule, from Karl XII's murder in 1718 to Gustav III's coup d'état in 1772.[21] But within Linnaean studies, only a few essays touch upon questions of economics. And since they do not link Linnaeus' science to the economic issues they discuss, they have been of limited value to my research.[22]

In reading my source material through the prism of "nature and nation," and in seeking to place Linnaeus within a history (or prehistory) of economics, I have inadvertently entered the vast and contentious area of modernization theory. And what has vexed me throughout my research have been the curious inroads and genealogies that lead from the subject I study to the modern methods most available for interpreting it. Indeed, the dominant discourses about early modern economics are often its latter-day ideological descendants.

This holds true for the limited historiography of cameralism itself, and for neo-Marxist development theory.[23] Thus, while classical economics was born out of a critique of British imperial mercantilism, and this critique in turn inspired the now standard history of mercantilism, the history of cameralism was written originally as an attempt to place the economic modernization of Wilhelmine Germany within a longer genealogy of German national independence.[24] Also, since the Nazis claimed the cameralists as their predecessors, German scholarly interest in cameralism declined after World War II. The topic of an economic *Sonderweg* became something of an embarrassment.[25]

Neo-Marxist development theory also derives from an essentially nine-teenth-century critique—Lenin's analysis of imperialism. Developed within the writing of history by Immanuel Wallerstein and Arghiri Emmanuel, it encouraged the former colonies to become economic autarkies. Thus cameralist historiography and neo-Marxist development theory differ from cameralism itself, that is, from the branch of early modern economics that this book investigates, only in that they overlay the same early modern economic assumptions with a Hegelian teleology, organized around race or class.[26]

"A Geography of Nature": Natural Philosophy

In 1778, the younger brother of the Swedish naturalist Carl Linnaeus (1707–1778) began to compose a *minne,* or memorial, of his famous relative who had died that year.[1] Writing from his pastor's farm in the forests of Småland (a poor and remote mid-Baltic province), the Reverend Samuel Linnaeus (1718–1797) could yet hope that his story might be read in Parisian salons. For in the year that he died, Carl Linnaeus was at the height of his fame. Plant catalogues, local floras, encyclopedias and dictionaries, botanic plate publications, natural histories of foreign countries, monographs on flowers, and even children's books and botanic primers, all used a Linnaean vocabulary.

Linnaeus' most fashionable treatise, *Philosophia botanica* (1751), had made botany accessible to amateurs and novices. It was reprinted ten times in Latin between 1755 and 1824. It was also translated into English, French, German, Dutch, Spanish, and Russian. One English summary of *Philosophia botanica,* James Lee's *Introduction to Botany* (1760), went through eight editions by 1811. Another, Philipp Miller's *Short Introduction to the Knowledge of the Science of Botany* (1760), saw fifteen editions.[2] Erasmus Darwin's *Botanic Garden* (1789), dedicated to "ladies and other unemploy'd scholars," even took on the chimerical task of versifying Linnaeus' principles of science.

Fifty years earlier, in 1740, in a pamphlet Linnaeus printed to support his bid for a professor's chair at Uppsala University (riotous students tore his competitor's dissertation into bits during his defense), Linnaeus noted that abroad he was compared to Newton, Leibnitz, and Galileo.[3] "Tell me," he demanded in a *vita,* "who else has done something similar."[4]

Linnaeus is remembered as a botanist, but he really was a jack-of-all-trades. He worked in medicine and natural history (except in experimental branches, such as anatomy, physiology, microscopy, and chemistry). His voluminous scientific writings are brilliantly illuminated with hasty, yet profound, insights. Most famously, he described many of the mechanisms of species interdependence, as Charles Darwin noted on reading his *Oeconomia naturae* of 1749. He also discovered the principles of dendrochronology, noting that tree-rings date individual specimens and record historical weather patterns. Yet Linnaeus never elaborated such insights. Indeed he could not have done so, given his ignorance of the experimental methods and evidentiary protocols that were being elaborated by his contemporaries.

Some of Linnaeus' schemata were never influential, such as his semi-Aristotelian zoology. Here he grouped mammals by toes and teeth, fish by fin bones, and birds by feet and beaks. Snails, insects, slugs and other molluscs, starfish, zoophytes, and crustaceans he shoved into a catch-all *Vermes*—an Aristotelian folk name akin to the English folk term "bugs" or "critters." And he classified hard-shell invertebrates twice: in the genera of Testacea for the shell, and in the order of Mollusca for the soft-bodied animal.[5]

Nor did Linnaeus' fame rest on his classifications of diseases or minerals (which this book will largely ignore), or on his medical or anthropological work. Rather, his reputation derived from his sexual system of plant classification, which was used across Europe from c. 1760 to c. 1800.[6] Linnaeus first presented this system in print in 1735, in *Systema naturae*. In this eleven-page folio pamphlet (it expanded to some 2,300 octavo pages in the twelfth edition of 1766–1768), he laid out a dichotomous key to all of nature's productions.

By means of repeated bifurcations, Linnaeus provided a five-tier botanic hierarchy. He laddered the plant kingdom downward from classes to orders, genera, species, and varieties. Phanerogam plants were divided into twenty-three classes, according to their stamens' numbers, length, degree of distinctiveness, and placement. He added a residual twenty-fourth class, encompassing various cryptogams such as fungi, mosses, algae, and ferns. In turn, he divided his plant classes into orders, according to the numbers and characteristics of their pistils.[7]

When Linnaeus devised this, he knew that the genera of his sexual system were not always natural kinds (although he wavered on the question of what natural kinds might be).[8] Despite its moments of contradiction and inexactitude, however, his sexual system made the ordering of floral collections less daunting both to the learned and to amateurs.[9] His lasting contribution to knowledge, then, was his patient labor to mechanize and standardize the science of botany.

Linnaeus invented a binomial nomenclature, designating each species of flora and fauna by a two-word code consisting of the name of its genus and a species epithet. As suggested by his name for this indexical practice, *nomina trivialia*, he himself only understood the importance of it toward the end of his life. Nonetheless, his species labels continue to this day to answer to the practical needs of the wider scientific community. The first edition of *Species plantarum* (1753, for flora), and the tenth edition of *Systema naturae* (1758, for fauna), remain the starting points for the modern codes of nomenclature for plants (1867), macroscopic animals (1906), bacteria and viruses (1948), field crops and garden plants (1953), and genetically engineered life-forms (in process).[10]

Linnaeus' practice of calibrating botanic names means that the Linnean Society of London, the custodian of his library, manuscripts, letters, and herbaria, is used by scientists as much as by historians.[11] Housing many of Linnaeus' holotypes (the type specimens that he based his species descriptions on), its archives function as the magnetic north for modernity's continued mapping of the natural world. Linnaeus' reputation, then, his place within the pantheon of Enlightenment naturalists, rests upon plain sorting devices, and not upon a single, towering achievement that sets him apart from other men and makes him into a glorified genius.

A man of charisma and drive, the Baltic naturalist was also a rude provincial—sentimental, superstitious, and devoid of general culture. His main patron and Sweden's chancellor, Count Carl Gustaf Tessin, himself a cultured man of the Enlightenment and Sweden's erstwhile ambassador to France, compared "our worthy Nature-Investigator" to a "hearty and energetic provincial Governor."[12] The French-speaking Swedish court, presided over by the sister of Frederick the Great of Prussia, regarded the naive and spontaneous Linnaeus as a boreal, overgrown Emile, and thrilled to his uncouth manners. Continental naturalists were shocked to meet in their "master" a greedy *gubbe* (old man). One described him as a "somewhat aged, not large man with dusty shoes and stockings, markedly unshaven and dressed in an old green coat from which dangles a medal."[13]

Apart from his intelligence, his charm, and his stinginess, Linnaeus' most salient trait was his pride. Adjusting reality to suit his self-image, he bragged that he "became the first president of the Academy of Science" of Sweden, implying that it was a distinction bestowed upon him singularly because of his matchless merit. In fact, the founders—democratically-minded men concerned about relations between commoners and nobles in their new academy—had cast lots for this position.[14] Linnaeus similarly considered the

Swedish Order of the Polar Star (*Nordstjärneorden*), awarded to him in 1753, as his greatest distinction. Yet this medal, pinned onto his "old green coat," proclaimed that honor resided at a provincial level.

In this chapter, I wish to explore that parochiality. I seek to locate Linnaeus within his culture, finding there a set of unexpected clues to his importance in the Enlightenment. To do so, I will return to the Reverend Samuel Linnaeus, writing his brother's *minne* in the spring of 1778. For alongside Linnaeus' *vitae*—the notes he composed for his funeral orators—Samuel Linnaeus' *minne* nicely captures the double way in which Linnaeus understood himself: as a civil servant and as a philosopher-naturalist. In turn, these roles express how Linnaeus triangulated society, self, and state, and how he introduced and conceptualized his natural knowledge in this calculation.

As a civil servant, Linnaeus thought of himself in static and traditional terms. Most immediately, he cast himself as a craftsman and a patriarch, cultivating a family patrimony. More widely, he conceived of himself as a state-builder. In its conceptual outlines this wider aim was also continuous with the past. Here Linnaeus projected onto the civil and economic realm the terms and concepts of governance that had structured Sweden's seventeenth-century military empire. Linnaeus fashioned his civil-servant persona as continuous with his environs in another way, too. In his joint performance as craftsman and bureaucrat, his reference points were larger than the individual self: they were the family, the state, and that confluence of the two expressed in his often-used phrase—"my dearest Fatherland" (*mitt käraste Fädernesland*).

As a philosopher-naturalist, however, Linnaeus cast himself in radical and individualist terms. Here his causative discourses aimed at making a decisive break both with the past and with his wider communities. Linnaeus posed as a lone innovator and as a founder of a new science. As we saw, he always wore his Order of the Polar Star and bragged that he was the first naturalist to have received this high national honor. And while he lived in an era that idealized leisure, he took pride that his knowledge was obtained through labor. This anecdote is typical: housed one June in the summer palace of Drottningholm where he dined privately with the Swedish royal family, a sour Linnaeus classified the queen's collection of shimmering sea-shells "with . . . incredible labor."[15]

The portraits he commissioned of himself also celebrated him as a naturalist at work. From the first extant picture, an anonymous pen drawing from the early 1730s, to the late, official portraits, Linnaeus instructed the artists to show him holding a *Linnaea borealis*.[16] As a personal attribute, this Nordic shadow growth was intended to draw attention to his 1732 voyage through Lapland. (It also emblematized his cameralist project, for he viewed *Linnaea borealis* as the botanic base for a domestic tea industry.) When Linnaeus was ennobled in 1762, he chose as his heraldic device the motto of a self-made man: "Win fame through deeds."[17]

Linnaeus' model for his "new science" was not Isaac Newton or Robert Boyle, however, or any of the Baconians and experimentalists of the seventeenth-century English scientific revolution. In dubbing himself "the Reformer," Linnaeus harked back to his own past and inheritance, or to Martin Luther. As in Lutheranism proper, Linnaeus' emphasis on the radical singularity of the individual self was quickly transmogrified into the practical drive to found a state of sorts, or a commonality of orthodoxy (although now in the realm of science rather than religion). For Lutheranism is most typically a religion of the early modern state—at once conjoined with the state, and making it an object of worship.

Carl Linnaeus was born in the village of Stenbrohult, in Småland, in 1707. He was the eldest son of Christina Brodersonia (1688–1733), a parson's daughter, and Nils Ingemarsson Linnaeus (1674–1748), a curate and a farmer's son. His self-images derive from the early modern state in which he grew up: the religiously orthodox empire of Sweden's warrior king, Karl XII (1682–1718). Linnaeus may even have been christened in honor of him. If so, it was an inauspicious choice. Two years after Linnaeus' birth, his namesake lost his Baltic empire to the Russian tsar (Peter the Great) at the battle of Poltava, in 1709. Nine years after that, in 1718, the king was murdered—leaving behind the Swedish mainland as a ravaged remnant of an erstwhile great power.

In its own small way, the Linnaeus family had linked its advancement to that of the empire, as the men took on parsonships and thus became, given the confluence of church and state in early modern Scandinavia, the most local representatives of secular authority. Linnaeus said he stemmed from "farmers or parsons," hence "simple and plain people."[18] Yet here, as often, he was falsely modest. Sweden's parsons and yeomen were local dignitaries, represented in local courts and the diet of national estates (*riksdagen*). They tilled inherited land, farmers owning it outright and parsons owning it by custom, in the form of church livings. They also both formed intermarrying castes—parsons since the Swedish Lutheran Reformation of 1527, farmers since time immemorial.

When Linnaeus' brother Samuel wrote his *minne* in 1778, he crafted it, with unselfconscious ease, as a *family* narrative. He began by discussing their first known ancestor, "my father's father's grandfather Ingemar."[19] The use of only a first name did not mean the family was somehow rootless, or known to itself only sketchily. The yeomanry of Småland had no need of such marks of identity as family names. They lived in close-knit villages, and passed their farms from father to son. In fact, most Swedes today have family names—many Icelanders still do not—only because in 1901 the Swedish state legislated patronymics into inheritable surnames, so as to track down tax-evaders, draft dodgers, and aristocratic pretenders.

Linnaeus' paternal grandparents were farmers—Ingemar Bengtsson (1633–1693) and Ingrid Ingemarsdotter (1641–1717). To flaunt his learning, their son took a Latin family name at university. Yet he signed himself Nils Ingemars*son* Linnaeus. His eldest son, even after he had been ennobled as "Carl von Linné," at times added an "N," for Nilsson, to his signature.[20] The vernacular patronymic lingered, a memento of a rural past, until Linnaeus' son Carl von Linné (1741–1783) more ambitiously tagged onto his last name a French *fils*.

Linnaeus' father, Nils Ingemarsson, had named himself Linnaeus to celebrate a triple-trunked linden tree growing next to the family farm, Jonsboda Östragård, in Hvittaryd parish, Sunnerbo hundred, Växsjö bishopric, Småland. So magnificent was the tree that two of his maternal uncles also gave themselves family names in its honor. According to local tradition, it was a magic growth (*vårdträd*), and its well-being was linked to that of the families who farmed its land—now calling themselves Lindelius, Tiliander, and Linnaeus.[21] Some hundred years later, in 1820, high winds tumbled its rotting trunks across a field and onto a Bronze Age cairn. The farm people let the "ruin" lie. As the romanticizing local parson contentedly noted, they still adhered to "that prejudice, which I am happy to forgive and do not want to exterminate, that it would not be good luck to remove the least splinter of that linden tree."[22]

Over the years, the Linnaeus family had established a customary right to one particular parsonship, the vicarage of Stenbrohult, in middle Småland. As a memorial to their triple linden clan, Nils Ingemarsson Linnaeus planted three linden saplings in his rectory garden.[23] Samuel Linnaeus, who was himself the vicar of Stenbrohult, explained in his *minne* of 1778 that the living had been held by Linnaeus' great-grandfather's father-in-law, great-grandfather, grandfather, father, and brother. As was common to Baltic Lutheran parsonships, the lineage weaved between the female and male side of the family, depending on family members' needs, skills, and powers.

It was typical that Carl's grandmother's father had married his predecessor's daughter, and that the death of Carl's grandfather in 1707 occasioned an enormous family fight. Should his curate son-in-law be promoted? Or (as was common practice in early modern Scandinavia) should his widow be married to the next incumbent? In the event, the bachelor candidate appointed to wed the relict of the deceased incumbent died unexpectedly. Her son-in-law then surreptitiously wangled a letter of attorney that appointed *him* to the vacant living: Karl XII signed it in a war camp in Polen.[24]

In 1748 Samuel Linnaeus and his three sisters gathered around their father's deathbed, while Carl was away in Uppsala. In his double capacity as son and curate of Stenbrohult, Samuel had "renounced all claims on Stenbrohult's parsonage in favor of a young, unsupported sister." The plan was that she would command the living by marrying the parson candidate.[25] Fortuitously, a local clerk then proposed marriage to this "unsupported" Emerentia (1723–1753). Linnaeus' two other sisters, Anna Maria (1710–1769) and Sophia Juliana (1714–1771), had already married parsons. Thus Samuel became the next vicar of Stenbrohult.

No wonder, given this family background, that Linnaeus thought of his science in terms of a family craft. He even imported concepts of the customary rights of female family members into an academic setting. After he was granted the right to dispose of his university chair in 1762, he long pondered whether he should save it for a potential son-in-law, or bestow it on his son.[26] And in 1746, when one of his parson students died on board an East India ship traveling to Canton, he lamented: "God help me with the widow who cries out for her right to marry a parson. I am married and prevented. There is no parson in the Academy. How shall we get one for the poor widow?"[27]

To entail positions was the rule not only in Linnaeus' family, which had passed down a parsonship for five generations, but a normal and typical procedure for the entire Swedish Lutheran clergy, which constituted an intermarrying civil servant caste following the Swedish Reformation. It was also the custom among Linnaeus' academic colleagues.

Linnaeus' Uppsala teacher, Olof Rudbeck the Younger, had taken over the chair of his father, Olof Rudbeck the Elder. Linnaeus also tutored his son Olof, so that he could succeed his father and grandfather. Olof Celsius the Younger, the son of another of Linnaeus' teachers, Olof Celsius the Elder, became an Uppsala professor too (and eventually a bishop). Indeed, a few closely intermarried clans dominated Uppsala—such as the Celsius, Aurivillius, and later Afzelius. They modeled their academic posts on their ancestral Lutheran parsonages, and used them in turn to establish dynasties within the higher clergy. Most famously, the Rudbeckius-Rudbecks produced one father and three sons who all became bishops, and the Benzelius-Benzelstiernas, one father and three sons who all became archbishops.

Like his kin and his colleagues, Linnaeus interpreted his natural knowledge as an inherited craft. All the same, science was not the calling to which he had been born. As the elder son, a *vita* laments, he "was condemned to become a parson by his parents."[28] After Linnaeus began preparing for a medical career at university, his father kept his change of studies secret from his mother for more than a year. If she had known that "Carl was *medicus*, it would have worried her more than if he had changed his religion."[29]

But then, if his parents guarded Stenbrohult's parsonship as a patrimony for him, why did Linnaeus pursue a medical career? When writing his *minne* in 1778, the Reverend Samuel Linnaeus found this puzzle of why his elder brother Carl worked in science so mysterious that he could only answer it with a piece of science in turn. He summoned the medical theory of the imprinting of the fetus: the belief, common in early modern Europe, that the expectant mother's experiences, acts, and mental states helped form the body and personality of her unborn child.

After the family genealogy, Samuel continued his *minne* by describing how their mother, the nineteen-year-old daughter of the Stenbrohult parson, together with her husband, "in the year 1705 moved to Råshult, which is the curate's residence for Stenbrohult." Around their new home (a grass-roofed cottage), they "laid out a little garden," which the eccentric curate adorned with a small and markedly homemade baroque folly.

"In this garden, father, blessed in memory, had with his own hands built an elevation as if a round table, around it beds and herbs or bushes that were meant to represent the guests, and flowers pictured the dishes on the table." Implying that this curious botanic feast formed the unborn child, Samuel Linnaeus added: "Our Mother saw this quite often: during that same time my Brother was conceived."[30]

Still innocently puzzling over why his elder brother did not become the vicar of Stenbrohult, the Reverend Samuel Linnaeus augmented his tale of fetal imprinting with images of an impressionable child who took his parents' floral pastimes to heart. As Samuel retold the well-rehearsed family legend (it described events that took place some seventy years earlier, before his own birth), Linnaeus' father "decorated the cradle with many flowers." "The following year the father took the little son outside with him, sometimes in the garden and sometimes in the meadows, often laying the child on the ground in the grass, and putting a little flower in its hand to amuse it."[31] A year later, in 1708, Linnaeus' father was promoted to Stenbrohult's parsonship. Laying out a garden around the vicarage,[32] he set one part aside as "Carl's garden."[33]

Perhaps, Samuel mused in a letter of 1781, there was an inherited botanic streak in the family (as the father's floral mania suggested). "I too have a natur. liking for botanical tasks," Samuel continued: "the inclination for bot. is inherited and is propagated in the children," citing his own four daughters as proof. At the same time he artlessly noted that female family members could not pursue the patrimonial profession. "My daughters regret that they didn't become the other sex and had been allowed to profit from their uncle's instruction, since they would have chosen the study of botany."[34]

Linnaeus himself, to explain his passion for flora, did not turn to the imprinting of the fetus, the environment of the child, or the heredity of the family. Instead, he radically reformulated the question of how to relate his family tradition and his own scientific career. (He was, after all, a more imaginative person than his worthy and plodding brother.) In his view, the tasks of the naturalist and the parson were properly the same. And he crafted his life's work in such a way that for him, this was nearly true. He understood his vocation as a religious calling (in the Protestant sense), and explained his teachings in these terms. He also structured his work according to the administrative models of the Lutheran parsonship.

Linnaeus ornamented his lectures and books with Lutheran homilies, warning against courtly and Catholic mores and urging his Scandinavian audiences to adhere to the ways of their "Gothic" forefathers. On a formal level, he framed his scientific writings as Bible verses, and as Lutheran sermons and catechisms. In principle, Linnaeus condemned all self-consciously literary forms of scientific writing. Yet in his own texts he drew on the writerly models of the Old Testament and its Lutheran commentaries. Indeed, he most sternly castigated a concern with style as contrived and decadent in his 1751 *Philosophia botanica*—a work that he artfully constructed in the form of a Lutheran almanac.

For his descriptions of living nature Linnaeus drew on his daily readings in Roman poetry, particularly Ovid and Virgil. He combined these citations with a rough-hewn poetic style that evoked how he experienced the natural world in a state of naïve ecstasy, and is enhanced by the synesthetic linkages between the senses that mark his literary genius. Mostly, he worked in his own, self-styled genre, a modified form of the Lutheran sermon—part rhapsody, part oral epic, part elegy, and part pericope exegesis. Read through the prism of later Romantic nature poetry, Linnaeus' nature writings are beautiful.

Generally speaking, in eighteenth-century Europe the observation of nature had become accepted as a source of sure knowledge. Linnaeus, however, grounded the veracity of his botanic claims in revelation. (He even announced this philosophy of knowledge in *Philosophia botanica*.)[35] He embraced the seventeenth-century doctrine of natural theology, too, borrowing it most immediately from the works of the English Puritan divine and naturalist John Ray (1628–1705). It was in this spirit that he read a work that was part of a Baltic country parson's library in the period, and a formative work for eighteenth-century Scandinavian pietism, *Vier Bücher vom wahren Christentum* (1606–1609), written by Johann Arndt. Arndt had developed his doctrine as a kind of armistice piety removed from the religious wars, sidestepping contentious issues such as the forms of church rituals, the nature of individual salvation, and the status of Holy Scripture. He took what was then still metaphorized as "the Book of Nature" as the prime object of religious contemplation.

In his sermons, pamphlets, and lectures, Linnaeus too elaborated how "to read Nature as any other Book."[36] Once, in a speech of 1759 addressed to the Swedish royal family, he audaciously concluded by subverting a Bible verse typically read as heralding the birth of the Messiah. His science, he explained, was "the light that will lead the people who wander in darkness."[37]

Linnaeus believed that he was one of the elect, called upon by God to reveal, Moses-like, the divine law of nature.[38] His acquaintances were struck by his arrogance. And his moods vacillated between euphoria, when he would work frenziedly, and a hideous melancholia, when he would become prostrate with fear that his half-hearted attempts to curb his maniacal hubris would not placate the fury of *Nemesis divina*, the god-figure he imagined ruling his destiny.

Linnaeus' understanding of his science and his personal importance thus drew on elements from Christian salvation history, with its principal *termini* of fall and redemption. He was dazzled—and frightened—by the fact that he had, as he saw it, "created a new epoch."[39] That epoch he understood in Lutheran (and not always Christological) terms, as the autograph manuscript *Progressus botanices* suggests. Indulging in his pastime of ranking botanists, Linnaeus penned in only two words next to his highest category, *reformatio— ego, mihi.*[40]

"No one," he added in a *vita,* "has so totally reformed an entire science."[41] The eminent physiologist and botanist of Göttingen, Albrecht von Haller (1708–1777), once famously complained that Linnaeus fancied himself "a second Adam."[42] While his own writings liken him instead to Luther, at least one image makes the same point: the frontispiece of the Lange edition of *Systema naturae* (1760) features the author in Eden, surrounded by fauna. Collapsing time past and time present, he at once names the animals and writes the *Systema.*

For his self-image Linnaeus took his cue from Genesis (a text he knew by heart). It tells how God brought "the fish in the sea, the birds of heaven, the cattle, all wild animals on earth, and all reptiles that crawl upon the earth" (Gen. 1:26) "to man to see what he would call them" (Gen. 2:19). "Whatever the man called each living creature, that was its name" (Gen. 2:19). Linnaeus, too, conflated fauna's and flora's "own names" with the names he had "written on" them.[43] But his inchoate search for the "true" or original names of flowers and animals stemmed not only from his Mosaic literalism, but also from his Swedish Gothicism.

It was commonly believed in early modern Swedish learned circles that Sweden was the world's oldest nation, settled by Magog, son of Japhet. This theory, rooted in the biblical allegories of the sixteenth century, was reiterated most powerfully by the famous polyglot Olof Rudbeck the Elder, the father of Linnaeus' Uppsala teacher.

This patriotic myth also held that Swedes had been too remote geographically to suffer the cacophony after Babel, and that they alone spoke an uncorrupted tongue, indeed the original language of Eden.[44] Unlike the family of his mentor, Olof Rudbeck the Younger, Linnaeus never engaged in philological research to prove this point. But the question of scientific inquiry he posed to himself remained Gothicist: how can we identify a primeval language of nature?

In the pursuit of this knowledge his biblical fundamentalism and his Gothicism merged. As Linnaeus knew, according to Genesis the denizens of ancient Babel had used the "single language" of Eden to forge "mortal men" into "one people" so powerful that "henceforward nothing they have a mind to do [would] be beyond their reach" (Gen. 11:5, 6). Read allegorically, as a lesson in statehood and in science, the story of Babel suggests that people's mastery of the natural world depends on large-scale cooperation, which a common language makes possible.[45] Such a language may have been what Linnaeus had in mind when, in a letter of 1737 to Albrecht von Haller, he described his botanical sexual system as a set of "laws . . . by which names could either be made or defended."[46] Thirty-five years later, toward the end of his life, he summarized the results of his career as a lawgiver: now "one can read nature's Book as easily as any other Tract."[47]

Linnaeus' Lutheran and Swedish Gothicist inheritance inspired his conception of a natural order that would end, rather than confront, the standoff between *res* and *verbum* after Babel. In turn, these local cultural references (as they were mediated in his scientific writings, and imaginatively reinterpreted by his continental readers) explain why the neoclassicists, primitivists, and Romantics of the later eighteenth century so admired Linnaeus. It explains why Jean-Jacques Rousseau and the French revolutionaries idolized this old-fashioned Baltic professor, and why Johann Wolfgang von Goethe claimed to have had only three teachers, Shakespeare, Spinoza, and Linnaeus—adding pointedly for the last, "not, however, in botany!"[48] It was the provincial naturalist's antirhetorical stance, and his appeals to virtue, that struck a chord in Europe's greatest men of letters. This is why Goethe owned no less than four Linnaean plant dictionaries, and carried *Philosophia botanica*—bound with a vernacular commentary and *Termini botanici*—when, in imitation of Rousseau's meditative botanic walks, he climbed Weimar's woody hills.[49]

"Russau"—as Linnaeus, who was hazy on who Rousseau was, spelled his name—even wrote to Linnaeus in 1771 to claim that he derived "a more real profit from your *Philosophia botanica* than from all books on morality."[50] That the French philosopher admired Linnaeus was so well known that when he arrived in Berlin, Frederick the Great is said to have asked him: "Would you not like to botanize a little in the park, in the conservatories? I hear that you are an admirer of the great plant knower over there [*da drüben*] by my sister of Sweden, the Monsieur Linné."[51] Rousseau's hugely popular *Letters on Botany*, written between 1771 and 1773 to educate a four-year-old girl, in turn molded the fad among women for Linnaean botany in the later eighteenth century.

Linnaeus' contemporary fame thus derived from what one might call his style of thought. The practicality of the Linnaean system, its simplicity, and its avoidance of rhetoric and complexity answered to a certain strain in eighteenth-century intellectual taste. The unadorned rusticity of Linnaean science contributed to its later success. Indeed, what appealed to Rousseau and his followers was that Linnaeus had constructed his botany in traditionally Lutheran terms—moralizing, popularizing, broadly iconoclastic, and in no need of scientific mediators. In the more sophisticated and Romantic late eighteenth century, the writings of the Swedish parson's son read like exercises in willful epistemological innocence.[52]

Linnaeus, after all, had pronounced himself a hater of rhetoric, that is, of the attempt to use language for persuasion and emotional effect.[53] Yet his preference for a plain style, even in scientific writing, was itself a form of rhetoric. It was bound up, for example, with his particular ethical perspective. Linnaeus regarded any attempts to use language as a vehicle for comparison and analogy as inefficient, obfuscating, and immoral. He thought this would lead to fuzzy and, in fact, false descriptions. *Philosophia botanica* explicitly disparages rhetorical tropes such as synecdoche, metaphor, and irony. Linnaeus pointedly described the same flower twice, once ornately, and once briefly.[54]

Since the eighteenth-century men of letters who lived after him were at once nostalgic, stern, and hungry for an unmediated language of authenticity, they could claim Linnaeus as their precursor, even though his attacks on civilization and his literary mannerisms were intended as a sort of Gothic protest against all things courtly and French, and even though the roots of his style and thinking lay in his Lutheran and austerely primitive childhood in a remote mid-Baltic province.

Linnaeus used old-fashioned terms not only for his scientific vocation; his science too was markedly antiquated. But in the eyes of his contemporaries, Linnaeus' sexual system of plant classification overshadowed his old-fashioned propensities and his many failed attempts at scientific theory.[55] Because of the invention of this new system, he was hailed as "the greatest Botanist that the world ever did or probably ever will know," as an English literary magazine put it in 1750.[56]

Nonetheless, he had critics. Some were important, others less so. In the 1730s and 1740s, a few aging naturalists, such as the German botanist Johann Georg Siegesbeck at the St. Petersburg Academy of Science, had suggested that Linnaeus' sexual system was immoral.[57] Others, such as

Johann Jakob Dillenius at Oxford University, argued he should return to a renewed study of Theophrastus. Still others, such as the wealthy collector Sir Hans Sloane in London, advised him to develop an alphabetical taxonomy. More importantly, at Göttingen University the distinguished naturalist Albrecht von Haller advocated bio-geographical criteria for classification. At Edinburgh University, the antisexualist school cast doubt even on the notion that plants had two sexes. And in Paris, Michel Adanson took it upon himself to demonstrate that, contra Linnaeus' sexual system, no single floral characteristic could satisfactorily distinguish plant groups. Another Parisian botanist, Antoine-Laurent de Jussieu, similarly used what we term claudistic methods—that is, he assigned different and varying weights to several plant characteristics.[58]

On a more fundamental level, some of Linnaeus' contemporaries could barely imagine functional and universal conventions of botanic nomenclature. Albrecht von Haller artlessly revealed this as he wrote to Linnaeus in 1740: "I wish we could have a European Flora written upon your principles. As to a universal System, it seems hardly to be hoped for, except from some man to whom every botanist would communicate his whole stock of observations, and all his dried specimens."[59]

In effect, Haller argued that naturalists were unable to describe plants so that they could be securely recognized by others in turn. The implication, which he attempted to escape in his letter by positing a single imaginary man of exceptional talent, was that even if one scholar *had* had access to all plants of the earth, the system which he then would construct could be used only by himself. A uniform and usable ordering of earth's diversity, Haller suggested, was impossible.

Nor was Haller alone in this pessimistic view. Already by the mid-seventeenth century, at least one natural philosopher had come to think that the entire enterprise of translating natural forms into human language was hopelessly corrupt. In 1668, the Puritan clergyman John Wilkins, Oliver Cromwell's brother-in-law and later bishop of Chester, published *A Real Character.* Inspired by the work he did on codes and ciphers in the Civil War (he even wrote a book on the subject) and by Jesuit missionaries' accounts of the Chinese script, Wilkins attempted to construct an all-encompassing order of nature. Aided at first by John Ray (for plants) and Francis Willoughby (for animals), Wilkins drew giant cross-grid maps of the world. He then issued to each species a homemade ideogram invented for the occasion. In this manner, Wilkins hoped, the natural order would be immediately experienced, rather than intellectually comprehended. Through his invented signs, an Adamite vision was to be re-created.[60]

By the early 1700s, the seventeenth-century craze for artificial languages had died down. When the professor of botany at Oxford, Dillenius, wrote to Linnaeus in 1737 to comment on his *Systema naturae,* he argued (like Haller) that "in my judgment, every scheme of classification offers violence to nature." He went on to prove that he had not grasped Linnaeus' essential point, which was to create an artificial, and therefore stable and robust, system. "I do not doubt that you yourself will, one day, overthrow your own system."[61]

In a later letter from the same year, Dillenius outlined an alternative botanical project: "The day may possibly come when the plants of Theophrastus and Dioscorides may be ascertained; and, till this happens, we had better leave their names as we find them." He added, on a more activist note: "That desirable end might even now be attained, if any one would visit the countries of these old botanists, and make a sufficient stay there; for the inhabitants of those regions are very retentive of names and customs, and know plants at this moment by their antient appellations, very little altered."[62] In one sense, Dillenius subscribed to the empiric apparatus of the "new science" of the seventeenth century, recommending in effect an anthropological field trip to antiquity's hallowed ground. Yet he also remained committed to a philological botany as founded on the Greek classics.[63]

The century's most famous naturalist next to Linnaeus himself, Georges-Louis L. de Buffon, vigorously mocked his competitor in the "Initial Discourse" of his celebrated *Histoire naturelle* (1749 f.). "This large tree which you see is perhaps only a bloodwort. It is necessary to count its stamens in order to know what it is."[64] Through Denis Diderot and L. J. M. Daubenton, Buffon's condemnation of Linnaeus became commonplace in the French salons of the Enlightenment, and in the *Encyclopédie* (1751–1765).[65] The famous materialist philosopher Julien Offray de la Mettrie even wrote a pornographic *L'homme plante* (1748), which was jeeringly dedicated to Linnaeus and described a "plant-woman" according to the sexual system.

Linnaeus retorted by dismissing Buffon's masterpiece as a work "in French" (a language he could not read). "Without pretty figures; wordy descriptions; *oratorice;* few observations, beautiful ornate French; much anatomy with sceletons; without any method; criticizes everyone, but forgets to criticize himself, although he himself has erred the most. Hater of all methods."[66]

Against his French critics Linnaeus found comfort in 2 Samuel 7:9: "And I . . . have cut off all thine enemies out of thy sight." This biblical verse he vaunted as prophecy of his eventual victory over that "Frenchman named Buffon," who "always wrote against Linnaeus" and "lived in the Botanical Garden in Paris, as Inspector."[67] (The professor always dismissed the count as a kind of gardener.) In a *vita* of 1774, he gloated (erroneously) that Buffon "now must arrange the plants after his [the Linnaean] System, nolens volens, since they have been so arranged by the Kings of France and England and in most Gardens in Europe."[68] The quarrel with Buffon stimulated the Romantics' accolades in favor of Linnaeus, and the French revolutionaries' celebration of Linnaeus was intimately bound up with his earlier and negative reception in Paris. The Parisian *Société d'Histoire Naturelle* (founded by "lovers of freedom" in 1790) even erected a statue of Linnaeus in the Jardin des Plantes, symbolically to close off Buffon's reign.

Yet Linnaeus did not live to see this later triumph. Instead, he saw his sexual system disregarded by other Enlightenment naturalists, who took Buffon's lead. A proclaimed goal of all eighteenth-century botanists was a "natural" order, or a taxonomy accurately reflecting how different plants are related to each other. This was true for Linnaeus' own "natural fragments" of 1738, for Jean Baptiste de Lamarck's projected "Théâtre universel de botanique" of 1778, and for Antoine Laurent de Jussieu's *Genera plantarum secundum ordines naturales disposita* of 1789.[69]

In 1737, Linnaeus wrote to Haller in an uncharacteristically pessimistic mode. "I think I shall be able to get together more fragments [of the natural order] than many other people can, though much remains behind, and I doubt whether I shall ever complete them."[70] A year later, in *Classes plantarum* (1738), he proposed sixty-five natural plant families, which he bracketed as *Fragmenta methodi naturalis*.[71] He also called for a natural order in the first edition of *Genera plantarum* (1737), in the dissertation *Cui bono?* (1752), and in *Philosophia botanica* (1751).[72]

Linnaeus suspected that the natural order was encrypted in the basic parts of fructification (calyx, corolla, pericarp, pistil, seed, stamen, and receptacle). But he never grasped the outlines of this natural order, botany's "final end" and "first and last goal." He hesitated even over its most basic metaphors: "the works of the Creator in a proper chain," "the countries on a map of the world," knots in a fishing net, or a grove of branching trees.[73]

In later life, Linnaeus gave private lessons in the natural order. These teachings too "always remained fragments." Addressed to a few foreigners, his seminars were intimate, as their speculative content warranted. Linnaeus welcomed his students in a red dressing-gown and a green fur hat, pipe in hand. On Sundays, he brought in a farmer to fiddle. While he looked on, his foreign students danced the minuet and reel with his four unmarried daughters, later memorialized by one of their dance partners as "good-hearted but coarse children of nature."[74]

In private tutorials that he gave in 1767 to a young doctor from Göttingen, Linnaeus claimed to know how plant families were related—and then refused to tell the student.[75] By that date, at the tail end of his productive life, he had examined and named some 7,700 plants and nearly 4,400 animals. Still he could not puzzle out his hoped-for natural order.[76] In a late *vita,* he admitted his failure. He also warned botanists against using his own natural fragments in their daily work: "He who can find the key to this, he has invented the natural method, but this will probably not happen until the squaring of the circle is invented; he who uses them instead of Method builds a house without a roof."[77] In 1756, a young naturalist, writing from Charleston, South Carolina, artlessly agreed: "The man who gives the natural system must be a *second Adam,* seeing intuitively the essential differences of things."[78]

Historians argue about how Linnaeus construed the relation between the artificial classification and the natural world of plants. Possibly Linnaeus believed that an artificial system and a natural order were not two distinct vegetable orders: rather, one might be inextricably linked to the other, as a provisional but necessary means to a more perfect yet remote end.[79] So in his later life Linnaeus could have rationalized his own failure to discover a natural order. He had already examined many plants, but he knew he had not examined them all.[80]

One scholar has argued that Linnaeus believed that "the exact delimitation of a natural character of a genus depends upon a complete survey of all species that fall under the genus."[81] If that is so, then the remoteness of this future, when all plants would have been collected and positioned within the artificial system, thereby revealing a natural order of things, would have opened up space in the present for boundless useful work, unencumbered by vaulting ambition or crippling doubt.

Certainly, from the beginning of his work in botany, Linnaeus rigorously limited the task of his science. We have seen that his work in the "natural" order amounted only to a few lists, a couple of seminars, and a handful of exhortations. In 1737, he instead outlined a new project. As he wrote to

Haller: "botanists seem to me never to have touched upon nomenclature as a subject of study, and therefore this path of their science remains still unexplored."[82] In another letter to Haller that year, Linnaeus granted the artificiality of his sexual system, even comparing it to an alphabetized list: "I am ready to agree with you that the stamens and pistils lead to no natural system, having adopted a method founded thereon as a substitute . . . for an alphabetical arrangement was always intolerable to me."[83]

Having compared his sexual system to an alphabetical one, Linnaeus again discounted any grander claims for his classificatory efforts: "If my harmless sexual system be the only cause of offense, I cannot but protest against so much injustice. I have never spoken of that as a natural method."[84] He went on to quote his *Systema naturae* as proof that he was aware his system did not mirror nature's actual kinships. He concluded his letter to Haller ingenuously: "Therefore, if you establish a natural method, I shall admit it."[85]

Linnaeus set out to create a self-consciously artificial system, designed for easy use. Again as he wrote to Haller in June 1737: "If you were to collect all the generic names that have been changed from the time of Tournefort to this present day, they would exceed a thousand, though insensibly introduced. What is the cause of all this innovation? I can perceive no other, than there having been no laws laid down, by which names could either be made or defended."[86]

In line with his own lawgiving inclinations, Linnaeus greeted with delight Gustav III's absolutist coup d'état in 1772, which closed off Sweden's Era of Freedom and ended its chaotic parliamentary rule. In his own realm of science, he saw himself as an enlightened despot. Espousing "laws," Linnaeus wanted to transform botany from an ungovernable living language with a multitude of provincial dialects, into a legislated code administered from a single center.

Such a botanical code necessitated a rigorous separation between essence and epithet. Before he introduced his binomial nomenclature, Linnaeus' plant polynomials also functioned as plant labels. As was common to all early modern botanical appellations, they were at once descriptions and designations.[87]

His refashioned teleology was less literal and philological, however, than was that of medieval and Paracelsian natural hermeneutics. Linnaeus did carry over the Renaissance project of decoding in nature the triangular relations among human use, divine intent, and natural law. He also continued to use its accompanying master metaphor of "reading" nature's "book." But he rejected the Paracelsian notion that each species was also its own ideogram, signifying its uses and symbolic meanings. To Linnaeus, final causes (which to him were the ways in which the natural world was part of salvation history) most obviously operated on the explanatory level of the globe itself, which he understood as a single self-regulating entity. He regarded nature as the revealed works of God in space, functioning as a geographic counterpart to God's revealed work in time, or salvation history.

Throughout his life, too, Linnaeus worked within the confines of a scholarly genealogy dominated by the Latin West's reception of Aristotle. On a rhetorical level and in prefaces, he indexed his botany as part of Aristotelian physics. But these multiple interpretive frames did not intrude on his classificatory endeavors themselves. As his contemporaries already sensed, on this separation the claim for Linnaeus' greatness rests.[88]

CHAPTER 2

"A Clapper into a Bell": Floral Names

When in the spring of 1731 the twenty-two-year-old Linnaeus first sketched a botanic order of his own, he only followed the routine procedure of early modern botanists. His best friend, who was also a parson's son and an Uppsala University student, had written a local flora, too. Like Linnaeus' more famous draft, it was "ordered after the very simplest and clearest Method"—namely, his own.[1]

Both Carl Linnaeus and his friend Petrus Artedi had graduated from provincial grammar schools. In these schools, which took students as far as admission to university, they had been trained in the logic and natural philosophy of Aristotle. Indeed, Linnaeus' first Latin reading at home was the *Historia animalium*. (It was a gift from his father.)[2] As a boy Linnaeus encountered the same handful of texts that a Roman boy might have. He studied the same few hundred Mediterranean plants. He learned the same systematic arrangements, the same lists of medical and economic uses, and the same Greek (and for Linnaeus also Christian) stories and tales and anecdotes encrusted around each species. Eight hundred years after Aristotle, in the fourth century, St. Augustine wrote that in the dusty Roman settlements along the African coast, boys were flogged for not knowing Aristotle. Twelve hundred years after St. Augustine, in the eighteenth century, Linnaeus wrote that in the snowy farm villages along the Baltic shores, boys were flogged for not knowing Aristotle.

Given their archaic schooling, Linnaeus and his friend Artedi must have perceived the natural world in the terms of Thomistic medieval logic, broken loose from their Latin readers. In a 1995 novel, the Swedish writer Magnus Florin has beautifully imagined their world view: "They imagined everything in the world divided into two parts. The hard in one part and the soft in another. The fixed and the mobile. The one-yearly and the many-yearly.

That without tail and that with. The quick and the slow. The two-legged and the four-legged. The hairy and the hairless. They imagined each such half in turn divided into two new halves. And so on in divisions to follow, never to end. By this they were enchanted and amazed."[3]

Linnaeus' and Artedi's botanic systems were both constructed, as was customary in Aristotelian natural history, by downward division, so they could not be adjusted to each other or amalgamated. That meant that when the two friends began sketching their flora, both Artedi, with his "very simplest and clearest Method," and Linnaeus, with his "proper method," undertook to re-categorize all of earth's plants. To these two boys, this was a less quixotic task than it seems to us. Five hundred plants were found in a work of botany still read in Linnaeus' youth, the *Enquiry into Plants* by Aristotle's student Theophrastus (370–285 B.C.). Working fifteen hundred years later, the sixteenth-century German herbalists—who were also read in Linnaeus' youth—had scarcely categorized more species. Even the more modern floras available to Linnaeus, such as Gaspar Bauhin's *Pinax theatri botanici* (1623), John Ray's *Historia plantarum* (1682), and Joseph Pitton de Tournefort's *Institutiones rei herbariae* (1700), were manageable in size. They counted about six thousand, eighteen thousand, and ten thousand species, respectively.[4]

Further bolstering that theoretical courage which empirical ignorance fosters was the limited access that the young Linnaeus had to exotics, herbaria, and botanic libraries. He lacked formal training in botany, and indeed in any science. As a schoolboy in Småland, his only zoological reference apart from Aristotle was Kondrad Gesner's *Historia animalium* of 1551. One of the first authors he read when he arrived at Lund University was Ulysses Aldrovandi.[5] Linnaeus only heard of Europe's most eminent botanist, Joseph Pitton de Tournefort, in May 1727, when he was twenty and was about to graduate from his grammar school in Växsjö, Småland—"where," as he impatiently put it in a *vita*, "he had tramped around for a full twelve years." "Only then" did his teacher "tell him that to know a flighty Latin word or name of a plant was nothing, but that all plants ought to be known by their flowers. . . . As it was said that a Tournefort had shown."[6] This aside refers to Tournefort's *Institutiones rei herbariae* of 1700, and it shows how in the Baltic provinces news from the learned republic took the form of vague and belated rumor.

Things hardly improved when Linnaeus went to the university in the autumn of 1727. At the time, North Baltic universities were sleepy hamlets.

They taught the rudiments of Lutheran orthodoxy to the future parsons and civil servants of the Swedish state, a Spartan war machine that was now, after the defeats of the Great Northern Wars (1700–1718), without purpose. News in the realm of philosophy and science were greeted cautiously. In 1732, for example, the president of Uppsala University, who was also a chancellor of the realm, urged his professors to instruct their students "to thoughtfully read Wolff's and Leibnitz' writings, and guard themselves against any news that may be found in them and could cause damage." In their response, the professors congratulated themselves that they had no students "that loved such dangerous news."[7]

Lund University, where Linnaeus spent his first year of study, had no public herbaria of any importance. One professor, though, showed him "something the boy had never seen." It was a collection of "Stones, Shells, Birds and Herbaria of pressed and glued herbs."[8] Uppsala University, to which Linnaeus moved in the summer of 1728, at least had a century-old herbarium of three thousand species, deposited at Uppsala as war plunder in 1666.[9] There Linnaeus saw examples of coastal Asian flora brought back by East India travelers. But he had probably never seen a live African plant, and only a handful of dried ones. As for the New World, by 1730 Linnaeus had examined nine American species.[10]

The libraries at Lund and Uppsala were meager and outdated.[11] As a student Linnaeus had "no books and no mon. to buy them." When he did scrape together a few coins, he had to purchase a second-rate botanic primer, "since," in the book-boxes of Baltic university towns around 1730, "Tournefort couldn't be had." Indeed, Linnaeus was largely self-taught. "From him," meaning Tournefort, "he himself [Linnaeus] taught himself his method."[12] In his quest for learning, the young student depended on professors' private libraries. He was overjoyed when, in the summer of 1730, he became a tutor in Olof Rudbeck the Younger's home and found at last "a pleasing Library open for himself daily in botany."[13]

Lacking botanic texts and dried specimens, the young Linnaeus also encountered relatively few live plants. He had never been abroad.[14] The only landscapes he knew were the taiga, the conifer biome girdling the northern hemisphere, and the European mixed broadleaf forests south of it. Nor had he seen them in a primeval state. Both at Uppsala (in middle Sweden), and at Lund (in south Sweden), Linnaeus botanized in ecologically impoverished landscapes, their few groves felled by land-hungry peasants.[15]

Around Uppsala, many now common indicator species, such as elk, forest hare, and roe deer, were exceedingly rare. Around Lund, the remaining forests bore the names of the aristocratic families protecting them for their shoots and hunts. They were small islands of diversity in a sea of grain, shaded by rows of coppiced willows. North of the Skåne plains, in poor and stony soils where goats grazed, the monolithic North Atlantic heather moors and even sand deserts were still spreading in Linnaeus' day.[16]

As Linnaeus ruefully admitted, in his first three university years he examined only "600 domestic" live plants.[17] Nor did he encounter many exotics. Lund's botanic gardens were puny. In November 1727, Linnaeus found it worthwhile to record that he saw artichokes there.[18] In part because of that dearth, but also following an academic tradition, his first flora—the manuscript *Spolia botanica* of 1727—began by describing those plants which "I have had imprinted with my mother's milk." "*Smolandium* I start with, as a dear fatherland."[19]

Uppsala's botanic garden was also small. It had been founded around 1655 by Olof Rudbeck the Elder. Under the dispirited tutelage of his son, and especially after the great fire of 1702, it had turned into a weedy, overgrown bushland. As a student, Linnaeus went over its meager plantings repeatedly. He complained that it "daily decays, so that there now are hardly two hundred in the whole botanic garden."[20] Altogether, he estimated, the garden housed 280 species in 1731 and perhaps fifty in the later 1730s.[21]

Rudbeck the Elder wrote three plant inventories of Uppsala's botanic garden, in 1658, 1666, and 1685. Linnaeus wrote three *Hortus Uplandicus* in a single year, 1730, and one more in 1731. During these two years, he shifted from Tournefort's method to his sexual system, his *methodo propria*. Like his 1731 *Adonis Uplandicus,* these crucial early manuscript floras centered on the "small and swampy Uppsala garden by the Black Creek."[22] They also drew on Olof Celsius the Elder's town farm (a semi-official annex to the public garden), the local apothecary's herb garden,[23] and Uppland county's estate gardens—stony oak groves owned by the grandsons of Gustav II Adolf's officer corps.

Linnaeus was taught to approach local flora within the context of a profoundly old-fashioned project. Indeed, his floral schooling was subordinated to another discipline altogether, Bible studies. In the summer of 1730, he worked as a plant collector for a "decent old graybeard," Olof Celsius the Elder.[24] By turns professor of Greek, Oriental languages, theology, and philosophy, Celsius had taken over the teaching duties of the formal holder of Uppsala University's chair in medical botany, Olof Rudbeck the Younger. In

1702, Rudbeck had abandoned botany for an immense and never published *Thesaurus . . . harmonicus,* proving that all languages derive from Hebrew. He did so after the great fire of that year consumed his botanic garden, his herbaria, and his many thousands of readied, but unprinted, woodcut blocks for the *Campus Elysii,* his family's illustrated edition of Bauhin's *Pinax* (1623).[25]

While the former Uppsala professor of botany labored on Hebrew philology, Celsius worked on a project no less biblical. During the summer of 1730, he and Linnaeus collected plants around Uppsala for Celsius's manuscript *Flora Uplandica* (1730–1732), a minor and never published preamble to his five-volume *Hierobotanicon sive de plantis sacrae scripturae* (1745–1747).[26] To classify this material, Celsius used Tournefort's method (adding his own modifications and mixing in the systemics of Ray). He also surrounded this local flora with a biblical commentary—searching Sweden's conifer forests for clues to Sinai's desert shrubs.

As Celsius explained in the preface to his *Hierobotanicon,* his "Biblical *Botanica*" had required the philological study of Arab, Greek, and Latin texts. It "also demanded some knowledge of Botany in general, so that many years ago I felt called upon assiduously to hunt the plants and trees growing around the academy."[27] Thus Linnaeus grounded his "proper method" in small, local floras and in Aristotelian and biblical epistemes.

"In this way Linnaeus lived from spring 1730 till the end of 1731, during which time he had to give public and private Lessons" and tend to "the small Rudbecks' information and education." In June 1731 he even wrote the doctoral dissertation of one of his "3 small Gentlemen," the twenty-year-old Johan Olof Rudbeck. This was a customary task for tutors. Yet Linnaeus felt ambivalent about being a ghost-writer. He announces in a title-page annotation on his draft manuscript: "This Dissertation I have put together in a day for 30 *daler* copper. For this another was honored."[28]

In this floral monograph, entitled *De sceptro carolino,* Linnaeus brilliantly mimicked the high-flown Gothicist oratory of the Rudbeck family. He also smuggled in an off-hand reference to a *methodo Linnaeana.*[29] For "apart from all of this [I] had already begun to doubt that Tournefort's was sufficient, therefore took on the task accurately to describe all flowers, to bring them into new classes, reform name and genera, in a completely new way, which took away all time and almost sleep itself."[30]

Yet Linnaeus derived most aspects of his "completely new" botanology from earlier naturalists. He followed Aristotle in inducing constant taxonomic characters from observed examples, and in layering classifications with cumulative differentiating characteristics. He used Ray's species concept and, like him, based his method on a principal part (such as roots, flowers, seeds, or leaves). He also borrowed Tournefort's use of single words as generic names. Indeed, he borrowed most of his plant genera.[31]

For the new botany Linnaeus appropriated the theory of plant sexuality as well. In 1729, he said, he came across a magazine review of a 1717 lecture by the French botanist Sébastien Vaillant. (He may have studied Vaillant's work already in 1726–27, however.)[32] Be that as it may, Vaillant's work fascinated Linnaeus. As a youngster at a mediocre and provincial university, he had not realized that plant sexuality had been investigated by seventeenth-century naturalists such as Ray, Nehemiah Grew, and Rudolph Jacob Camerarius.[33] Here, he felt, was the key to building his system.

Linnaeus never met Vaillant, though he met his widow and inspected his herbaria when he visited Paris in 1738. Involving himself in a priority dispute, he never properly recognized his predecessor.[34] In *Classes plantarum* (1738), Linnaeus—who liked drawing up lists, and would tabulate his colleagues, his soldiers, and his cows—enumerated all botanic methods he had come across.[35] Alphabetizing the improbable with the important, this botanic bibliography only complicated his obligation to Vaillant.

In his Uppsala home, Linnaeus admittedly hung a portrait engraving of Vaillant alongside sixteen other "great *Botanici.*"[36] Vaillant also found a modest place on his handwritten "botanic calendar," where he arranged the death dates of botanists through the year, to form a memorial wall calendar.[37] But in a vernacular manuscript Linnaeus dismissed Vaillant as having only "touched upon" plant sexuality.[38] In a Latin *vita,* intended for the French Academy of Science, he was more careful. "The sexuality of plants, which in reality Vaillant discovered, he made indisputable and built his plant system upon it."[39] In other *vitae,* Linnaeus claimed that earlier "Botanici" had never noticed stamens and pistils, or that he was the first to bring plant sexuality "out into the open day," "so clearly that all his adversaries grew quiet."[40]

After reading the magazine review of Vaillant's work in 1729, Linnaeus summarized its main points in a handwritten pamphlet, *Praeludia sponsaliorum plantarum.* This he gave on New Year's Day of 1730 to his patron, Olof Celsius, instead of the laudatory poem which was the customary gift on that day. He also read it aloud in the Uppsala Science Society. It so impressed the professors in the audience that they appointed him to the post of curator of Uppsala University's botanic garden.[41]

In his derivative "little tract," Linnaeus addressed "the great analogy, between Plants and Animals." Citing vessels, fibers, and other anatomical similarities, as well as diseases, hibernation, and other likenesses, he elaborated the ways plants, too, are sexual beings.[42] As we saw, a few months later, in the summer of 1730, Linnaeus began to "doubt that Tournefort's was sufficient." A year later, in May 1731, in a *Hortus Uplandicus*, he first delineated his "autopsy" of nature, the sexual system of plant classification.[43]

In his old age, Linnaeus noted about himself, in the third person: "he had conceptualized everything, before he was even 23 years old." It is a measure of his naivete that he meant this damning confession as praise. He added that he had "worked out everything, before he returned to his fatherland" from Holland in 1738. Linnaeus traveled to the Low Countries in 1735 to take a degree in medicine, which at that time was not granted in Swedish universities. He knew, too, that his botanic project could only be completed in a European center of learning. As he put it, he went there hoping "the Germans" would print his scientific manuscripts—his "unhappy children," "night-works," "cockroach feed."[44]

In 1737, Linnaeus graphically acknowledged his dependence on patronage and on intellectual networks in one of the copies of Martin Hoffman's full-length oil portrait of him as a Lapp, or Sami. This copy (three are extant), Linnaeus had painted for a Dutch friend and mentor, Leyden senator and naturalist Johan Fredrik Gronovius.[45] He made the artist depict a pile of neatly labeled books (projected and actual) that Gronovius had helped to finance: *Systema naturae* (1735), *Fundamenta botanica* (1736), *Musa Cliffortitiana* (1736), *Critica botanica* (1737), *Genera plantarum* (1737),[46] *Flora Lapponica* (1737), *Corollarium generum,*[47] and *Methodus sexualis.*[48] As these ambitious and multiple titles announce, Linnaeus took it upon himself to apply his systems all-encompassingly. He intended this to be an empirical project and also a personal one, because he positioned himself as a final single arbiter of flora. As he explained in his seminal *Species plantarum* (1753), "Plants NOT SEEN by me I have excluded here, since so many times I have been fooled by authors, this so as not to mix the doubtful with those entirely surely known."[49]

As we saw in brief, Linnaeus' fame in the eighteenth century rested in the democratizing accessibility of his achievement. The virtue of his classificatory system consisted neither in its faithfulness to the natural order (it was patently artificial), nor in its inherent logic. Rather, its workaday usefulness appealed to both learned men and novices. An index of this is the appearance across Europe in the 1750s and 1760s of vernacular primers explaining Linnaean terminology. Linnaeus' *Termini botanici* of 1762, which explicated 673 botanic terms for use on the Linnaean field trip, was reprinted twenty-seven times before 1811.[50]

Linnaeus simplified his botanology through guides and handbooks and encouraged its use by devices such as floral names honoring discoverers of new species. This practicality in turn explains why, although some eighteenth-century naturalists rejected Linnaeus' sexual system for the orthodoxy of classical philology or for the Holy Grail of a natural system, most of them embraced it in their work.[51]

In 1738, Johan Fredrik Gronovius described his thoughts on Linnaeus' systems (note his cartographic metaphor): "Last winter we had a very excellent club, or society, which met every Saturday . . . Linnaeus being our president. . . . We have made so much progress, that by his Tables we can refer any fish, plant, or mineral, to its genus, and, subsequently, to its species, though none of us had seen it before. I think these Tables so eminently useful, that every body ought to have them hanging in his study, like maps [*Tabula Geographica*]."[52]

In his most famous theoretical work, the 1751 *Philosophia botanica*, Linnaeus similarly analogized the five levels of the sexual system to the administrative map of the early modern state, as "kingdom, province, territory, parish, village."[53] In the preface to his *Libellus amicorum*, a souvenir and autograph booklet began in 1734 for his European travels, he uses the same metaphor of mapping when he introduces a working title for *Systema naturae*: "Because, as I hope, shortly one is likely to see all three of nature's kingdoms depicted in maps or paintings, printed under the title *Geographia Naturae*."[54]

When Linnaeus made botany easy for people without schooling or wealth, he acted compassionately. He himself had been a destitute, self-taught youngster. One of his earliest works, a handwritten guide to the botanic teaching collections at Uppsala University of 1730, he wrote to help his fellow students avoid "the great inconvenience of copying all names with pen flying, in the Garden under the open sky, which after all seldom can be done without errors in the name or the citations."[55]

Linnaeus wrote handbooks brief enough to be read with ease, and small enough to be carried into the field. In their format, each divided into twelve chapters and 365 aphorisms, they were reminiscent of the Lutheran almanac. As a fifth-generation parson's son, Linnaeus probably imagined his followers learning an aphorism a day, just as his father's parishioners drilled their daily catechism. He also pointed out that his sexual system, which relied on a few, easily observable features, made hand-colored or copper-engraved images unnecessary. Thus poorer people (students, women, and common folk) could become botanists without an expensive library.[56]

Linnaeus insisted that *Philosophia botanica* (1751) be so cheap that students could afford it. He labored on its language, a straightforward, unornamented Latin that incoming university students could read. In a sermon, he argued that his system "is by no means more difficult than any other European Language; it is just as unavoidable here first to get to know the Letters, which, however, are not as many as ours. Then one learns the Syllables; and, at last, at most 100 Vocabulary items. When one is a little used to this, it is as quick to read Nature as any other Book; yes even for Women themselves."[57] In another sermon, Linnaeus likewise promised that after one year's study, anyone "at first glance can distinguish any plant whatsoever, even if it came from farthest India."[58] Invoking the female gender as a hyperbole for a wide audience, he bragged that this was true even for "that sex that more loves darkness."[59]

Linnaeus also lowered the entrance cost to botany by spelling out—indeed legislating—botanic practices. *Philosophia botanica* ends with a series of one-page instructions on how to set up a herbarium, organize an excursion, plant a garden, and embark on a voyage of discovery. Linnaeus added ten full-page diagrammatic line drawings elucidating plant parts, as well as several indexes. He even instructed the reader how to use *Philosophia botanica* itself. Nicely, the pamphlet closes with this loop of thought.[60]

The book described Linnaeus' actual botanic practices, rather than some ideal notion of what could be done. His provincial journeys, undertaken between 1733 and 1754, provided the model for the methods of natural history that he formulated in the back pages of *Philosophia botanica*. Linnaeus' lectures and courses in practical botany also helped him formulate its instructional maxims. Already in 1733, he guided students—including "many noblemen and barons"—on floral excursions outside Uppsala. The participants paid him with coins, hats, gloves, half-sleeves, socks, books, and buttons.[61] In the summer of 1739, he combined his natural history lectures to the Swedish House of Nobility with floral ramblings around Stockholm's islands.[62] During the summers of the 1740s, he led students and auditors on twelve-hour natural history walks twice a week. Sometimes as many as three hundred people participated. Women as well as men took part, and town people as well as "several foreign Gentlemen and Gentlemen from Stockholm."[63]

During these hikes, Linnaeus developed in practice what some years later he would put in writing, in *Philosophia botanica,* as the proper regimen of botany. He listed field equipment such as vasculas, field microscopes, magnifying glasses, notepapers, butterfly nets, insect pins, and pocket knives. He suggested reading lists (*Systema naturae,* and material on local flora and fauna). He set down times for departure (seven o'clock in the morning), lunch (between two and four in the afternoon), and demonstrations (once every half-hour). He itemized natural objects worthy of collecting: stones, plants, insects, reptiles, fish, and "little birds that are shot."[64]

Showing his interest in the organization of his science, Linnaeus devised military rituals and routines for these excursions. Sharp-Shooters killed the birds. An Annotator recorded the results. The participants paid Linnaeus, and a Fiscal, or policeman responsible for "the troops' discipline," punished late arrivals and unexplained absences.[65] For men a "uniform" was mandatory, consisting of a short jacket, wide and loose sailor's trousers, a wide-brimmed hat, and an umbrella to shield them from the sun.

In this theatrical manner, the professor led his "couple of hundred auditors that collected plants and Insects, worked at observations, shot Birds, kept a Protocol."[66] As one participant wrote: "An army of botanists was formed. He himself was general; companies were subdivided, captains and second lieutenants appointed. A drill book was set up, according to which people spread over all meadows and fields around here. The uniform was a sweater and sailor's trousers; the weapon, insect nets and pins. The trophies: flowers, garlands, and butterflies, pinned with these pins to hats."[67]

At twenty-one hundred hours in the evening, the uniformed "army of botanists"—waving banners, bearing their skewered trophies from the field—marched back into the hamlet of Uppsala, past the imposing burial mounds of its Viking kings, the half-ruined medieval city walls, the royal castle perched high on its hill, and the burghers' wooden town houses and kitchen gardens clustered around its great cathedral. Linnaeus walked at their head, accompanied by a band of musicians.

"They returned into the town with Flowers in their hats, and with Kettle-Drums and Hunting Horns followed their leader to the Garden through the entire Town."[68] Last-minute botanizing was done in "the professors' hop gardens" outside the city gates, "the house roofs in Uppsala, which are covered with green growing sods of turf," and the burghers' "fenced garden patches, planted with vegetables, in particular cabbage."[69]

In 1748, the rector of Uppsala University ordered Linnaeus to abandon these excursions. He objected to the "introduction of a uniform and a new way of life that turns the youths' minds from all other duties and tasks." As he laboriously explained (proving his point by making it), "we Swedes are a serious and slow-witted people; we cannot, like others, unite the pleasurable and fun with the serious and useful."[70]

Linnaeus had earned a substantial income from these tours, however. As he wrote plaintively in a *vita* (referring to himself in the third person as was his habit), the rector's letter "almost killed him; because of it he couldn't sleep for two months."[71] Still he continued to arrange botanic outings, though now circumspectly and shorn of their theatrical and military aspects. Yet it is tempting to speculate that when he described them in *Philosophia botanica* in 1751, he also memorialized his "botanic army" now passing into oblivion.

Linnaeus' fame rests on his greatest invention, the binomial nomenclature. Scholars have argued that Linnaeus' binomials originated in scattered precedents in Renaissance and Greek herbals, in folk names, or in his own bibliographic work.[72] None, however, explains why Linnaeus developed his binomialism *when* he did, or why the inventor was Linnaeus and not some other Enlightenment botanist—all of whom, after all, inherited the same traditions of botany.

Here, I propose an alternative hypothesis, which accounts for the dating of Linnaeus' invention. It is based on *Philosophia botanica*'s last, 365th aphorism, instructing the reader to investigate the "economic use" of plants.[73] I argue that Linnaeus' binomials resulted from his attempts to practice science as an auxiliary branch of economics, and from his efforts to create a simple language for it. Linnaeus' students—young men indebted to him for their entire botanic learning—found his polynomials difficult to use as proper names, both on their voyages of discovery and in their collaborative economic botany. He developed his binomials as a stopgap response to this problem, to make his students more effective as support staff and collaborators.

We saw earlier that Linnaeus developed the sexual system in his early twenties, between 1729 and 1731, and without access to the libraries, herbaria, or natural habitats that would encourage him to challenge the traditional view that the earth's species were few and stable.[74] Indeed, he constructed his flora at the tail end of the classical era of the reception of Aristotle.

Since he worked with a local empirical base, and within the context of old-fashioned theory, he was content to let his plant diagnoses serve as proper names and to follow the two thousand-year-old Aristotelian principle of *genus et differentiae*, that is, of recounting what the species has in common with its genus, as well as what makes it different.

True, in old age Linnaeus repented of the 162nd aphorism of *Philosophia botanica* ("the number of species is constant"),[75] and of his famous assertion in the preface to *Systema naturae* ("We count so many species as there were in the beginning"). As he pondered the mysteries of generation, he came to feel that hybrid cross-breeding explained the earth's variety of life forms. In the sixth edition of *Genera plantarum* (1764), he theorized that modern species stemmed from semi-Platonic, Edenic Ur-animals that embodied the qualities that the taxonomic rank of order contained in the present.

It was a difficult view to hold at a university devoted to educating parsons, and dominated by an orthodox Lutheran faculty of theology (empowered by the cabinet and the national estates to censor heterodox texts). After Linnaeus' death in 1778, his son asserted that "never was my Father, Blessed in memory, an atheist; no, on the contrary, he couldn't bear listening to those kind of people, who spoke in that direction," as a way to preface that "he probably did believe that animal and plant species also the genera emerged through time but that the natural orders were the Creator's deeds."[76]

Although Linnaeus did come to assume a historical formation of species, until the late 1760s he never doubted that it was possible to tabulate all forms of nature. As he saw it, this principle of change had either ceased or was slow enough for naturalists to keep up with it. In the 1756 dissertation *Cynographia,* he distinguished eleven types of dogs, such as "Knee-dog," "House-dog," "Shop-dog," "Partridge-dog," and "Naked dog." He was certain, however, that he described varieties that did not need to be integrated in a Linnaean scheme of nature. (His contemporary Buffon would have sensed the budding species in these niche pets, reminiscent of Darwin's Galapagos finches.)[77] Linnaeus held to his belief in species stability even when he was tempted to abandon it as an economic improver, when, after Sweden's great famine of 1756, a craze for botanical alchemy and grain transmutation swept the country.

In one early letter of 1737 to Albrecht von Haller, Linnaeus admittedly claimed that naturalists can only classify fragments of earth's flora, given its boundless diversity.[78] Nonetheless his polynomials, denoting as they did their species bearers' place within a universal kinship hierarchy, assumed taxa were limited in number. And at other times Linnaeus—who never valued consistency over imagination—took a more optimistic view on man's ability to inventory nature. "If according to gross calculation we reckon in the world 20,000 species of *vegetables,* 3,000 of *worms,* 12,000 of *insects,* 200 of *amphibious animals,* 2,600 of *fishes,* 2,000 of *birds,* 200 of *quadrupeds;* the whole sum of the species of living creatures will amount to 40,000."[79] Linnaeus never grasped magnitudes. By "40,000" he meant "a big number."[80] But he did not mean "a number so large that I can classify only a fraction"— even if by the late 1760s he occasionally wondered if flora and fauna could be indexed into a *Systema naturae,* no matter how bulky.

Indeed, Linnaeus puzzled more over the fact than over the number of species. The real problem, he stated in a 1763 lecture, was why there was *any* natural diversity. "I have often asked myself, why this power, who has created everything in the most simple and wise way, did not create the globe as a cheese, which we worms could have gnawed while we grew up, lived, and multiplied."[81]

By "names" Linnaeus meant diagnostic phrases, not arbitrary references. This is why he could claim that his "names" enabled everyone to "distinguish any plant whatsoever at the first glance, even if it came from farthest India, in that the plant itself informs him about its name, its taste, its smell, its properties, powers, and use, yes points him with a finger as it were to all that one knows about her, for the good of mankind."[82] Linnaeus' diary entry for April 1729 illustrates that species names were generally system-dependent in this period. Here he recorded his burgeoning botanic learning, noting how one day, as he was puttering in the Uppsala botanic garden, he was noticed by an Uppsala professor, Olof Celsius the Elder. When this professor quizzed him on the names of individual plants, he "answered them all with the names after Tournefort's method."[83] Name followed method.

In a sermon of 1773 Linnaeus repeated this point, again meaning by "name" at once a description and a designation: "The thing is, that each Stone, Plant, Animal itself shall tell the ignorant its own name so that it will be understood by everyone who has learnt the language."[84] But the process that Linnaeus presented as automatic demanded that the naturalist memorize a word string for each species. Also, since Linnaeus' diagnostic phrase names adhered to the Aristotelian principle of downward logical division and were formulated to distinguish its bearer from other species within the genus, each time a new species was discovered, all the congeners' phrase-names would also have to change.[85]

Eventually, Linnaeus' binomials, his brief and arbitrary names of natural forms, would shortcut the problem of how diagnosis (description) and reference (name) mingled in his polynomials.[86] But at first Linnaeus and his students experimented with other solutions of how nature should be "tidied" (*i ordning stält*). They turned to vernacular names, bibliographical references, relative number systems, and—as a brief experiment with a different form of true names—absolute number systems. Over time, however, they found that all these naming practices had disadvantages.

To turn first to the use of the vernacular, Linnaeus recognized the value of folk names in his Swedish economic works, which took the form of translated dissertations, sermons, broadsheets, newspaper articles, tracts, and almanac entries, and were intended as items of popular education. In his Latin works he listed vernacular plant names. He made his students do the same, arguing that local names reflected unknown uses or properties of plants. His student Johan Peter Falck, for example, was no linguist, yet as he traveled through the Russian empire, he listed botanic names in Russian, Armenian, Circassian, Turkic, Kalmyk, Voyak, Ostyak, Tanguit, Mongolian, Georgian, Chechen, Ukrainian, Persian and Ossetian.[87]

Common folks knew no Latin. In the preface to the *Narrative of Domestic Plants* (1757), a list of edible native flora inspired by the Swedish famine of 1756, Linnaeus explained that to help "the poor farmer" use his pamphlet, "I have also added most Swedish names of the plants." But as he observed in a letter to the Swedish king, folk names "are often different for plants in different provinces."[88] Advising the court on wild native plants that could be eaten in that great famine year, he listed twelve dialect words for *Epilobium angustifolium*, such as "Weasel-Milk," "Heaven-Grass," "Elk-Food," "Calf-Ass," "Milk-Grass," and "Fox-Ass."[89]

Since the vernacular was not codified in the eighteenth century, Linnaeus found the line between spellings and words difficult to draw. In his letter of 1756, he told the king that *Epilobium angustifolium* was "called *Allmocke* in Västerbotten, *Almycke* in Ångermanland, *Almecke* in Medelpad."[90] At a more personal level, too, Linnaeus was aware of the fluidity of early modern vernaculars. At the outset of his lectures, he apologized for his obscure dialect terms and his "difficult pronunciation."[91] His Småland pronunciation and vocabulary carried over to his Swedish writings, too. As Linnaeus saw it, to write in the vernacular meant to imitate sound. No wonder, that in the index to a 1748 catalogue of the Uppsala botanic garden, he artlessly distinguished between "Swedish Names" and "Botanic Names."[92] More broadly, he feared that the use of the vernacular would split the European learned republic into insular communities of science, locked into separate languages. (This in fact happened in the early nineteenth century). He mourned the decline of Latin,[93] and admitted that he "learnt neither English nor French nor German nor Lapp, yes not even Dutch, even though he stayed a full three years in Holland."[94]

Beginning with *Flora Lapponica* (1737), Linnaeus numbered species sequentially throughout the book. In later works, he denoted them by a trinomial, consisting of their generic name, an abbreviation of the title of the book where he first described them, and their sequential number in that primary work. This practice incorporated the printed source of a plant's first mention in the Linnaean corpus into the name itself, elevating that source, no matter how occasional, into a crucial reference point. By extension, it became impractical to include newly discovered species in newer editions of older classificatory works. For while plants needed to remain grouped according to Linnaeus' sexual system, inserting new species in a consecutive list disrupted older number references that had taken on a life of their own in derivative works.

To solve this problem, Linnaeus at times used two number references. In the *Narrative of Domestic Plants* (1757), "at each plant I have added the Numbers of the first and second Edition of *Flora Svecica*, from which he who doesn't know the plants can get more information."[95] In his *Öland and Gotland Journey* (1745), he instead used a novel trinomial, consisting of a single-word generic name, a book number, and a single-word individual epithet.[96]

Species plantarum (published in 1753) first introduced a consistent use of alphabetical binomials. Happily, there is extant a little-known first draft of this global botanic species list, now kept at the Linnean Society in London. This draft, which was written between 1746 and 1748, contains only a few scattered binomials.[97] Thus 1748 marks the earliest date for Linnaeus' binomial nomenclature. *Philosophia botanica*, which first legislates the use of his binomials, has the imprint date 1751, though it was printed in the autumn of 1750.[98] But this book espouses the binomial nomenclature only halfheartedly. More likely the year 1752, when Linnaeus commenced his second draft of *Species plantarum,* is the latest date assignable to his invention. To learn how Linnaeus came to use true binomials, it is these four years, 1748 to 1752, which we must investigate.

Why just then? In 1745, in the *Öland and Gotland Journey* and in *Flora Svecica,* Linnaeus used a trinomial system, assigning to plants a genus name, a species epithet, and a sequential book number.[99] Four years later, in 1749, when he began drafting *Pan Svecicus,* he did not start a fresh numbers list, unique to that publication. Instead, he re-used the species numbers from the 1745 Stockholm edition of *Flora Svecica.* In the preface he explained that he did it "in order to save paper."[100] Since *Pan Svecicus* makes note of species that are already listed in *Flora Svecica,* Linnaeus believed that even if he had started—as was ordinarily his custom—a fresh numbers list for *Pan Svecicus,* he would still have had to list the species numbers of *Flora Svecica.* In turn, this indicates that *Flora Svecica*'s number references had become established among the Swedish amateur naturalists and improver landlords who were its targeted readers.

Perhaps, then, it was by observing the actual, local use of his economic botany that Linnaeus grasped the practical opportunities of true names, which is to say universal and brief ones. In recycling *Flora Svecica*'s plant numbers, Linnaeus really envisioned a numerical binomialism. More generally, he established the principle that each plant would be assigned a permanent and arbitrary sign. Three years later, in *Species plantarum* of 1753, he shifted from a numerical to an alphabetical binomialism. He now extended and systematized the use of a species epithet and a generic name (it was a practice he had employed occasionally for plants in the index for *Flora Lapponica* [1737] and for the *Öland and Gotland Journey* [1745]). But the first work in which Linnaeus introduces true alphabetical binomials, which still appear alongside a numerical sorting system (itself *also* in the process of becoming arbitrary and absolute), is again the 1749 *Pan Svecicus.*

Pan Svecicus is only an unassuming pamphlet, immediately translated into the vernacular, which addresses a circumscribed problem, the domestic production of imported cattle feed.[101] Why would a brief tract on animal husbandry inspire the use of not only one, but two, arbitrary reference systems (the numerical one quickly disappearing again, and the alphabetical one becoming the foundation for later nomenclature)? The answer, I believe, can be found in the circumstances of how it came to be written.

Pan Svecicus was assembled jointly by Linnaeus and his students. It is a table correlating fodder plants and domestic animals, and thus providing a guide of how to fatten these beasts. To assemble this table, Linnaeus farmed out tasks to his students. Each young man was assigned a test animal (cow, pig, goat, horse, sheep). Clutching goose quills, scrap paper, and ink wells, the students tracked their animal subjects as they foraged the meadows girdling Uppsala, freed from foot hobbles, snout irons, chains, and muzzles. Throughout the day, the students wrote down the plant species their animal subjects ate.

Linnaeus' students listed 856 plant species, while watching the botanic specimens disappear at the moment they realized that they needed to identify them. Tracking the hungry beasts as they swallowed representatives of new and yet newer plant species, the students needed to scribble their botanic names at great speed, in the field, and without reference works. Those tense moments, which tested at once memory and hand, were replicated throughout the day, and were experienced by a cohesive group of young, semi-educated field naturalists. Although I now rely on conjecture, I suspect that at such times short trivial names seemed more practical than long phrase-names.

Linnaeus' binomial nomenclature might also have been stimulated by his students' voyages of discovery from the late 1740s on. These men were the first Linnaeans. They were young. They came from modest homes in the Baltic provinces. They had little scientific grounding or general culture, apart from what they had absorbed at Uppsala University. Their troubles in coping with novel flora may also have inspired their teacher to create binomials.

Both Linnaeus' botanic practice in the 1740s and its program, as expressed in *Philosophia botanica,* required immediate, deductive field observations and an eventual synthetic analysis. (In this way the work foreshadowed the modern division of labor.) "In *one* day," Linnaeus wrote of his botanic excursions, "as many natural objects are found as otherwise in as many days as there are participants."[102] The collector no longer needed to synthesize his findings. And the synthesizer, rather than being the field worker, was, as Linnaeus described himself in a list of "Officers of Flora," the "General."[103]

This division of labor assumed commonly understood and uniformly followed reporting routines. But archival sources suggest that these routines proved elusive, at least in the everyday practice of the student report. Consider, for example, Linnaeus' collaboration between 1750 and 1752 with one typical first-generation Linnaean, his student Pehr Osbeck. Osbeck was the son of landless peasants (*torpare*). A clever boy who had been noted by a country parson, he got a scholarship to study at Uppsala University under Linnaeus. In 1750, Linnaeus secured for him a post as ship's chaplain for the Swedish East India Company, to journey to Guangzhou (Canton). He had never been abroad before.

Osbeck lead morning and evening prayers, preached on Sundays and holy days, taught the ship boys and cadets to read, comforted the ill, and buried the dead.[104] Apart from these pastoral duties, he collected foreign plants. This he and his teacher regarded as the first step toward determining their uses and incorporating them into a Swedish economic policy. Linnaeus was proud and excited by Osbeck's journey. Now, he enthused, he had finally begun his global and collaborative botanic project. His only previous student-traveler had been a middle-aged parson, Christopher Tärnström, who died on his way out to China.

Osbeck wrote his first letter to Linnaeus on 26 February 1751, from the port of Cadiz in southern Spain. In it he characterized the local Spanish flora. As a floral shorthand, he used the relativistic binomial system that Linnaeus himself used in the early 1740s. That is, Osbeck referenced plants by the book in which they were first described in the Linnaean manner. Osbeck's abbreviations of book titles varied; for *Flora Svecica* he might use "Flor. Sv." or "Fl. Sv." But then, he only owned eighteen natural history books (thirteen of them were written by Linnaeus).[105] His little library was easily sorted. Indeed, so small was his botanical canon that at times he neglected to put in a book reference altogether, writing, for example, about "Lichen 945."[106] In those cases, he implicitly referred to his "master tome" of plant description, the Stockholm edition of *Flora Svecica* (1745).

Osbeck also used plant numbers (in the case of Linneaus's publications) or page numbers (in the case of most other authors).[107] At times he also added a Latin abbreviation or a vernacular name. For example, when he found mallow on the streets of Cadiz, he called it "Malva Fl. Sv. 581."

Linnaeus' students were helpless without their travel libraries. As Daniel Rolander wrote Linnaeus from Amsterdam on his way to Surinam in 1755: "I have no books, so I cannot identify any plants correctly, since our luggage, which traveled by sea from Stockholm, has not yet arrived."[108]

Osbeck and Linnaeus, the field worker and his teacher, shared a reference library of a dozen-odd books and a self-evident "master tome." Even so, Osbeck's on-the-spot, homemade names confused even their inventor. As he explained in a later letter from his China travels: "that which in the letters from Cadiz was quoted from Hort. Clifort. is taken from Geuttardi observat. and that which was called C.B. was taken from Boerhaven's index plantarum, which I borrowed while in Cadiz."[109]

At Cadiz, Osbeck only discussed plants he knew from the botanical literature. During the sea journeys from Spain to the Cape, and then to China, he studied "only 2 Elements and 2 Naturalia namely air and water, Whalefishes and Seagulls."[110] The real challenge came when he returned to Sweden in the summer of 1752 with his Chinese and African plants. To his knowledge and within his library, they had not been described. His letters make clear that he felt overwhelmed by his work. "Drying and care-taking gives me plenty of work." Helplessly, he embedded the fruits he had collected in sand. "I do wish they wouldn't rot during the travels."

At the moment of picking, his South African plants had been torn apart by goats, and the Chinese ones by sheep. During the voyage, many plants "were lost, and many rubbed to pieces." Osbeck had been allotted one "chest, which served me as a clothes chest, a bed, and a natural history cabinet." On arriving at Gothenburg, the chest was sealed by customs officials, suspecting contraband.[111] Facing his rotting, half-eaten plants when they were released at last from the Swedish customs, Osbeck devised an alphabetical list of his travel collection. Abandoning that list, he began a numerical list. This in turn he jettisoned for a fresh numerical list. He then immediately confused the two.[112]

Of course, a preliminary plant numbering was a routine procedure for eighteenth-century botanists. This is what Linnaeus imagined his student Daniel Solander was doing on his return to London from Captain Cook's 1768–1771 circumnavigation of the globe, a fantasy that outlines how Linnaeus regarded the ideal organization of a herbarium. "I have every day been figuring to myself the occupations of my pupil Solander, now putting his collection in order, having first arranged and numbered his plants, in parcels, according to the places where they were gathered, and then written upon each specimen its native country and appropriate number."

Linnaeus continued his classificatory elegy by imagining a seamless transition from a temporary number system to a proper Linnaean classification: "I then fancied him throwing the whole into classes; putting aside, and naming, such as were already known; ranging others under known genera, with specific differences; and distinguishing by new names and definitions such as formed new genera, with their species."[113]

Yet Osbeck's pleas to his teacher, as he sat in his rooming-house in Gothenburg harbor trying to sort his China collections, were more typical of the Linnaean novice than his teacher's imagined notion of how the perfect student "put his collection in order." As Osbeck worried in one letter: "Here [I] enclose herbs with a most humble request to learn their names, since the last Numbers didn't work."[114] Another letter from Osbeck to Linnaeus, dispatched from Gothenburg sometime in the autumn of 1752, contains a plethora of worries. He sent "a Chinese that I guess I have sent up before, since I don't have any number that matches it, it too is enclosed." A long list of flora follows, most of it accompanied by wavering and uncertain comments. "That I believe is 124 by me." "By me this is the Chin. plant 5," while this "is probably Nr 69." Other plants "are probably different numbers." For yet other botanic specimens Osbeck simply confessed: "I can't find it any more under that number."[115]

Linnaeus established a true binomial nomenclature in the early 1750s, after seeing students like Osbeck haphazardly mingling and confusing their own homemade abbreviations. Another impetus might have come as early as 1748, when, as we saw, Uppsala University's rector curtailed Linnaeus' botanic excursions—public events where up to 300 people, both lay and learned, took part. In *Philosophia botanica* (1751), a work which first legislated his binomial nomenclature, Linnaeus writes about these excursions. Perhaps the ban on them inspired him to philosophize the conditions necessary for a collective data-gathering, such as he had practiced in these outings.

Be that as it may, from the early 1750s on Linnaeus issued to each plant and animal a permanent and arbitrary two-word name, denoting its genus and species. That he had thereby initiated a revolutionary change was not clear to him. The newfangled binomials seemed to present instead a fresh and arduous task. Linnaeus often strained to think of new epithets. For example, he romantically undertook to name butterflies only after persons and deities in classical mythology. By the time he was entering new butterfly species into the twelfth edition of *Systema naturae,* he had depleted the Elder Pliny's *Natural History* and Hyginus's *Fabulae,* and had to turn instead—and less romantically—to a practical German handbook, the *Grundliches mythologisches Lexicon.*[116]

Though Linnaeus set out the rules for generating binomials in *Philosophia botanica* (1751), under the heading "Differentiae" he only added the phrase: "Trivial names can perhaps be allowed in the way I have used in *Pan Svecicus*."[117] Even in *Philosophia botanica*, the first text to regularize both his sexual system and his binomialism, the technique was scattershot and at the mercy of hurry and chance. Crippled with gout, he dictated the text to a student, "from bed, as quickly as the Printer could typeset."[118]

Linnaeus even hesitated whether to publish it at all. For he feared that it would delay the appearance of his Skåne travel diary (also published in 1751). It was only when his publisher announced that he had decided to reissue *Fundamenta botanica* of 1736 that he hastily added—and even then, only at the request of his students—descriptions of plant parts, and a few lines suggesting a binomial nomenclature. Thus haphazardly, and in competition with economic projects in which Linnaeus was more immediately invested, did the most influential systematic text of the Enlightenment come into being.

In *Species plantarum* of 1753, Linnaeus similarly recommended trivial names in an offhand manner, merely by noting that he found them practical when he worked on *Pan Svecicus*. He did not legislate them, and indeed wrote in the preface that species epithets could be changed—which would destroy what we today see as their essential function.[119] He positioned them on the page in such a way that they read like marginalia of sorts even in print. "Simplified names (trivial names) I have put in the margin, so that without elaborations one can denote a plant with one single name; these I have, however, put there without any strict selection, this can be provided at another time."

Linnaeus worried that botanists might use binomials as a license to skip plant diagnoses (descriptions). He immediately turned his discussions of binomials into an admonition: "All sensible botanists ought to most strictly restrain themselves from ever introducing a trivial name without completing a diagnosis, so that science will not decline into the same barbarity as previously."[120]

For fauna, Linnaeus first used binomial names in a work directed not to university-trained naturalists, but to a courtly audience, a 1754 vanity printing of the Swedish king's natural history collection, *Museum Regis Adolphi Friderici*. He only introduced a consistent binomial nomenclature for animals in the 1758 edition of *Systema naturae*. In a footnote, he revealed that the function of species epithets still eluded him: "I have changed a few trivial names that had previously been assigned without plan in *Fauna Svecica*."[121]

In 1753, Linnaeus also issued a doctoral dissertation, *Demonstrationes plantarum in horto Upsaliensi,* which listed the plants he had shown to his Uppsala students that year.[122] Next to *Species plantarum*'s first edition, published the same year, it is his first true binomial work, though his most brilliant student, Pehr Löfling, used binomials already in November 1749, in his dissertation *Gemmae arborum.*[123] Seeing Osbeck's troubles when left to his own devices, Linnaeus perhaps thought it wise to choose at first a didactic work to illustrate how to use binomials. At the same time, his use of a demonstration manual for Uppsala's botanic garden to introduce a scientific novelty recapitulates his introduction of the sexual system in an earlier such manual, of May 1731.

Regardless of his own underestimation of his invention, Linnaeus' binomial nomenclature succeeded almost at once, and despite some early protests. It mattered little that in 1754, Peter Collinson, a London wool-draper and a fellow of the Royal Society, complained to Linnaeus that he "perplex the delightful science of Botany with changing names that have been well received, and adding new names quite unknown to us." He added: "None now but real professors can pretend to attain it."[124] The Earl of Bute was more direct: "I cannot forgive him the number of barbarous Swedish names."[125]

Linnaeus' first generation of students, who were less concerned about Swedishisms, grasped the usefulness of binomials more quickly. Osbeck, the China traveler, was anxious to settle on stable names and wrote Linnaeus in 1757 concerning his floral voyage: "Since You, Sir, are so gracious and cite my Travels, I won't change the names, but will simply keep quiet about all the printing errors."[126]

Yet it was only toward the end of his life, as he summarized his life's achievements, that Linnaeus himself came to appreciate his binomials. Now botanists could name plants "as easily as one names a person."[127] "Previously," he wrote, "in order to determine a given plant, a whole *differentia* had to be recited, with great trouble to the memory, the tongue, and the pen."[128] "Trivial Names were unheard of before." "Linnaeus introduced them everywhere. That was the same as putting a clapper into a bell. Two names are easy to remember, easy to say and write."[129] Even so, Linnaeus' understanding of his binomial nomenclature was formally expressed only in 1772, when he published his *Nomenclator botanicus.* A page-reference guide to *Systema naturae, Generum plantarum, Mantissa plantarum,* and *Species plantarum,* it turns on the use of the binomial.[130]

*   *   *

At some point during the eighteenth century, as ever more flora was placed within the sexual system's frame, some kind of arbitrary botanic names would doubtless have emerged. In that sense, Linnaeus' economic project was a contributory cause, and not the causal factor, of his binomial nomenclature. Yet it explains the timing of, and immediate impulse behind, his greatest invention.

In themselves such names were only technical tools, of course, and not scientific inventions or discoveries proper. What Linnaeus contributed to knowledge was not a true map of nature. Indeed, over the last two centuries, the theories informing Western maps of living nature have shifted paradigmatically, from a mediated and fragmented reception of the ancient writings of the Greeks and the Jews, to a unified theory of evolution.

Linnaeus provided an arbitrary index, making obsolete the synonym lists or translation guides between different floras, which were common in the seventeenth century. His greatness does not rest upon his fragmentary natural orders or his economic botany, but on his filing cabinet of nature. Linnaeus developed this tableau of names incrementally, through his daily work practice rather than through a preconceived theory, and within a context of local economic problems. Between 1748 and 1752, he undertook his first large-scale collaborative work with his students, and was thus able to observe how these first-generation Linnaeans—the human testing ground for the efficacy of his botanic research program—themselves classified (or failed to classify) flora. Thus Linnaeus' trivial names first appear in works of economic botany, where he demanded instant practicality and a broad readership, and where he and his students most immediately experienced the lack of brief names as a problem.

Afterwards, binomials spread from these more occasional works into Linnaeus' systematic Latin tomes. By the later 1750s, they had become his preferred cross-reference. By the later 1760s, he bragged, in an idealistic inversion of the facts, how he had arrived at his binomial nomenclature as an idea and insight. But it was, rather, his belief in economic utility that fueled his floral innovation. Linnaeus' classificatory botany was not a complex theory. It was a useful technology.

CHAPTER 3

"The Lapp Is Our Teacher": Medicine and Ethnography

Carl Linnaeus was a doctor. He had studied medicine at university, and his doctoral dissertation of 1735—a hasty affair of some thirteen pages submitted for the degree granted during his eight-day stay at the Dutch mail-order university of Harderwijk—was on a medical topic (the etiology of malaria).[1] Between 1738, when he returned to Sweden from Holland, and 1741, when he was appointed professor of medicine at Uppsala University, Linnaeus worked as a clap doctor in Stockholm.

Specializing in syphilis, Linnaeus gained a lucrative aristocratic clientele (as well as contempt for the nobility whose ranks he still yearned to join). Trading on newfound connections, he was appointed chief physician to the Swedish navy. His *vitae* boast that he could cure scurvy, tumors, smallpox, and syphilis. (But he could not heal his own rotting teeth, gout, migraine, and melancholy. Nor could he reverse the many small strokes that eventually killed him.)[2]

Linnaeus did not see himself as only a healer, however. He spent much time classifying diseases, and pondering their origins. Most famously, he diagnosed and described miner's lung.[3] He also believed that epilepsy is caused by washing one's hair; leprosy by eating herring-worms; and ergotism by eating black radish seeds.[4] To explain his acute septicemia of 1727 (to cure him, a field surgeon slit open his pus-filled, blackened arm and rubbed the knife wound with excrement), Linnaeus invoked a mythical insect, *Furia infernalis*.[5] And his mystical "double key to medicine," a thirty-page pamphlet that divides nature into male and female principles, he believed was "a masterpiece and one of the greatest jewels of Medicine."[6]

Linnaeus' most important contribution to medicine, however, was his work on diet and nutrition.[7] During his 1735–1738 European journeys and stay in Holland, he publicized himself as a physician who had mastered an unknown ethno-medicine. In *Flora Lapponica* (1737), in his dietary manuscripts, and in other writings, he fashioned the Lapland indigenes—the Sami—as an moral and medical microcosm complementing his economic macrocosm, the ideal of a self-sufficient principality.

To this end, he presented his Lapland travels of 1732 in the light of Ovidian literature, seventeenth-century Gothicism, early modern anthropology, and neo-Hippocratian environmental medicine. Arctic reindeer herders may seem unlikely candidates for cameralist citizens. Yet Linnaeus thought of these nomads as *exempla* of virtue and health. Their health, he believed, augured the rewards that awaited farming folks if they would only reject imports like coffee, sugar, and salt. The Samis' dietary and medical customs thus lay at the heart of Linnaeus' medical *and* economic philosophy. On them hinged, in his view, the improvement of both Scandinavia's economy and the health of its inhabitants.

The West's first anthropological monograph on a single people, Johannes Schefferus' *Lapponia* (1673), is in fact about the Sami. Authored by an Uppsala professor who never himself visited Lapland, it conflates Lutheran missionary reports with the *Historia de gentibus septentrionalibus,* a travel account written by Sweden's last Catholic archbishop, Olaus Magnus, and printed in exile, in Rome, in 1555.[8] Despite these two learned tomes, well into the eighteenth century the Sami were little known in Scandinavian learned circles.[9] In 1729 the Uppsala Science Society even discussed whether they were a New World people, speaking "the American language," or a lost tribe of Israel, speaking Hebrew. Other Scandinavian scholars held they were pygmies, or Scythians.[10]

Farther afield, in France, Pierre-Louis Moreau de Maupertius suggested, as had Schefferus before him, that the Sami were "Banish'd or Expell'd" Finns.[11] Jean-François Regnard compared "this little animal that is called a Lapp" to apes. George-Louis Leclerc de Buffon, recalling perhaps the ancients' coupling of Scythians and Ethiopians, lumped together Arctic Sami and African San as dwarfish degenerates.[12] In England, Oliver Goldsmith, in his 1774 *History of the Earth,* ranked the Sami as the lowliest of his six races.[13] Linnaeus' own canonical tenth edition of *Systema naturae* (1758), which has the dubious honor of pioneering a global race order, places the Sami in the freak category of "Monstrosus," and as supposed dwarfs, alongside mythic Patagonian giants.[14] (Much later, in the Nazi era, the Sami were classified by German race anthropologists again as a "patologische Rasse" of degenerate Finns.)[15]

Schefferus, at least, had denied German propaganda claims during the

Thirty Years War that the Sami were actual devils, working as army war-locks for the Swedish king, Gustav II Adolf.[16] Yet, like his contemporaries, he mostly saw them as cunning, lowly pagans. In the Enlightenment, Euro-peans began regarding the Sami more benignly. A few treacly "Lapland songs" originally from *Lapponia* were reprinted in the *Spectator* in 1712. Sub-sequently they reappeared along with James Macpherson's Ossianic frag-ments in 1760, Thomas Percy's runic poetry in 1761, an anthology of Welsh bards in 1764, and Johann Gottfried von Herder's *Volkslieder* of 1778–79.

These sentimental outpourings were far removed from empirical knowl-edge or even ethnographic curiosity, however. Indeed, Schefferus' so-called Lapland songs and their reception bear no relation to Sami "jojk" or oral poetry. The *Spectator* even mistook a geographic place-name for a pastoral heroine, and had the lover sighing plaintively for the love of "the heath of Orra" (Orramor).[17]

In Scandinavia Olof Rudbeck the Elder, a colleague of Schefferus' at Uppsala and the main architect of seventeenth-century Swedish Gothicism, was the first to reject the fiction of the Sami as a vicious race, founding in its place the myth of them as innocent children of nature. Rudbeck the Elder so objected to the inherited view that his habitual catchall phrase of sarcastic disbelief became: "It's as true as if it had been written in Schefferi *Lapponia*."

In turn, Linnaeus followed his son and Uppsala teacher, Olof Rudbeck the Younger, and announced in *Flora Lapponica* (1737): "I swear on the Bible that I have not seen a more innocent people than the Lapp."[18]

Around the same time as professors at Uppsala University proclaimed the natural innocence of the Sami, missionary zeal drew other Swedes to Lap-land. By the 1690s, Lutheran missionaries had rooted out the last indige-nous pagan culture of Europe. We know only in fragments the erstwhile spiritual practices of the Sami, such as their bear cult, runic calendars, sha-manist divinations, spirit worship, and syncretist creation narratives (blend-ing Viking paganism, Roman Catholicism, and local animism).

The Swedish Lapland mission was made permanent in 1632 with a semi-nary in Lycksele. Later it was centralized into the Lapland Ecclesiastical Bureau of 1739. It aimed at full-scale assimilation, and it modeled itself on continental efforts to preach to the Jews.[19] The missionaries introduced compulsory church attendance (a special problem for nomads), tithes, cate-chism exams, and inspection tours of Sami households. Sami shamans were fined, tortured, imprisoned, and burned at the stake.[20] By the time Rudbeck the Younger arrived in Lapland in 1695, the Sami had become tourist guides of their own culture's destruction, showing sightseers their former cult places. Only, as the Younger Rudbeck warned Maupertius in 1736, these guides did not like it when the visitors laughed.[21]

In the half-century between Schefferus' *Lapponia* (1673) and Linnaeus' *Flora Lapponica* (1737), Baltic naturalists produced only minor scientific works on Lapland. Rudbeck the Younger traveled to Lapland as part of an astronomical expedition. The astronomers journeyed to the Torne marshes on the Norwegian border, about 125 miles (200 kilometers) north of the Arctic Circle.[22] Rudbeck, however, remained behind in more southern latitudes along the Baltic shore. What is supposed to be his Lapland description, *Nora Samolad sive Laponia illustrata* (1701), ends even further south, at the border of Uppland.[23]

In 1711, the Uppsala Science Society sponsored an expedition by Henric Benzelius to correct the measurements that had been done in 1695 of the sun's refraction at midsummer at the Arctic Circle. Benzelius, a theology student and a grandchild of Lapland homesteaders, also sought to determine the specific density of the Arctic air; analyze Arctic plants for salt and sulphur; record Sami animal names; and determine if kitchen vegetables grew on the tundra. In a project important to a salt-importing nation, he tried to extract salt from seawater by means of cold winds and melted snow.[24]

Benzelius never published on Lapland. Later he became a professor of Greek and Oriental languages and, eventually, like his father and two of his brothers, archbishop of Sweden. He thus became the *ex officio* chair of the Lapland Ecclesiastical Bureau, the agency responsible for converting the Sami. Even when traveling in Lapland, though, he hardly encountered that people, and he showed little interest in the indigenes his forebears displaced.

In 1732, the Uppsala Science Society again sponsored a Lapland voyager, Carl Linnaeus. It seems that the Society gave him no written instructions. He himself understood his task broadly, as involving mineral prospecting, botany, zoology, ethnography, and medicine. Linnaeus set out "from Uppsala city on May 12, 1732, it was a Friday, at 11 o'clock, when I was twenty-five years old within half a day." He wore "trousers quite elegant in leather" and a "braided wig," and he carried a gun and an épée. In his saddlebags this aristocratic pretender had packed a spare shirt, two nightgowns, a passport, letters of recommendation, a microscope, an ink pot, goosequills, and prefolded papers for pressing plants. Against mosquitos, he brought a hat with netting, leather trousers, and knee-high boots. Neither his diary nor his published accounts mention it, but Linnaeus also carried maps of Lapland, and the Younger Rudbeck's Lapland diary of 1695. He admitted, however, to taking four manuscripts, all written by himself: a flora of Uppland county, an early sketch at a sexual system of plant classification, an outline for a work of ornithology, and "this here Protocol," or travel diary.[25]

For some two weeks, Linnaeus rode northward through unkempt forests along the Baltic shoreline and the Bothnian Sea. In his diary, he noted signs of the Arctic: the last elm, oak, and hazel, the first iced-over puddle, and the first glimpse of Sami and the midnight sun. "Who among foreigners wouldn't want to see this?"[26] Like the Younger Rudbeck, Linnaeus marveled at how spring seemed to last forever as he wandered north in May. "This summer we had no summer, but a steady spring that at last closed in a dark blue autumn."[27]

Once Linnaeus arrived in Lapland, he encountered Swedish farm villages with hay-growing "new-builders," and Sami lands with reindeer-herding "Lapps." Led by Sami guides, he walked in their undulating hills. He saw lowland forests of Norway spruce, at higher altitudes dwarf birch groves and thickets of dwarf willow, and on the hilltops fields full of rocks, dotted and streaked with flecks of last year's snow, and guarded by a huge silent sky. Linnaeus sensed with wonder that he was "led into a new world, and when I came up into it, I didn't know if I was in Asia or Africa, for the soil, the landscape, and all plants were unknown to me."[28]

Like all voyages of discovery, Linnaeus' journey was mediated by indigenes. Sami sold him food, shelter, and pack animals. They were his guides and rowers. He rode on hired horses and boated on craft rowed by Sami. When he walked, it was along trading routes and accompanied by interpreters and porters.[29] One of his *vitae* nonetheless declares, "Linnaeus must walk by foot through roadless Lapland, and everywhere with unbelievable work crawl around and snoop about for plants."[30]

At Linnaeus' memorial service in the Swedish Academy of Science in 1779, the speaker suavely glossed his Lapland vision. Linnaeus' travels were so arduous that "the Lapps themselves pitied him."[31] In his Lapland diary of 1732, though, when he described a fast-paced march, Linnaeus claimed the reverse: "The Lapps, who are born to suffer, as birds to fly, moaned that they had never been in such a state; I pitied them."[32] In his *Iter Lapponicum* Linnaeus enthuses over how he was "led" into this "new world" (by his largely unmentioned Sami guides). He makes less of how he scurried back at nightfall to the homesteaders' settlements—huts huddled on south-facing hillsides. There he was greeted by his colonial hosts: parsons, curates, schoolmasters, masters of mines, quartermasters of regiments, bailiffs, burgomasters, governors, and judges.[33] These Arctic superintendents ruled over 40 percent of the country's landmass—and 0.2 percent of its people.[34]

Initially, fur-bearing animals had drawn Swedes to Lapland, just as they attracted the French and English to Canada and the Russians to Siberia. The Sami also hunted commercially. By the 1730s, however, hunters had depleted the brown bear, lynx, ermine, otter, wolverine, marten, and wolf, as well as the red, black, and Arctic fox. Now Lapland earned money for the Swedish crown from iron, alum shale, copper, and silver mines, such as Nasafjäll (1634), Kedkevare (1659), and Gällivare (1742), worked partly by forced Sami labor. Lapland also exported tar and pitch, salmon, cloudberries, ptarmigan, black grouse, hazel hen, and reindeer meats and skins.

Homesteaders too settled in the region. These were of two kinds, Finnish slash-and-burn corn-growers and Swedish cattle farmers and hay-harvesters. The Lapland Settlement Act, in force between 1673 and 1873, granted colonialists crown land leases, shooting and fishing privileges, and freedom from extraordinary taxes.[35] Nonetheless, the settlers, who were living at the northern edge of crop-growing, barely survived.

Altogether, Linnaeus made three attempts (lasting two, three, and sixteen days) to pass into the "entirely different" and "entirely foreign" lands beyond the Scandinavian settlements.[36] To find the elusive migrant bands of highland Sami, he ventured about six miles (ten kilometers) into the Ume regions. He discovered only wastelands (a decade before, the Sami had fled well-armed colonialists). Ordering a fire to warm him as he lay on a reindeer skin, he sent off his Sami guide to find a Sami camp. His scout, however, found just a destitute old Sami woman. After gobbling her last reindeer cheese, Linnaeus boated back to civilization.[37]

In the Lule Sami lands, Linnaeus crossed the highlands along the Norway trail and stayed overnight in a Sami camp. But in the Torne Sami lands, as in Ume, his river track ended with a U-turn on the second day, and without coming across a single Sami.[38] After some weeks of rest in the Swedish settlements of Torneå and Kalix, he returned to Uppsala through Finland. He was home on 10 October 1732.

Apart from the time he spent traveling to and from Lapland and in homesteaders' settlements, Linnaeus' Lapland journey lasted eighteen days. He never passed the sixtieth degree north latitude, which marks the Arctic Circle. While he did cover an impressive 2,000 miles (3,500 kilometers), in his 1732 report to the Uppsala Science Society he doubled that distance to 4,500 miles. Not coincidentally, perhaps, his contract stipulated that he be paid per mile.[39]

To bolster the claims of his invoice, Linnaeus submitted a map of his travel route to the Science Society. In the Sami lands of Ume, he drew in a lengthy, winding, and entirely fictive path toward the southwest. In the Lule lands, his pen meandered toward the uncharted northeast. (Actually, he walked the well-worn trade route to Norway.) In Torne he added, with an easy flourish, a colossal eastern detour, and penned in a rapid raft ride to the sea-coast along imaginary rivers.[40] He also wrote the Science Society that in the lands of Lule he had marched the equivalent of some 500 miles (800 kilometers) in four days.[41]

Understandably, the Science Society hesitated to pay for such miles. Linnaeus ironically remarked that it only "admired all the pains I endured, and condoled me for the dangers to my life that I escaped." A month later the Society refused to publish his travel report. In a phrase defining how he understood his voyage of discovery, Linnaeus fumed: "That's the way *oeconomica* is encouraged in Sweden!"[42]

Throughout the winter of 1732–33 he sent invoice reminders. He also angrily scribbled onto the margin of his travel report: "I now wouldn't want to travel that same even if I were offered a thousand copper plate coins."[43] In the self-congratulatory poem that introduces his Lapland diary, and which he added during this payment dispute, Linnaeus tripled his mileage: when he visited "the end of the world" he "traveled in one year on land" some 6,200 miles (10,000 kilometers).[44]

Linnaeus presented his Lapland travels as an empirical enterprise and an adventure upon a *tabula rasa*. He claimed to dismiss older texts, and set himself apart from both Gothicist tomes and ethnographic literature. In fact, however, he fashioned his Sami savage from international ethnography on indigenes of America, India, and West Africa,[45] as well as on Ossetians, Samoyeds, Inuits, and Kalmyks.[46] His main model was a perennial favorite of the Enlightenment, Jean de Lhery's *Histoire d'un voyage fait en la terre du Brésil* (1578). He also read literature specifically on the Sami, including Magnus (1555) and Schefferus (1673).

At the same time, Linnaeus' travels of 1732 imitated the Younger Rudbeck's journey of 1695. He followed a truncated version of Rudbeck's route, and brought along Rudbeck's travel diary on his journey. He shared Rudbeckian and Gothicist sentiments about nature and the Sami, too. Linnaeus copied into his diary Rudbeck's remarks on such matters as Sami longevity, the beauty of their feet, and the medicinal qualities of glacier meltwater.[47]

In regard to its aim, however, Linnaeus' voyage differed from Rudbeck's. In 1695, Rudbeck had been commissioned by an absolute monarchy with the vague task to "observe and draw this and that, which might convey glory and honor to the fatherland."[48] Accordingly, his Lapland descriptions emphasized the merely representative, for example, the naming and symbolic value of *sceptrum carolinum*, a stately meadow flower. In 1732, Linnaeus was commissioned by a learned society to find objects of *oeconomia*. He focused on the economics of colonization and the appropriation of indigenous medicine.

As we saw, Linnaeus never admitted that he had brought along two maps (both from 1660) as well as the Younger Rudbeck's Lapland journal. He openly dismissed the classic in the field, Schefferus' 1673 *Lapponia*. Instead, he mythologized his Lapland journey by claiming it for the cult of the unmediated eyewitness report. This imaginary rupture with the past found an ironic comeuppance in his eventual inability to publish his 1732 Lapland diary. According to his own criteria for empiricism, it fragmented into fantasy. Even Linnaeus' *Flora Lapponica* (1737), though more guarded, is careless with facts and attributions. Its descriptions of fungi, which Linnaeus later transferred to his canonical *Species plantarum* (1753), derive from Rudbeck's 1695 field sketches from southern Finland.[49]

Flora Lapponica was a published work intended for the learned republic. In it, Linnaeus cautiously exaggerated measurements of time and unverifiable past events, rather than measurements of space and traceable geographic facts. He inflated his fifteen-day ramble in Lule into forty-eight days of "hunger, thirst, sweat, rambles, cold, rain, snow, ice, rocks, mountains and the Lapp language." Spinning out one peckish afternoon, he asserted he went without bread for over four weeks.[50] To Albrecht von Haller, Europe's foremost specialist on Alpine flora, author of the poem "Die Alpen," and an intrepid Alpinist with eleven ascents behind him, Linnaeus magnified his two-week saunter into "several years."[51]

Only once, in a late *vita*, did Linnaeus truthfully report how long he had stayed in the Sami lands. Even this statement he coupled with a wistful claim. "Half a month I lived with the Lapps, as a Lapp." Yet his field notes reveal that he regarded his Sami guides and interpreters as lowly servants. He bullied them, at one point using an épée to force a Sami bearer to move along.[52] Sporadically, he sympathized with Lapland's indigenes. For example, he pitied "the poor Lapps" who were driven away from their rivers by homesteaders yet continued to be taxed by the Swedish crown for their fishing privileges.[53]

But mostly Linnaeus lacked empathy. When he heard that Lutheran missionaries forced Sami to convert by opening their forearm veins and threatening to bleed them to death, he laughed aloud. Again, he laughed—with the same curious spasmodic bark—when he heard that to make the Sami burn their shamanist drums, the missionaries frightened them with mechanical toys.[54]

Linnaeus also exaggerated his knowledge about the Arctic. He bragged to Haller that for years he had been "conversing with Laplanders, Finmen, and Norwegians."[55] In fact, he knew only a few words of Finnish. On returning to Uppsala through Finland in 1732, he fretted that the peasants had not learned the language of their Swedish lords. A local schoolmaster drew up his Samish glossary of about two hundred (mainly botanical) words. Insulting both the helpful homesteader and his Sami informants, Linnaeus entitled the list "voyagers' names" (*nomina perigrina*).[56]

No high-altitude plants are described in *Flora Lapponica* or found in Linnaeus' extant herbaria. Yet in *Flora Lapponica* Linnaeus made much of how in Lapland he climbed "the sky-high mountain overhangs, the terrifying rocks, the threatening boulders, the precipitous stone slabs."[57] He clearly kept to the valleys. But his travel report to the Uppsala Science Society describes how he fumbled about, blinded by fogs, on the cloud-covered rockfaces of what he argued were Europe's highest mountains.[58] *Flora Lapponica*'s frontispiece, the design of which Linnaeus oversaw, piles up jagged alpine peaks—far removed from Lapland's pudgy, worn-down hills.

For his tour of Germany, Holland, France, and England, Linnaeus assembled a Sami costume. He brought his Lapland diary, and prepared to publish the account of his voyage. Already in Uppsala, he had begun to embroider his travel tales. Over time, and in Holland, Linnaeus' Lapland invention passed beyond self-aggrandizement. It became governed by a central fantasy: the colonizer masquerading as the colonized.

A comparison of Linnaeus' frontispiece sketch for the projected *Oeconomia Lapponica* (executed sometime between 1732 and 1735) with the *Flora Lapponica*'s frontispiece (1737) demonstrates how Linnaeus developed his

Sami persona during his stay in Holland. The autograph sketch for his pro-jected *Oeconomia Lapponica* is a clumsy *Streubild*, with anecdotal scenes scat-tered before a mountain backdrop. Sami milk reindeer, travel about in rein-deer sleds, pose with flowers, and (in a more somber vein) shoot European travelers. Visitors appear less lucky. An elk (having wandered northward beyond its usual forest range) is swept down a waterfall. Another elk tum-bles into a glacier crack. And at the same time as one European traveler plunges down a mountainside, another tourist is shot by gun-toting Sami.[59]

The naïve drawing is almost comically autobiographical: Linnaeus' three greatest frights during the Lapland journey were a fall down a hill, a slide into an ice hollow, and a Sami letting off a gun nearby (in order, Linnaeus always claimed, to kill him).[60] Linnaeus intended his sketch as high drama. It reveals him as a fumbling foreign fool—allegorized even as an errant elk.

Like the earlier sketch, the published frontispiece of 1737 presents minia-ture vignettes in a wild and jagged mountainscape. Sami row about, portage boats, tend a storehouse, and drive reindeer sleds. Here, however, Linnaeus does not portray himself as a hapless traveler in the midst of disaster. In-stead, he projects himself into a fantasy that he had hinted at in the manu-script *Diaeta naturalis* of 1733: "I wish that I had been a rich farmer, if I had been used to that since childhood, but even more, if it wasn't so cold, a wealthy mountain Lapp."[61]

In the 1737 frontispiece to *Flora Lapponica*, the Lutheran parson's son ap-pears outside his tent in a Sami costume, cross-legged and smiling, banging a shaman's drum. In the foreground, the artist has added vignettes of the changeling's drum, his floral talisman *Linnaea borealis*, and an intelligent-looking reindeer (see ill. following p. 78).

Linnaeus brought to Holland the Sami dress he wore in *Flora Lapponica*'s frontispiece, and he probably showed it to his Amsterdam engraver. (His own clumsy hand could not have fashioned a sketch usable for the copper engraving's fine details.) Like Maupertius, Europe's other famous Lapland traveler at the time, Linnaeus dressed as a Sami at Amsterdam parties. He also wore Sami clothes in Falun in 1734, when he courted his wife-to-be, Sara Lisa, "*absentibus parentibus*."[62]

Or at least, Linnaeus dressed as what he vaguely understood as a "Lapp." The Sami themselves were meticulous and elaborate dressers, who used the considerable profits from their reindeer exports on conspicuous clothing. Their costumes originally had been designed in the sixteenth century, when they had completed their turn to large-scale reindeer transhumance, and

they translated Renaissance cuts and ornaments into a folk style that defined the gender, marital status, and family of its wearer.

At the same time, the Sami dress jangled with evidence of trading contacts
and social stratification.[63] Rich highland herders favored brightly colored
Lübeck wools, "whimsically loaded with ornaments,"[64] or hung with silver
coins, glass beads, and "an abundance of Brass Chains and Rings."[65] The
metal weight of a wealthy Sami woman's dress, adorned with a huge silver
collar, could exceed twenty pounds (ten kilos). Sami men also wore the
feathered skins of seabirds as hats. As already Schefferus had noted, these
headpieces, with the bird's head artfully stuffed and its wings drawn down
around the wearer's ears, were "not unbecoming."[66]

Next to the Sami in their wondrous garments, Linnaeus in his so-called
Lapp costume cut a poor figure. His beret, which a Swedish tax collector had
given him, was part of Ume women's summer clothing. His reindeer fur liv-
ery was a Torne man's winter garment bought in Uppsala after the trip. His
reindeer leather boots were of a type the Sami manufactured for export and
did not wear themselves. And his shaman's drum—an artifact illegal to own
in Lapland itself—had been presented to him by an Uppsala professor as he
packed for Holland.[67]

Still, his ad-hoc assemblage of souvenirs made a nice display. (He also
dangled from his belt a runic calendar, a knife, birch-bark boxes, and
pouches made of reindeer fur.) The only trouble was that the garb was, perhaps, a tad reminiscent of the "Lapp" costumes worn by other Sami pretenders, such as the wayward son of a Swedish archbishop who had put
himself on public exhibit in Leyden in the 1660s, the two "Lapp" sisters (actually daughters of a north-Swedish grocer) whom Maupertius brought to
Paris in 1737,[68] and the "prince of Lapland" who in the 1710s had toured
the Continent and even married a German blue blood, before he was un-
masked as a Swedish shoemaker's son and had to flee.[69]

The display of "savages" was a practice Linnaeus knew well.[70] The Swedish king Adolph Fredrik had African servants, and the Swedish queen Lovisa
Ulrika, sister of Frederick the Great of Prussia, even attempted to rear an African Emile. (She abandoned the experiment when the boy became too un-
ruly: he grew up to become a courtier and a Stockholm burgher.) In 1758,
Linnaeus himself would attempt to buy an albino girl exhibited in London,
"born in Jamaica of black parents."[71]

The Danish court kept Greenland Inuits, who "pined away and died"
from—in Linnaeus' diagnosis at least—"Gnostalgie." This illness, Linnaeus

and his contemporaries held, was particularly apt to strike Sami (and their reindeer). Linnaeus noted that "the Lapp girl who stayed with queen Ulrika Eleonora couldn't be happy but wanted to go home to Lapland."[72] Here, as so often, he echoed seventeenth-century wisdom. In *Atlantica,* Olof Rudbeck the Elder described a broken-hearted Sami in Holland.[73] Perhaps this story inspired our Sami impersonator's claim that he, too, nearly perished from "Gnostalgie" in Amsterdam.

Linnaeus, then, is part of a long history of Scandinavians parading as Lapps on the Continent. Indeed, he immortalized himself as such, not only on the frontispiece of his Lapland flora, but also in a more formal portrait painted in 1737, "on the occasion when he showed himself in a Lapp costume to display that People's dress for a Company gathered at Mr. Clifford's."[74] (This is the painting reproduced here.) George Clifford was a banker and Linnaeus' main Dutch patron. Between 1735 and 1738, Linnaeus was superintendent of the botanical gardens and herbal collections of his country place, De Hartenkamp, outside Leyden.[75]

The portrait was inscribed, in Latin: "Carl Linnaeus returned from Lapland. Thirty years old 1737." It collapses into one moment the three years, 1732 to 1735, that Linnaeus spent in Uppsala upon returning from Lapland. Probably Linnaeus kept these years to himself. (That same summer he wrote to Haller that he had lived in Lapland for "several years.") Depicting a smiling Linnaeus wearing his Sami woman's fedora—rakishly, at an angle—the portrait reflects the problems of veracity for a struggling explorer of the Enlightenment, forced to transform *himself* into a noble savage, since he could not afford to bring a real one home with him.

Linnaeus' anthropology, then, valued knowledge of self over knowledge of others. And it exercised this in typical Enlightenment fashion, through the perspectivizing device of contemplating non-European peoples. At the center of this self-reflection lay the queasily compelling fantasy of being contemplated in turn. "The Lapp marveled at my clothes, as I was about to describe him."[76]

Informing Linnaeus' reflexive exoticisms were notions of the Golden Age that he had inherited (via his Latin grammar school) from classical antiquity. In Uppsala he reframed his *Iter Lapponicum* in imitation of Ovid's poems from exile, and in vernacular fragments from that period he experimented with rhetorical devices from Virgil's *Georgics.*(He also used them to structure *Flora Lapponica.*)[77]

Linnaeus ornamented such general cultural inheritances with specific accounts of the Sami's health. "O happy Lapp," he enthused in *Flora Lapponica*. "In all innocence you spend your years up to even over 100 with a carefree old age and excellent health; the thousands of diseases that are common among us other Europeans are entirely unknown to you." He concluded, more weakly: "You are not overwhelmed by scurvy, fevers, bloatedness, or corns."[78]

The Sami, Linnaeus thought, were the most representative of "natural" beings—such as "apes, Turks, Persians, Chinese, Siberians, Americans," "children, cats, wild predators," "feral Floridans," biblical patriarchs, republican Romans, wolf children, and "heathens."[79] This disparate, almost Borgesian group all lived, Linnaeus believed, according to the dicta of nature and hence in perfect health.

At the same time, the 1732 Lapland voyage, and as much his encounter with the homesteaders as with the Sami, presented Linnaeus with what he came to understand as a key research problem. In modern terms we can shorthand it as preventive medicine.

Linnaeus' home province, Småland, lay about 950 miles (1,500 kilometers) south of Lapland. There farmers wintered on turnips, rye bread, and pickled herrings. Turnips and rye cannot grow in Lapland, however, and corn and oats ripen only sporadically. Lapland homesteaders instead ate "extremely bitter and detestable" breads made from "chaff without any addition, except now and then a handful or two of barley."[80] "Rye bread is kept aside, to give to visitors."[81] As Linnaeus noted in his Lapland diary, they baked bread from fir bark, pine needles, moss, finely chopped straw, dried fish, wild roots, seaweed or—if "in the greatest poverty"—potatoes.[82] On returning from Lapland, Linnaeus wrote a manuscript treatise of "Arctic inhabitants' stepmother grains," or "famine breads." It describes *tunnbröd*, flatbread, "usually of rough *materia* like grain stalks, chaff, bark, etc." It even includes a recipe jumbling wild greenery for "when there isn't enough bark bread."[83]

As a condiment to their grainless breads, Arctic homesteaders ate half-rotted whitefish and herring. The "smell alone" (outside observers agreed) "made one nauseous."[84] For they had too little salt (a highly taxed commodity) to pickle their catch, and too little sunshine to dry it. As Linnaeus commiserated, "even though their work is extremely hard," they "live from rotten so-called *luns*-fish" buried in the soil in spring, and dug up in winter.[85]

In 1711 the Uppsala Science Society, wondering how Scandinavians survived in the Arctic, wrongly guessed that the homesteaders subsisted on Iceland moss. Linnaeus even advised them to, although such moss is in fact inedible. The Lapland settlers did, however, feed reindeer lichen to their cattle. A dwarfish, ferocious breed, the Lapland cow wintered on straw, bark, and dried fish. It even ate boiled horse dung, "instead of hay, the latter being a very scarce article."[86]

Like the erstwhile Greenland Scandinavians, the Lapland Scandinavians lived on the northern edge of their eco-culture. Those who kept to their own ethnic ways, instead of learning from Sami and fur trappers, suffered from what we now know were deficiency diseases such as rickets (from lack of vitamin D) and scurvy (from lack of vitamin C). Scurvy, a winter illness endemic to early modern Scandinavia, was widespread in Lapland in the summer as well. "In every town," Linnaeus marveled, "large groups of people ran around seeking help, with swollen, wound-infested legs."[87]

In his Uppsala inaugural address of 1741, Linnaeus proclaimed Lapland scurvy to be part of his medical research program.[88] This program in turn hinged upon his "discovery," referred to above, that "the Lapp is ignorant of countless diseases."[89] Yet in 1732 Linnaeus had observed among the Sami diseases that we regard as illnesses of colonized people, such as malnutrition, alcohol dependency, and susceptibility to Europe's novel infections.[90] Indeed, the tundra's low population density may have kept the Sami—and eventually the Scandinavians homesteading in the region—immunologically unprepared for Europe's infectious diseases long after the colonialists first appeared.[91]

In 1732, some 150 years after homesteaders arrived, Linnaeus observed that "the smallpox, when it comes with the settler, gets [those] over seventy and eighty years old and [they] die as if [struck by] the plague." He added: "measles *idem*." That year Linnaeus also diagnosed Sami with what we recognize as tuberculosis, leprosy, epilepsy, and rheumatism.[92] He also lamented their "harrowing, repulsive, and woeful illness" when frogs, accidentally imbibed as tadpoles, spawned in the gut (this is how he explained the etiology of the disease). This *Colica Lapponum* he compared with labor pains. "The poor Lapp wiggles like a worm on the ground."[93]

Later in life, Linnaeus remembered instead that the Sami's legs were straight, and not bowed; that their skin was smooth, and not wound-infested; and that their babies were chubby, and not emaciated. He explained these nomads' health by their customs and manner of life. Especially, he noted that the Sami breast-fed their babies.[94] The Lapland homesteaders, by contrast, "have no time to breast-feed their children, thus they suckle horn all day," Linnaeus wrote, referring to the regional custom of feeding cow's milk to infants from hollowed-out cow's horns. "From this," he continued, "the wives bear more children than I have seen anywhere, but still [they] are of a lower quality than others, for they die away in their youth. The mothers are their own children's murderers, in fact if not in law."[95]

Nor did the Sami use wet nurses. Together with his Uppsala colleague Nils Rosén von Rosenstein, the author of a seminal early handbook in pediatrics, Linnaeus opposed this common practice. As he saw it, not only might the child imbibe the wet nurse's personality with her milk but she might also "shake, whip, torture, and toss" her little charge. He also noted that some wet nurses (and some mothers) stunted or even killed infants by guzzling vodka. At times, they poured spirits down their babies' throats. And "how will those mothers answer to God, who let little children, three to four years old, drink aquavit?"[96]

In all ways, Linnaeus felt, Sami babies were better cared for than European infants. He was astounded that at four months, a Sami baby could "turn its eyes and its head." In a telling commentary on the health of his smallest patients in Uppsala, he added that he had never seen a "European" baby manage this feat.[97]

Linnaeus was devoted to children, and was himself a tender and affectionate father. In his writings, he mourned over how Scandinavian children were "gruesomely beaten." In the 1752 dissertation *Nutrix noverca*, Linnaeus described the many battered children he saw in his medical practice, "lame, hunchbacked, or covered with runny wounds."[98] He sadly noted children were bullied and frightened (for adult amusement, he felt, disguised as discipline). In the same tone of grief, he described a local custom to deprive boys of sleep, to make them more clever at Latin.[99]

Linnaeus also found it "quite painful to see babies tightly swaddled, since they can do nothing but cry all the time. Our Lapp never swaddles children." Here was an adjustment of reality typical of Linnaeus' use of the Lapp as foil. In 1732 he had admired the patience of well-swaddled Sami babies: "They mostly are quiet, even though their hands are tied."[100]

Linnaeus noted, too, that Sami babies and adults slept naked in moss, safe from "lice, fleas, bedbugs, scabies, syphilis."[101] Later the Sami child, or so he told his Uppsala students, dressed in skins "which animals all wear. . . . It is therefore most natural also for us to wear skin clothes." At other times,

Linnaeus suppressed all memories of indigenous apparel, and used as a generic name for the Sami the phrase "the naked Lapp."[102]

At least the Sami wore loose clothes and lacked knowledge of corsets. (In his *Systema naturae* of 1758, Linnaeus classified corseted women as "monstrosities"). The Sami also walked barefoot or in reindeer slippers, "pigeon-toed according to the instructions of God, and not of the dancing teacher."[103]

Similarly, Linnaeus honored the Sami for not wearing wigs, rouge and powder, swords, or silks. He abhorred such civilized luxuries, which he regarded as French, courtly, and decadent. In the 1758 *Systema naturae,* he included in the polynomial for the "European" race the phrase "ruled by fashion." He wrote labyrinthine lists of wicked delights that included almond cakes, oysters, raisins, wall paper, music concerts, chests-of-drawers, maid servants, oil paintings, plaster, and "large windows."[104]

The Sami bachelors' custom of carrying on their person a pleasant-smelling willow growth inspired him to a typical tirade linking women, imports, and luxury: "O you ridiculous Venus, who in foreign countries have at your service coffee and chocolate, jams and confections, wines and lemonades, jewels and pearls, gold and silver, silk and pomade, dances and parties, music and theater! Here you must be content with only a dry fungus."[105]

Linnaeus interested himself in preventive medicine, and this he understood largely as a matter of diet.[106] Here, too, he presented the Sami as a moral and medical ideal, and as a last living avatar of virtues that once ruled among Scandinavians. Like the ancient Goths, he argued, the Sami were ignorant of "inflaming alcoholic beverages, tobacco, tea, coffee, sugar, silk, most spices."[107] Therefore, they enjoyed perfect health. At the heart of Linnaean medicine thus lay a call to moral rejuvenation. And here, as he once scribbled on an undated scrap of paper, "the Lapps are our teachers [*docentibus lapponibus*]."[108]

Why, the Sami even kept their teeth! This proof of health particularly impressed Linnaeus, whose own teeth were rotting, and who remembered his father as a handsome but toothless man. He himself enjoyed what he termed the "new invention" of sugar.[109] But he recognized its links to cavities. "The wild nations in canada," he noted, had begun drinking sugared tea. Their teeth turned "carial, loose, aching, and rotten, while otherwise they would be healthy and white, like dogs'."[110]

"Nowadays," he noted wonderingly about his own country, "among the richer people this sweet substance is eaten more than salt."[111] The "ugly and carial teeth" of wealthy Skåne farmers Linnaeus associated with their *kavring*, a syrupy bread.[112] As a guest at the court of the Swedish queen, Linnaeus could personally "vouch that aristocrats, men and women alike, have nasty-smelling mouths and blackened teeth."[113]

In contrast to them (Linnaeus claimed), the Sami refused to eat sugar and cakes. They preferred cloudberries mixed with reindeer milk, and "fresh, almost liquid, or transparent fir resin." He also claimed that they would not drink beer, or even "the most delicate wine"—only water, mixed, possibly, with birch sap.[114] Elsewhere, dwelling as always in a state of contradiction over Sami reality, Linnaeus describes their hangovers.[115]

In 1732 Linnaeus had observed that Sami used salt to preserve meats and fish, and to season reindeer butter, cloudberry sauce, and coffee.[116] But the Sami had also offered him fish, "unsalted and crawling with vermin," and boiled beaver, "very insipid, for want of salt."[117] Editing his memories, he concluded in *Flora Lapponica* that "salt is the only cause of scurvy and not a lack of vegetables as other physicians believe. This is clearly seen by our Lapps and courtiers; for neither of them have scurvy, since they don't eat salt."[118]

Linnaeus also praised the Sami for eating only twice a day, as he claimed, and living solely on cold meat. "The Lapps are entirely carnivorous animals."[119] (At times, comparing the teeth of orangutans and humans, Linnaeus regarded natural man as a fruit-eating ape. Mostly, however, he classified him as a flesh-feeder.) The highland Sami staple was indeed the meat, offals, blood, and milk of the reindeer. Linnaeus noted they also ate "wolverine, squirrel, marten, bear, beaver, yes, all except wolf."[120] Although he does not write about it, Linnaeus knew that the Sami bought cow milk from homesteaders and rye flour (for hot-stone flatbreads) at Norwegian fairs. In his cameralist tracts on native wild plants, he does describe how the Sami gathered wild herbs, barks, berries, and roots.[121] With the incredulity of a farmer wondering at nomads' opportunistic snacking, he artlessly distinguished between plants that are "edible" and those that are "eaten by Lapps."[122]

Linnaeus admired Sami indigenous medicine, too. He recommended their reindeer cheese for frost bite, their birch bark for wounds, and their milfoil flower for intestinal parasites. Already in a letter of 1731 to the Uppsala Science Society, he had announced his plans to study Sami "home cures."[123] And in May 1732, the government-operated news sheet *Posttidningen* announced that he was traveling north to investigate the Samis' "strange way to live, and the health benefits and drawbacks that arise from it."[124]

As a medical anthropologist, Linnaeus emphasized native remedies. He ignored the main Sami medicinals, since they were also shared by Scandinavians: tobacco, aquavit, beaver glands, and moxibustion. This last cure involved, in its Sami version, applying a birch resin clump to the patient's face, and then setting it on fire. Its frequent use, Linnaeus placidly noted, left adult Sami with "the face, the forehead, and the chin disfigured by a mass of severe scars."[125]

Alongside his vision of Lapp "herders following the customs of Biblical patriarchs,"[126] Linnaeus knew the Sami were long-distance traders.[127] In *Flora Lapponica* of 1737 he wrote of the Norwegian fairs at which the Sami sold their goods: wild furs; reindeer skins, meats, and cheeses; dried whitefish and salmon; handicrafts, such as gloves, shoes, and "most elegant" birchbark baskets; cloudberries and red whortleberries; and dried breasts of ptarmigan, capercaillie, black grouse, and hazel hen.[128]

In the unpublished *Iter Lapponicum* of 1732 he annotated what they bought in turn: porcelain "holland vessels," ox hides, salt, flour, cloths, silver ornaments, needles, knives, and "pots, sometimes of brass, sometimes copper."[129] He knew at first-hand, too, that Sami scouts expected pay in tobacco, vodka, and silver coins. At most he could comfort himself that their foreign vices had a local twist. The Sami took their tobacco with Arctic herbs, their coffee with salt, and their vodka with gunpowder. And "all wealthier Lapps sprinkle beaver glands on their Snuff."[130]

In presenting the Sami as living in Edenic bliss, Linnaeus followed Olof Rudbeck the Elder, who located erstwhile Eden on the eroded slopes around the royal mines of Kvikkjokk, in Lapland. There, throughout the eternally dark winters, conscripted Sami stoked the blast furnaces and Sami reindeer dragged burnt timber great distances: conscription bound both the individual nomad and his semi-wild herds to the engines of industry. In winter, the Sami were ravaged by cold and smothered by snowstorms. In summer, they lived in the perennial smoke from the rotten tree stumps they left to smolder—a measure against the midges and mosquitoes, which "fly up from their marshes in such numbers that they fill the air like a sun mist."[131] At Uppsala, Linnaeus lectured on how cattle, reindeer, and babies perished in insect attacks.

Indeed, the Linnaean fiction of the "happy Lapp" is an especially strained variant of primitivism, for in the eighteenth century the Sami were a thoroughly colonized people. They suffered from smallpox, measles, and alcoholism; as nomads crossing state borders, they labored under double and triple taxation; they were conscripted into Lapland's mines; they were driven from their hunting grounds, fishing creeks, and grazing lands; and the Lutheran churches burdened them with tithes, catechism exams, and compulsory church attendance.[132]

This image of the "happy Lapp" also spectacularly dismissed the complexities of Sami cultures and economies. On his 1732 journey, Linnaeus met Sami at five days' journey north of Uppsala. Revealing how he saw the Sami as more radically other than they were, he noticed they "spoke good Swedish" and asked them what they were doing so far south. They replied that they lived there. Perhaps offended by Linnaeus and his obvious surprise, they added that they intended to die there.[133]

Yet these forest-dwelling South Sami (who kept reindeers as dairy animals, in stables) were a people in the process of being displaced. Where Finns settled, Sami either moved or assimilated. For unlike Swedish fodder growers, who worked the marshes and watery lowlands, Finnish slash-and-burn corn farmers destroyed the dry lands where the reindeer grazed for lichen.[134]

In his writings Linnaeus ignored transcultured Sami—the parsons, fur traders, farm hands, and domestic servants, as well as the beggars and hangers-on living by the mission stations. Thinking within the categories of seventeenth-century taxation, he divided the Sami into "Highland and Wood Laplanders"[135] (reindeer-herding nomads and reindeer dairy farmers). He may not have known that "the Lapp language" comprises three mutually incomprehensible tongues, and he was probably ignorant of Lapland's creole mix of Finnish, Swedish, and West Sami (*nybyggarmål*).[136]

Linnaeus invested his ideal construct of the Sami with near-cosmic significance. Essentially, he conceived of a scale or great chain of health, reaching downward from Arctic nomads at the top, through Scandinavian farmers somewhere in the middle, down to the lowest group of all, French courtiers with their vile venereal diseases.

Linnaeus never distinguished the moral from the natural. He conflated illnesses of the body and the mind, classifying both as moral failings. This principle structures his manuscript *Nemesis divina* (late 1750s–1765), which is his testimonial to God's workings in history. It also governs what he believed was his most important medical study, the manuscript *Diaeta naturalis* (1733). Cartesian "physicians took man to be a little clockwork, a machine, and didn't know he was an animal."[137] Now, as the title page of *Diaeta naturalis* announces, Linnaeus had "shown and admitted that man is an animal and ought to live like an animal; thereafter the possibility is presented that traces of man's natural and almost unspoilt state exist" among the Sami.[138] Writing to the Uppsala Science Society in 1733, he elaborated: "When I saw the healthy Lapp in Lapland, I discovered some principles through which man could double his age, without illness, according to natural principles."[139]

Linnaeus felt the Sami lived in harmony with their environment, and therefore with their essential self, because they followed the customs of their forefathers. Conversely, he mocks his countrymen as *Suecus simia.* "He eats like an Englishman, drinks like a German, dresses like a Frenchman, builds like an Italian, smokes like a Dutchman, takes snuff like a Spaniard, and guzzles vodka like a Russian."[140]

The Samis' health also derived from their lack of property. Linking property to slavery, Linnaeus argued that "the Lapps all live alike, without *luxuria, hinc* no one is the slave of anybody else."[141] *Absque Cere & Baccho* (without grain and wine), he "doesn't have to sweat at a laborious burden of work as our farmers do."[142] "The Lapp gets from his Reindeer herd almost all his needs; lives content and happy in his cold and sterile land." In an outline to his projected *Oeconomia Lapponica* (c. 1732–1735), Linnaeus similarly described "a naked Lapp . . . who seeks his livelihood in a herder's life . . . thereby winning a long life and few illnesses, which he seeks to prevent in his own manner by living in a simple way."[143]

Already in his 1732 *Iter Lapponicum,* copying the Younger Rudbeck's 1695 *Iter Lapponicum,* Linnaeus had gushed that the Sami "inhabit *Campos Elysios,* and Ovid's golden age is their economy."[144] The claim that "Ovid's description of the age of gold is still true among the Lapps"[145] ultimately became to Linnaeus an article of faith. Mingling an Enlightenment notion of the noble savage with older Christian concerns for salvation and classical antiquity's nostalgic ruralism, he believed that having identified a people living in natural innocence, he had proved that man could live in harmony with nature. The Sami were Edenic humans.

It followed that Original Sin was not universal, and that neither Lutheranism (salvation by faith), nor Calvinism (salvation by grace), nor Catholicism (salvation by good works) were strictly necessary. Perhaps empirical studies of "the Lapps' customs, economy, diet, etc." could eradicate sin among Europeans. As Linnaeus mused, "all theologians assert that nature is fallen; that we always have an appetite for evil. This is the case, for we are taught by parents and upbringing and conversation to desire unnatural things, many-colored clothes, gluttony, stinginess, etc. But are the poor Lapps, Ossetians, Americans?"[146] In Linnaeus' hands, the noble savage became a token of—and ultimately a vehicle for—salvation. That, too, is the meaning of his stray annotation: "The Lapps are our teachers."[147]

More immediately, Linnaeus' "happy Lapp" helped him come to terms with his nation's backwardness. In 1735, at the age of twenty-eight, he arrived in his first continental city, Hamburg. He had never been abroad before. That first night, scribbling his impressions in his diary, he artlessly revealed how shocked he was by the riches of this foreign place—or rather, by how poor Sweden was. But at least he could take comfort in his "happy Lapp."[148]

"O You great God, when I must think about what difference there is between Swedes and foreigners, what gardening there is in Sweden, what in Hamburg . . . what setting, nature, climate, human skill . . . Sweden is worth nothing; no one, however, happier than the Lapp."[149] At the moment of his deepest realization of the material and intellectual poverty of his homeland, it was to the imagined memory of the Sami that Linnaeus turned to be comforted in his patriotism.

In a more practical way, Linnaeus' "happy Lapp" helped him overcome his marginal position among Holland's naturalists. In Amsterdam during the same year, he turned up on the doorsteps of the homes of physicians and naturalists dressed in his Sami costume, shamanist drum and runic calendar dangling from his belt. To a modern observer, his snuffbox and gunpowder pouch detract from his claim to have visited true children of nature.[150] To judge by his success in Amsterdam, however, his hosts must have been less suspicious.

Perhaps Linnaeus' inability to speak any modern language besides Swedish, and his schoolboy Latin mouthed in a Nordic singsong, served to authenticate his exoticism. Also, the eighteenth century typically criticized itself by observing itself through the prism of other cultures. Then as now, concepts of "wild nations" punctuated high moral discourse.[151] Linnaeus' noble savage skillfully drew together the well-worn themes of the Enlightenment's benign exoticism. The image also powerfully demonstrated the inability of the colonizer to approach the colonized, even as he stepped—in this case, literally—into their shoes.

Over the years, then, and with the help of ethnographic readings, a Gothicist inheritance, and the primitivist and nostalgic *Zivilisationskritik* typical of the

Enlightenment, Linnaeus honed his brief glimpse in 1732 of highland Sami camps into a vision of the "happy Lapp" "untainted with the Luxury of softer Regions."[152] At the same time, however, he regarded their ancestral lands as Sweden's inner colony, offering a means to reestablish Sweden's seventeenth-century glory and to challenge the Atlantic seaboard's economic hegemony. Both views were conventional in learned Baltic circles. Already in the late 1600s, one Swedish noble had proclaimed that "in Lapland we have our West Indies."[153]

Linnaeus was the first, however, to devise in detail economic projects to render the Arctic frontier profitable for its European overlords. These projects nicely express the limitations and contradictions of the Linnaean knowledge formation and his imperialism alike. As he noted, the Sami were long-distance traders, "very Expert and Cunning in exchanging their Commodities." Yet to Linnaeus they exemplified moral austerity and economic autarky. And though he approved of this, he also complained that they produced too few goods "for Exportation," and that their wealth, such as it was, did not sufficiently enter the national economy through trade, taxes, and rents.[154]

Linnaeus first presented his Lapland strategy to the Uppsala Science Society in 1732. Speaking from a sheet of notes (*index observatorium*), he listed over two hundred points of facts, such as alum shale discoveries, tips on pearl culturing, recipes for bark bread, and descriptions of dye plants and Sami botanic compasses. Yet the meeting was a failure. The members of the Science Society interrupted, to ask him to present only his mineralogical observations. Linnaeus, furious, felt treated like a prospector. As to the *collecta curiosa in re naturali* which he exhibited, the meeting protocols dismiss them as "some insects" and "the skin of a mink." Linnaeus' rage grew when, that winter, the members twice declined to publish in the Society's annuals a couple of longer Lapland write-ups he had penned for the purpose.[155]

At first glance, Linnaeus' *index* seems extraordinarily paratactic for a "scientific" talk. Yet it constituted a coherent economic geography. As for Linnaean lists generally, minerals were followed by plants, animals, and local technologies and ethnography. On a conceptual level, and within the master plot of import substitution that governed his natural history, Linnaeus had four Lapland strategies: to improve reindeer herding; to harvest natural resources; to support dairy and grain farming; and to introduce exotic cultivars.

In undertaking to make the Arctic tundra support more people, produce more goods, and pay higher taxes, Linnaeus urged naturalists to map its resources. "Today when all of Europe's, yes, almost all of the world's, natural history is explored, this still lies as if in the most cruel barbarism."[156] Linnaeus also sought to involve Arctic residents (indigenous and colonial alike) in a permanent project of exploration. Might not the Sami, he asked, perform thermometric experiments when shooting snow grouse in the highlands? And could missionaries not "collect observations, deduce proofs to be sent to the R. Aca. of Sc. and therefore serve the public and their Lapland?"[157]

To start with, these missionaries must begin attending Linnaeus' lectures in natural history when they studied theology in Uppsala. "They often don't understand any more than farmers what nature's Master so wonderfully has placed in front of their eyes."[158] On Linnaeus' recommendation, in 1747 the Lapland Ecclesiastical Bureau ordered prospective Lapland parsons to learn how to "improve and cultivate the lands . . . next to their main task of teaching the Lapps."[159] Seven years later, in 1754, clerics were more practically engaged in natural history when, on Linnaeus' suggestion, the Academy of Science commissioned two parsons to plant saffron crocuses on a north Lapland mountain.[160] The next year, in 1755, the Academy awarded Linnaeus a gold medal for his "useful and elegant proposal" to "plant on these mountains other useful trees and plants that grow on foreign mountains, and that therefore probably could as easily grow on ours."[161]

At other times, Linnaeus acknowledged that in their own way of life the Sami were adapted to the Arctic tundra. "In my opinion," he wrote in *Flora Lapponica,* "considering the land's natural conditions the Lapps' economy is the best."[162] "These lichen-covered moors, which a stranger would consider cursed, are the Lapps' fields and most fertile meadows."[163]

Upon a visit to a hill region in the middle of Sweden in 1734, Linnaeus even voiced his "opinion that it would be better if this place was populated by Lapps." Calculating how this would increase crown revenues, he noted that the reindeer were migrant lichen eaters that wandered into the unpopulated mountain chain in the summers. They "by no means would drive away" dairy farmers, "for in the summer the reindeer do not harm the meadows and in the winter they do not need hay."[164]

Yet even as he celebrated the Sami as they were, Linnaeus could not resist giving advice. He urged the forest-dwelling southern Sami to raise elk.[165] And he reproached "the Lapps' negligence" for not collecting horsetail and reindeer lichen in the summer, in case the winter snow glazed into crusty

VIRO NOBILISSIMO ET CONSULTISSIMO
D: GEORGIO CLIFFORTIO J.V.D.

Figure 1. The frontispiece of Linnaeus' *Flora Lapponica* (1737), ed. and trans. by T. M. Fries (Stockholm, 1905), with Carl Linnaeus featured as Sami in a fanciful Lapland landscape. Courtesy of the Gray Herbarium Library, Harvard University.

Figure 2. Portrait of Linnaeus "in a Lapp costume," by Martin Hoffman (1737). Courtesy of Linnémuseet, Uppsala.

sheets of ice (as sometimes happens), and the reindeer were unable to nose the snow aside for lichen.[166] Voicing the farmer's ancient complaint against the herder, Linnaeus also asked why the Sami should "lie about the whole day in the greatest comfort, when he is not taking care of the reindeer."[167] In his father's parish and in the poor, forested region of Småland generally, the subsistence peasants loathed the rich agricultural-export farmers of the plains nearby. Linnaeus linked the two enmities when he compared a "man from Skåne and a Lapp: both are lazy dogs."[168] In a more conciliatory mood, he likened these nomads to Jesus' ideal disciple. The Sami "often does not know from what he will live the day after, as the bird in the air lives in the wild wastelands, harvests nothing, seeds nothing."[169]

For Linnaeus, then, one aim was to rationalize the traditional relations between Lapland's indigenous people and their ancestral lands. He also hoped to exploit the region's natural resources. Its metals had been mined since the 1630s: Linnaeus urged continued prospecting for Lapland gold "and other expensive ores."[170] Searching his mind for other exploitables, he hit on "our Ordinary Falcon that they fuss over so abroad." In the summers, Dutch and French falconers congregated in northern Sweden. Linnaeus himself had visited their camps, during his voyage through Dalarna in 1734. "It would be better if someone of our own nation would have that profession, and catch falcons, and sell to the foreigners; then the gain would be a profit for the country." Ever optimistic in his cameralist plans, Linnaeus instructed his Uppsala theology students on how to snare these northern birds of prey.[171]

In a similar vein, he counseled Swedish apothecaries to sell pulverized reindeer horns as a substitute for the aphrodisiac *praeparata* ground from deer antlers.[172] He urged them to cultivate medicinal herbs on Lapland's highland tundra, and to open spas around the abundant wells.[173] In 1740, as a test for a plan to work its lowland forests, he introduced a "Lapp tea" brewed from the delicate shadow growth *Linnaea borealis*. (Attempting to capitalize on the cult of colonial wares, he called it "Cape tea.")[174]

Linnaeus also hoped to introduce cattle and grains to Lapland. "There is no doubt but that over time most of the Lapp lands will become Farming country."[175] Naturalists, he believed, could further this colonial process in two ways. First, they could train temperate plants to tolerate the climate. Second, they could domesticate Arctic species, substituting them for temperate cultivars and domestic animals. In 1733, for example, Linnaeus was queried by the governor of one Lapland county, Västerbotten, whether the region could "be seeded and bear fruits and be of use like the southern places." In response, Linnaeus proposed turning a native wild plant, lyme grass, into a "Lapp wheat."[176] This would "earn the fatherland a couple of Barrels of gold," and "feed millions."[177] In 1734, Linnaeus wrote to the same Lapland governor with a new scheme, for growing root vegetables in Lapland's mountain chain. "Oh, if in this way I were able to travel through all provinces in Sweden, namely one each summer, how much would not be discovered for the benefit of the country!"[178]

By the end of the 1740s, Linnaeus recognized that lyme grass, a saltwater green, was a poor substitute for wheat.[179] But if Lapland was not a breadbasket, perhaps it could be a timber producer. In 1732, he had advised expanding the peasant industry of making tar and pitch, in all ways to use the spruce forests along the Baltic coastline. "The woods are filled with large pine trees, completely in vain, because nobody makes houses of them and," he added quixotically, "they are eaten by nobody."[180]

In 1754, Linnaeus also proposed that Lapland's tundra be planted with spruce, Siberian cedar, and "the well-known Cedar of Lebanon."[181] He especially recommended larch trees, claiming that they grew above the treeline, provided ship timber and Venetian turpentine, and attracted sables. These rare Siberian fur animals would now, in their pursuit of larch trees, flock to Lapland "in great numbers."[182]

Linnaeus envisioned Lapland plantations of nutmeg, mace, and cinnamon and especially hoped to grow saffron. (At the time, saffron was used as a dye, spice, and simple. He himself rubbed a tincture of saffron, mercury, and sulphur into the eyes of smallpox victims.)[183] As he saw it, saffron "is an alpine plant and tolerates our climate; it would be quite profitable to plant in Lapland . . . because of the considerable amount of money that leaves [Sweden] every year" for its purchase. These crocus flowers, he predicted, "alone could accrue more interest for the Lapp than all his Reindeer herds."[184]

Linnaeus projected his Lapland spice plantations in typically cameralist ways, modeling them on the tobacco manufactures he had admired in Småland in 1741, where the workers were "many small Children." In Lapland, too, "small children, cripples, and the paralyzed could just as well work on these plantations."[185]

That Linnaeus' peers doubted the feasibility of his Lapland spice farms is subtly signaled by the 1754 index to the Acts of the Swedish Academy of Science. This exhortative list follows Linnaeus in recommending that "Cedar," "Spruce," and "Larch tree" all "ought to be planted on the *fjeld* mountains" of Lapland. But it suggests that "Saffron, a dye plant, ought to be planted on Öland," a sunny Baltic island situated about 950 miles (1,500 kilometers) south of Lapland.[186]

At the same time, Linnaeus' medico-moral fiction of the happy Lapp was intended to persuade citizens to accept a subsistence standard of living (by means of a promise of improved health). If domestic consumers gave up luxuries such as cinnamon and tea, then the state elite's need to produce these goods within the boundaries of the nation would be less pressing. Baltic cedar groves and nutmeg plantations could still generate "a considerable profit for the whole country."[187] These projects could use marginal land, such as Lapland's "wild wastelands" which now existed "completely in vain."[188] They could still employ marginal labor, such as "small children, cripples, and the paralyzed" as well as those Sami "lazy dogs."[189] But if the country's inhabitants rejected the goods of these state-governed ventures (*without* turning to smuggled goods), then they could be exported, instead of being—as this economic philosophy saw it—uselessly consumed at home.

If Linnaeus' economic Lapland reforms had been—or could have been—implemented, they would have destroyed the Sami transhumance, which in this fragile Arctic region was more productive and sustainable than Scandinavian homesteading. Thus Linnaeus plotted to destroy the basis for his admiration of the Sami: their (supposed) preternatural well-being and health. His construct of the "happy Lapp" is an extreme example of how prescriptive Linnaeus' "natural" knowledges really were. Pestering travelers for tobacco and vodka, trading what few cult objects their Lutheran overseers had not robbed them of, the small, sad figures of the Sami inhabit the margins of Linnaeus' economic *and* medical thinking. They map at once its only possibility and supreme irony.[190]

CHAPTER 4

"God's Endless Larder":
Theology

In 1739, Carl Linnaeus delivered the first public lecture at the newly founded Swedish Academy of Science. He chose to discuss the "economy of nature," a topic he addressed by examining "curiosities among insects," and which he put in the form of a thematic sermon on some verses in the Book of Job.[1] Thirty-three years later, upon resigning as rector of Uppsala University, he returned to the topic. In the interim, he wrote many briefer statements, speeches, and treatises on the subject, such as *Oeconomia naturae* (1749), *Politia naturae* (1760), and *Deliciae naturae* (1773). This theme was also the focus of his preface for the Swedish king's vanity catalogue of his natural history collection, *Museum Regis Adolphi Friderici* (1754).[2]

Linnaeus derived his concept of a divine economy of nature from seventeenth-century British natural theologians such as Robert Boyle and John Ray. Since Ray was also a great natural historian and classifier, he was Linnaeus' special hero. In 1748 one of Linnaeus' students was invited to meet the eighty-eight-year-old Sir Hans Sloane. (Sloane told the young Swede he was ninety-four.) The student wrote Linnaeus that Sloane was "now the only one who is still alive, of the friends of the great Ray." To Linnaeus, this fact counted for more than all of Sloane's natural history collections.[3]

What Linnaeus meant in his treatises on the "economy of nature" was that nature has self-regulating properties. To express it in modern terms, his model of nature was that of a feedback-governed equilibrium.[4] This cybernetic model he in turn imported into his notion of the household economy. As he wrote in a 1763 sermon: "If dead animals were to lie around everywhere, stinking, the earth would look dreadful and the air would be fouled by deadly plagues. But, as it is, wild animals, eagles, owls, ravens, and meat-eating flies hurry there and in a couple of days finish the lot."[5]

"In a well-appointed household, nothing superfluous ought to be left over, but all food and everything else ought to be turned to some purpose. This is also the case in the divine household, where the animals eat superfluous things."[6] Traveling through provincial Sweden in 1746, Linnaeus noted that people used churchyard soil for their cabbage patches. Human heads, he mused in his travel diary, thus turn into cabbage heads, which turn into human heads, and so on. "In this way we come to eat our dead, and it is good for us."[7]

In this context, Linnaeus discussed what he called a "war of all against all." "There are some viviparous *flies,* which bring forth 2,000 young. These in a little time would fill the air, and like clouds intercept the rays of the sun, unless they were devoured by birds, spiders, and many other animals."[8] Much later, Charles Darwin found this aspect of Linnaeus' thought fascinating. It even foreshadowed Darwin's own naturalization of humankind. For Linnaeus considered wars as a necessary regulatory mechanism of the divine harmony of nature.[9]

At the same time, by "economics" Linnaeus meant the science of how people can cooperate with, rather than battle, nature. Only by adapting ourselves to our environment, he believed, could humankind make use of nature, since in it everything is so complexly interdependent. In his Uppsala lectures on zoology, Linnaeus told his students that the king of Prussia taxed his crown peasants sixty dead sparrows per year, "so that year by year they may go extinct, but perhaps his land will be much more tortured by grasshoppers which come from Tartary. . . . In this way one sees that nature does not allow anyone to master her."[10]

A poem by one of Linnaeus' students, written in 1762 and lamenting deforestation, asks the modern question, "why should we treat with contempt / and plunder forests and meadows / an inheritance given to us." "I deeply pity those who will be born a hundred years from now," in a world bereft of forests, glades and groves.[11]

In his university lectures, Linnaeus exhorted his students to emulate his own alternative approach to the natural world: "Until now no one has thought about exterminating insects with insects. Most every insect has its lion which persecutes and exterminates it; these predatory insects ought to be tamed and taken care of, so they can purge plants." He especially wished to find the "lion" of bedbugs. "Nowadays they infect almost every house."[12] Using such "lions," farmers could "easily drive away snails from the fields, butterflies from the meadows, ants from the garden."[13]

Yet Linnaeus never questioned humankind's moral right to assert its power over nature. He took it for granted that most human activities improved the environment. In 1749, for example, he endorsed an invention that purportedly would "exterminate all wolves in Sweden within a year."[14] And he preferred bucolic culturescapes to pristine nature. His favorite Swedish province was Skåne, the southern peninsula of the Swedish mainland. He especially admired its southwestern tip, around the little university town of Lund. There, every vestige of the primeval broadleaf forests had vanished and had been replaced by grain fields, shaded by a few coppiced willows and hornbeams. In 1749, as Linnaeus climbed a hill and gazed over this windswept plain, he described it as "a land of Canaan, covered with the most wonderful fields and the most beautiful grains, as far as the eye can see." As he later elaborated, it was "the most wonderful [region] in the world and looked mostly like Flanders, for it was a plain without mountains, hills, stones, rivers, lakes, trees, or bushes."[15]

Despite his cautionary anecdote about the king of Prussia's campaign to exterminate sparrows, Linnaeus could not imagine a fundamental conflict of interest between nature and humankind. Nature remained a servant of humankind, which was "the miracle of nature and the lord of the animals for whom Nature has created all other things."[16] An occasional poem of 1754, written by one of his students, innocently elaborates this view: "It is as if all [plants] have sworn to be faithful to Adam's grandsons, / and to stay by their homes, under their supervision."[17] The poem goes on to describe how Linnaeus worked to domesticate native wild flora.

> When it was noticed that [wild plants] were no less useful than tame
>   plants,
> . . . he took them home with him from rain showers, swamps, and
>   crowds;
> put them in peace and quiet within barriers and walls and fences.
> To serve the children of men in many ways.[18]

Since nature was created for "Adam's grandsons," domesticating the wilderness meant restoring it to an Edenic state. Linnaeus called Uppsala's botanic garden "my Rhodos . . . my Elysium."[19] One of his friends described it as "a little picture of the first Pleasure Garden." It always remained a showcase of exotics. As a fancier of flora, Linnaeus preferred the spectacular and rare—a banana in bloom, say. His "Rhodos" housed numerous subtropical and tropical plants, as well as monkeys, peacocks, and parrots.[20] And it was while he was himself living in this garden (his Uppsala town house lay within it) that he coined this artless definition: "Paradise. Where all kinds of foreign plants are tended."[21]

In Linnaeus' case, "paradise" also meant, at least by 1743 and after spirited battles with the university, a fresh "professor's house," with five rooms and a kitchen, two offices, and plaster ceilings throughout.[22] In 1750, one of his students, Fredrik Hasselquist, writing from Smyrna, promised his teacher a cage of Mediterranean tortoises "to decorate your Paradise; they are rather easy to send home alive, and I shall arrange it; in particular it is pleasant to see how they make love."[23] A few years later, Linnaeus in his domestic "Elysium" could delight in two goldfish. Shipped in 1759 from London to Uppsala, they fed during the sea journey on "sugared biscuits and flies."[24]

Since Linnaeus believed that nature was created for man, "the lord of the animals," he could hardly entertain the possibility that an ecosystem (as we understand it) could suffer the extinction of life-forms by human agency. There are inklings in his writings of such a "tragic world" in which "the earth would be as if extinct" and in which people would be the only remaining zoological presence in a denuded culturescape. But these are more like Linnaean nightmares from which the dreamer always awakes.

Thus, in a 1763 eulogy to animals, Linnaeus experiments with a dystopia of a world without beasts. Yet he concludes by rejoicing in their obvious and continuing presence. Describing what he heard and saw as he stood at the doorstep of his Uppsala town house, he wrote: "If animals did not exist, how tragic the world would be. Now oxen bellow, sheep bray, horses neigh, frogs croak, and birds sing in thousands of ways in the treetops. The cuckoo calls out his coo-coo, the thrush babbles, the nightingale sings at night. Without animals, the earth would be as if extinct, but now dogs hunt hares, flies oxen, falcons pigeons, grebes fish, storks snakes, eagles hens, and everything rapidly moves about."[25]

Like his English precursor, the naturalist John Evelyn, Linnaeus did ponder northern Europe's deforestation. His travel diaries narrate how southern Swedish goatherders battled heather moors and sand dunes, while northern Finnish slash-and-burn corn growers seared lichen moors and birch forests down to the rock face. He even describes a rainy Atlantic countryside that was turning into a man-made desert due to overgrazing, where he walked on sand dunes so large and so fast-moving that the tops of still green and leafy tree canopies protruded from the dune ridges.

Yet he understood even such vast destruction as temporary and reversible blunders.[26] Linnaeus' student Daniel Rolander captured this sentiment when he wrote to his teacher from Surinam in South America in 1755. Insects, he noted, were "the only [creatures] that seem to be able to devastate these extensive forests, which neither humans, fire, nor other things can do. If they do not do it, then I am sure that this huge Hothouse will stand unmutilated as long as the earth itself."[27]

Linnaeus and his students regarded the densely peopled plains of northwestern Europe as most typically reflecting God's intention for nature. But they also declared more primeval regions, inhabited by horticulturalists or hunters-and-gatherers, to be Edenic. In a letter to his student Carl Peter Thunberg, a traveler to Japan, Linnaeus called southern Africa "your Paradise."[28] He described the Orinoco delta's jungle swamps (on the coast of present-day Venezuela) as an "Earthly Paradise, where the Land is like the rarest Botanic Hothouses and overflowing with Plants."[29] Only then did he go on to tell how his favorite student, Pehr Löfling, while working as a botanist for the Spanish crown, perished at the age of twenty-seven from the fevers of this "most wonderful country under the sun."[30]

Even the mildly exotic hills of the Mediterranean excited these Baltic-born naturalists. Another Linnaean student-turned-discoverer, Christopher Tärnström, in 1746 declared Cadiz in southern Spain to be "an Earthly Paradise . . . along the road everywhere there grew spontaneously all kinds of plants rare in Sweden, that hardly are found even in our Orangeries, there were also entire big fields drooping with lemons, the trees being quite yellow covered with these fruits." It was a spring marvel to a scurvied Scandinavian. When the same student reached Java, he declared it a "masterpiece of nature." As he naïvely noted, it "far surpasses the islands" of the Baltic archipelago.[31] Three months later, the tropical fevers of this "masterpiece" killed him.

In 1755 Linnaeus' student Daniel Rolander wrote his teacher one last letter from Surinam, just before he went insane. "I now am allowed to wander in Surinam, an earthly paradise, and see the great god's greatest miracles in Nature." After genuflecting at Linnaeus' altar of natural theology, the young traveler continued in a different vein: "I am pretty sure that even you, Sir, on neither the first nor the second day, would dare to walk into the forests, for you would see big Snakes, Lizards, Insects, and other animals . . . [who] with open mouths and terrible roars throw themselves at the passerby. I don't want to mention the thorny trees and bushes that everywhere catch one, so that one can't escape . . . fallen and rotting fruits spread a stench that can kill." Rolander then hazarded a forbidden thought: "For here not every-thing seems to be made for the sake of man, at least not everything serves him in an immediate way."[32]

Yet this suggestion was taboo, for it undermined the founding assumption of Linnaean cameralism. It also rendered meaningless Rolander's stay in what he—now at the edge of genuine madness—perceived to be a deadly jungle. The insanity to which the young student succumbed as he wrote these lines reflects this dilemma of Candidean cameralism in the heart of darkness. Indeed, it took what could be called a classic Linnaean form. Im-poverished, in broken health, and without any "oeconomic" finds to show his teacher on his return, Rolander began obsessively to fantasize about how he had found a Surinam bush bearing real pearls, and how these pearls were the elixir of life.[33]

Another aspect of Linnaeus' natural history, which was also at odds with his sunny prelapsarian view of nature, was his belief that humankind was a spe-cies of monkey. In his famous letter of 1747 to a German explorer of Russia, Johann Georg Gmelin, Linnaeus asked whether he should "call man ape or vice versa."[34] Seven years later, he added: "even to this day scientists search in vain for any distinguishing mark by which the Apes can be separated from humans."[35] He himself could only suggest canine teeth.[36]

Linnaeus classified humans in that new zoological group he himself estab-lished, as primates (and as relatives of the sloth). True, he designated hu-mankind as "wise man" (*Homo sapiens*). But this was an arbitrary, trivial name, and not a definition. In any case, he hesitated between that binomial and "day man" (*Homo diurnus*). In manuscripts of *Systema naturae,* he crossed out at times one, at times the other.

"Day man" had a counterpart, the "night man" (*Homo nocturnus*). Follow-ing the Elder Pliny, Linnaeus also termed this second human species "troglo-dyte man" (*Homo troglodytes*). Relying on second-hand reports, he identified albino Africans as such "troglodytes." (In eighteenth-century Europe these unfortunates, who were diaspora slaves, were exhibited by their owners as freaks. Voltaire wrote an essay on one of them, a boy put on display at the Hôtel de Bretagne in Paris in 1744.)[37]

In his Uppsala zoology lectures, Linnaeus also told his students that exotic apes shaded into humans, playing chess and worshipping God, and that ex-otic humans shaded into apes, growing tails and crawling on all fours. As a counterpart to *Homo sapiens,* he described a *Simia sapiens* that played backgammon. He argued, too, that in their societies apes execute criminals, wage "formal wars," and sing in choirs.[38] On a more personal level, Linnaeus cherished his guenon monkey, Diana. He recorded her first two "words"— "grech" and "hoi." He also spoke of her "mild eyes" and her endearing passion for almonds and raisins. Even in his scientific description of her he kept her proper name "Diana," and hence her claim to personhood.[39]

As Gunnar Broberg has shown in his important work on Linnaeus' anthropology, Linnaeus' classification of man was original in the realm of scientific thinking and among contemporary scholars. It might, however, have reflected lay ideas in his circle. Broberg's research has unearthed a discussion in 1735 between one of Linnaeus' sponsors, a county governor, and one of his friends, an Uppsala student. (That night, the boy recorded the whole debate in his diary.) The student argued over dinner that "apes are also a race of humans, rather intelligent, like certain people. . . . If they had been brought up as humans to begin with, there wouldn't have been any difference between these animals" and humans. The governor agreed. After telling a long and involved story about a family he knew in which all had tails, he concluded triumphantly: "Also there are people in North America who are furry all over their body with hair. What difference is there between them and apes?"[40]

On the one hand, then, Linnaeus believed that generic man is the master of the cosmos. Nature's complexity, variety, and interdependence proved that God created earth as a "theater"[41] for people to contemplate. As he summarized the religious task of humanity, "you consider the world so that from this work you shall come to know the almighty Creator, the all-knowing, all-powerful, and eternal God."[42] On the other hand, Linnaeus doubted that humans differed from other animals. He conceptualized both animals and humans as mortal and subordinated components of an immortal and harmonious whole, nature. To return to one of the examples above, he held that nature retains its balance by wars, which cull surplus people.[43]

In his most peculiar manuscript, the spiritual diary *Nemesis divina* (1750s-1765), Linnaeus theologically justified the agonies of true mortals by displacing the Christian concept of hell onto earthly suffering. He insisted that people are mortal: "I conceptualize man as a wax candle. . . . Some people he has made into large candles, some into candle ends. They burn for as long as they remain, and when they have burnt out, God puts others in their place. . . . Just as a candle can't say that the castle [which it illuminates] is made for its sake, so man can't say that the world is made for his sake, but everything is for God's majesty."[44]

Linnaeus assumed that people possess eternal life only in the displaced sense that they have blood descendants. Writing to his mother-in-law, he elaborated his notion of the mortal individual: "If the umbilical cord remained, the children would hang together, like a tapeworm, and then it would be more obvious that they are one. Brother Petrus and the other children are, then, nothing else than Mrs. Mother yourself with Father-in-law in blessed memory."[45] To illustrate his point, he drew a long thin tapeworm in the letter's margin.

At the same time, Linnaeus maintained that beasts had souls. (His love for his monkey Diana was equalled only by his passion for his raccoon Sjubb.) "Theologians legislate that man has a soul, and that animals are only mechanical automatons, but I think they would be better advised that animals have souls, differing only in degree of nobleness."[46] Mostly, however, Linnaeus ascribed immortality solely to an impersonal global entity, which he called "the eternal fire," "the world soul," or "Nature." He understood it as an animated Creation, put in place by a distant sky god he named "the Wisdom" or "the Intention."[47]

Unsurprisingly, the theology faculty at Uppsala University complained that Linnaeus conflated God and nature. Linnaeus was actually not a pantheist. But his colleagues correctly sensed that he was also not a Lutheran (in the technical religious sense). Arguably, he was not a Christian at all. He did obey the law to attend church on Sunday—although he trained his dog to leave during overlong sermons, so that he could go off in ostensible search. But he did not believe that Jesus Christ rose from the dead, that by dying on the cross he took on the sins of humankind, and that thereby we gained eternal life. Central Christian concepts such as grace, mercy, and salvation are absent in his theology.[48]

Linnaeus famously announced in the preface to *Species plantarum* (1753) that humankind "has been put here as a guest, who while enjoying these gifts shall announce the Lord's greatness."[49] But he also regarded humankind as an indigenous part of nature. This naturalized humanity retained certain mediating and declamatory duties vis-à-vis the supernatural. Yet it had no corresponding benefits or privileges—except the rather dubious one of being uniquely aware of its own condition and mortality. In a funeral oration over Linnaeus, his best friend summarized this philosophy: "man, who is so proud, has no other privilege over a clod of earth or an earth worm except that of knowing Nature's Lord from his works, and to praise his endless wisdom."[50]

Late in his life, Linnaeus wrote a fragment of a sermon in free verse, addressing the relation between man and God consequent on this theology.

> We are, then, like . . . candles, with which God
> has
> illuminated
> this, his theater.
> We stand here like the ocean in quick illumination,
> like
> snow crystals
> . . . we glitter
> toward the sun.
> We have the honor to be candles in God's palace
> We thus reflect the creator's shining majesty,
> duplicating the light.
> When we have burnt down, when he doesn't want to keep
> us,
> we are removed.
> [He] lets others be put in our place.
> Thus nature dooms us, contra Theology.[51]

Yet Linnaeus bitterly mourned his children—his "flower buds," "snow crystals," or "candle ends"—when they died. In one *vita*, or aide-mémoire for his funeral orators, he sandwiched between botanical annotations a note on the death of his smallest boy.[52] Using turns of phrases that suggested this long-past night had just occurred, he noted, with the exactitude which mourning can take: "1757. March 7. My little son Johannis, who had just begun to say a few words, caught whooping cough, which turned into a steady fever and thrush, and after eight days' illness this night between 12 and 1 says farewell to this world, who had not yet turned his three years."[53] It was as if Linnaeus hoped that the funeral elegy in the Swedish Academy of Science would commemorate also a little boy who "had just begun to say a few words."

Yet the toddler's death was commonplace. Linnaeus' student Osbeck reared four of his nine children. Carl and Samuel Linnaeus together had seven sons: six died in childhood. Thirteen years before losing Johannis, in 1744, Linnaeus had lost a girl, Sara Lena. She lived fifteen days.[54] These children's sufferings are hinted at in Linnaeus' classification of toddlers quintessentially as beings who suffered from scabies, thrush, intestinal worms, excoriated raw skin, and teething pains. One rule of thumb for marking the passage from toddler to child, he suggested, was that "the lice multiply."[55]

Linnaeus came to fear the coming of children into his family, writing in 1758: "I am a child of misfortune; if I had had a rope and English courage, I would have hung myself long ago. I fear my wife is pregnant again; I am old, gray, emaciated and the house already filled with children; who will feed them?"[56] Linnaeus' vision of his orphaned children was not hyperbolic. In his time and place, even adults could not confidently expect life. Of his sixteen classmates at grammar school, six had died while still at university.[57]

No wonder that in the preface to the tenth edition of *Systema naturae* (1758) Linnaeus warned his reader that "*pathologically,* you are a swollen bubble till you burst, dangling from a single strand of hair in one brief moment of fleeting time." "Nothing," he lamented, "is frailer than human life, nothing so vulnerable to so many diseases, so many troubles, so many dangers. . . . Old age is filled with pain, then the senses darken, the limbs grow numb; sight, hearing, movement, and the teeth—the tools for eating—die before you." He went on to state that "*naturally,* you. . . . are a crying, laughing, singing, speaking, teachable, experimental, wondering animal, but frail, naked." Yet "*theologically,* you are the final goal of creation. You are the almighty's masterpiece, placed on this globe."[58]

What, Linnaeus worried, did these extremes of "pathological," "natural," and "theological" man tell us about God? In an undated autograph fragment, he mocks us—and God—despairingly:

> Oh what kind of marvelous animals are we, for whom everything else in the world is created. We are created out of a foaming drop of lust in a disgusting place. We are born in a canal between shit and piss. We are thrown head first into the world through the most contemptible triumphal gates. We are thrown naked and shaking on the earth, more miserable than any other animal. We grow up in foolishness like apes and guenon monkeys. Our daily task is to prepare from our food disgusting shit and stinking piss. In the end we must become the most stinking corpses.[59]

Throughout his life, Linnaeus wondered why God created us "more miserable than any other animal," when we live in the midst of a nature he considered prelapsarian. Why had not the divine "Oeconomy of Nature" occasioned a perfect societal "Oeconomy"?

This riddle structures all of Linnaeus' economic science. In a pamphlet of 1762 on edible native flora, *Plantae esculentae patriae,* he stated that the human animal has at its disposal "an infinite larder. . . . In short, all inhabitants of the earth, the air, and the sea are at his service." He dedicated that same pamphlet, however, to "those among the countryside Dwellers, who in famine years are forced either to take to unnatural means to survive, or to die entirely because of a severe hunger."[60]

Similarly, Linnaeus began a pamphlet that addresses crop failures, the *Flora oeconomia* of 1749, by grandly stating: "The final goal of Creation is man; this the Bible teaches; this the natural sciences teach. Everything is thus made for the use of man." Yet the pamphlet's index of berries has headings such as "too sour," "unpleasant, are eaten by some people," "sour raw," and "black and unpleasant." A few are bracketed by their consumers: "children ate," "only greedy children eat these berries," "is eaten by the Lapp." A terse "edible" brings together the rest.[61]

At heart, Linnaeus agreed with his insane student in Surinam that God had designed nature for his own enjoyment, and not for people's material benefit. "I once heard him say," a former student remembered in 1820, that if the world was made for man, "God would have made the globe into a cheese and us to worms in the cheese."[62]

If the world was made for humankind, how could Scandinavia's famines be explained? In a letter to the Swedish king, written during the great famine of 1756, Linnaeus mourned "this present time, when so many thousands of poor citizens [*medborgare*] . . . not only must themselves wither away, but also, what is even more gruesome, must listen to their little children's whimpering, suffering and death agonies."[63] In a similar vein he puzzled over human suffering in this best of all possible worlds. "I don't know that any other country in Europe has more access to healthy and nutritious plants than our dear Fatherland, so it seems strange to me that the common people among us, at least in the summer and when native plants are available, should ever die of hunger."[64] He then invoked a crucial qualifier, natural knowledge, by concluding: "If they knew and could select between healthy and dangerous" plants.[65]

Linnaeus' actual solutions to starvation were less than practical. On a theoretical plane, however, he took comfort in believing that (to put it in modern terms) human suffering is caused by technological underdevelopment, and not by ecological necessity or divine will. To rephrase it in his categories,

our secular economy is underwritten by nature's sacred economy. But in order to husband nature, we must become *oeconomi botanici*.[66] Linnaeus' natural theology thus informed his belief in material progress, at the same time
as it induced the spiritual anguish that made him turn toward that creed of
improvement.

In the preface to the 1753 *Species plantarum,* Linnaeus summarized this
thought loop: "the WORLD is the Almighty's theater . . . we must research
these creations by the Creator, which the Highest has linked to our well-
being in such a way that we shall not need to miss anything of all the good
things we need." This is why "each object ought to be clearly grasped and
clearly named, for if one neglects this, the great amount of things will nec-
essarily overwhelm us and, lacking a common language, all exchange of
knowledge will be in vain."[67]

Linnaeus defined his natural knowledge as a proximate theodicy also because he struggled against superstition. Though he was steeped in south
Swedish folklore, and believed in much of it, he protested, for example,
against the folk belief that water can turn into blood.[68] When, during a provincial journey in 1741, he visited a rural courthouse in his home province
of Småland, he mocked its assembly of magic tools (originally collected as legal evidence by magistrates conducting witchcraft trials). "We blew on the
magic hornpipe without devils appearing, and we also used the milking rod
to no avail."[69]

Ten years later and during another provincial journey, he saw a farmer
use a divining rod to locate a purse of coins he had lost in the grass. He was
proud to report that "I still don't believe the rod."[70] In 1749, after Stockholm
newspapers had reported mermaid sightings, he had admittedly urged the
Swedish Academy of Science to launch a hunt and "catch this animal alive
or preserved in spirits." But he had also warned that it might be "fable and
fantasy."[71] In addition, Linnaeus opposed the Paracelsian doctrine of "signa-
tures" of nature: the notion, that is, that each plant has written upon it instructions for its human use or allegorical meaning. Once, in 1747, when he
came across a medicinal grass in a botanic excursion outside of Uppsala, he
even warned his students: "Some people use this against fevers, but is probably taken from the ancients' signature."[72]

Linnaeus wrote his most moving declaration of faith in his scientific theodicy in 1759, when the royal family visited Uppsala University. Reading his
speech today, one is struck by how linked its sentiments are to the fact that

he spoke from Europe's northernmost university (excepting Åbo), separated from the Arctic tundra only by a drawn-out conifer forest. His talk was infused with that anxiety which comes from sensing, within and without, a proximate wilderness.

One imagines Linnaeus spoke slowly at first. His speech was in Swedish, and French was the language of his courtly audience. He began with a homely simile: "Only the Sciences distinguish Wild people, Barbarians and Hottentots, from us; just as a thorny sour Wild apple is distinguished from a tasty Renette only through cultivation."[73] Warming to his topic, he projected how natural history would reproduce imperial economies within the home country: "Lesser knowledge of science by us still cause . . . Tea, Quinine, and Cochineal to be bought yearly for great sums of money, that could be planted in Europe as easily as ever . . . Rhubarb." Moreover: "Without science our Herrings would still be caught by foreigners, our Mines be mined by foreigners, and our Libraries be weighted down by foreign works."[74]

Probably frightening princess Sofia Albertina (the sister of the future Gustav III was only six years old), Linnaeus went on in a heightened pitch: "Yes, Demons of the forest would hide in every bush. Specters haunt every dark corner. Imps, Gnomes, River spirits, and the others in Lucifer's gang would live among us like gray cats, and Superstition, Witchcraft, Black Magic, swarm around us like Mosquitoes."[75]

The pastor's son exorcised these river and forest spirits, as much a part of Scandinavian rural life as "gray cats" and "Mosquitoes," by citing a biblical image that is usually read as heralding the coming of the Messiah: "The sciences are thus the light that will lead the people who wander in darkness."[76]

CHAPTER 5

"A New World—Pepper, Ginger, Cardamon": Economic Theory

In his *Culina mutata* of 1760, a tract on the changes in European foods over time, Carl Linnaeus described some ingredients from "a new world: pepper, ginger, cardamon, cand. nutmeg, and whatever all the kinds are called." "Not to mention sugar," he added thoughtfully.[1]

Linnaeus (born in 1707) had grown up in an impoverished northern Baltic province, Småland, at a time when the Carolingian empire was collapsing, and its heartlands were swept by famines and plagues. In 1760, ensconced at Uppsala University as Sweden's first ennobled naturalist, he noted with wonder that global trade now supplied European city dwellers with once fabled glories—Asian spices, cloths, drugs, medicines, and manufactures. As *Culina mutata* and similar tracts such as *De potu chocolatae* reveal, however, Linnaeus deeply worried over the natural cornucopia of the "new world" of Asia and the Americas.[2] He even urged Scandinavians to return to the old "Gothic foods," such as acorns, pork, and mead.

To Linnaeus, the word "economy" meant both an eternal natural order and a new human science.[3] This chapter turns from his conception of the economy of nature to his view of the economies of man. In particular, it looks at how, from his assumption of an Edenic nature operative in the present, he derived the more stringent hypothesis that every country possesses all the natural resources necessary for a multifunctional economy. Linnaeus' natural theology underwrote his prophesied Swedish autarky: his assumed prelapsarian nature provided a broad cosmological frame supporting his more narrow cameralist argument that nations prosper in commercial quarantine.

Linnaeus' applied science—a new natural knowledge for the state—was inspired not only by the instrumental utilitarianism general to the early Enlightenment, but also by his adherence to the older economic doctrine of

95

cameralism. At the same time, he framed cameralism less as a fiscal doctrine and more as a recipe for human welfare. For example, in contrast to most Swedish economic writers of the period, Linnaeus did not espouse the classic slogan of seventeenth-century German cameralism, as it was received in Sweden: "a poor people, a rich country" (*fattigt folk landets rikedom*).[4] In this rejection of workhouses and subsistence wages and his pursuit of schemes for general prosperity, Linnaeus' personal Enlightenment, such as it was, rests.

Yet if we address how Linnaeus thought that the human economy actually operated, we must return to seventeenth-century fiscal theory. Indeed, his understanding of German cameralism, while largely second-hand, was entirely orthodox. Like the German cameralists before him, he was suspicious of money. He exclaimed that the Aztecs—who used cacao beans as a currency—were "innocent people. If they didn't have anything to eat, they ate their money."[5]

Like the German cameralists, too, Linnaeus was obsessed by gold. "Does it not make all things into slaves? And where it is missing, is not everything missing?"[6] He loved handling gold coins, and was fond of displaying his hoard to his penniless students. Being innumerate, he counted great sums by one measure only: "a barrel of gold."

As a unit of account, Sweden used copper until 1766. The Hat party (*hattarna*), which formed in the second half of the 1730s at the same time as the rival Cap party (*mössorna*), and ruled Sweden between 1739 and 1765, financed Sweden's wars and its industrial subsidies by borrowing abroad and by printing paper money. The government thus engineered hyper-inflation. By the early 1760s, and despite Sweden's great copper mines, there was no metal money left in the country. People reverted to barter and to local, home-made moneybills.[7] Under such circumstances, Linnaeus' ambitious yet prudent metal measure of value, gold, makes sense.

No wonder, too, that Linnaeus felt that his worst fears about economic mismanagement by the state were confirmed when his first student traveling abroad wrote from Cadiz in 1746 that seventy silver chests were carried aboard his East India ship to pay for Chinese tea.[8] Indeed, Linnaeus "considered no thing more important than to close that gate through which all the silver of Europe disappears."[9] In classic cameralist fashion, he celebrated Sweden's one gold mine, which yielded two pounds of metal a year, but he never understood the importance of her iron industry, which represented seventy-five percent of the value of Sweden's exports.[10]

Linnaeus was a state interventionist, too. Without pondering the matter deeply, he supported tariffs, levies, export bounties, quotas, embargoes, navigation acts, subsidized investment capital, ceilings on wages, cash grants, state-licensed producer monopolies, and cartels. To use modern analytical terms, he supported legislated market imperfections favoring domestic producers over foreign competitors and local consumers.[11] This was so even when such laws affected his science. Thus in 1757, together with an entomologist, Count Charles de Geer of Leufsta, Linnaeus pleaded with the Swedish king to be allowed to import yearly 30 pitchers of *spiritus vini*, or else to distill spirits themselves. The problem was that the monopoly on aquavit production and the consequent ban on home distillation (*hembränning*) threatened to ruin these men as they bottled their zoological specimens.[12]

Eighteenth-century economic thinkers regarded the trade balance as "the infallible centre of interest for politicians."[13] They differed, however, on how to manage it. At times, Linnaeus embraced the English mercantilists' goal of a positive trade balance even if arrived at by means of international trade. Broadly, this was also the Hats' goal. Their 1739 table of custom dues encouraged imports of raw materials and exports of finished goods. In turn, however, this *tulltaxa* was overlayered by import and export licenses. Here more cameralist turns of thoughts came into play. The Hats typically banned several hundred types of goods entering the country. More often, like the German cameralists in their uncompromising moments, Linnaeus felt that states should be autarkies, withdrawing altogether from the commercial bonds tying them to peoples and places not politically subjugated to them. He artlessly elaborated his reasoning on this to the Academy of Science in 1746: "Everything that we buy from abroad is therefore more expensive, since we must fetch it from far away, and pay others who harvest it."[14]

On a philosophical level, too, Linnaeus argued that international trade was superfluous. He even posited a kind of divine geographic distribution of equivalent goods. In his most intricate and ornate program statement of his natural theology, the *Oeconomia naturae* of 1750, he noted that though "foreigners" owned goldfish, lemons, peacocks, and gold, Scandinavians possessed herrings, cloudberries, black grouse, and iron. In the same breath, this Northern naturalist conceded that Southerners had more "green things," as well as "beautiful turkeys" and other useful animals. Yet this was only because God so compensated them for their diseased air, putrid waters, ill health, and "snakes, lions, tigers, crocodiles, etc."[15]

As Linnaeus saw it, this divine replication of functional types of *naturalia* across political boundaries supported him in his goal to maximize national economic self-sufficiency. In the 1740 program of his science, he admitted that "Nature's Master has given each and every country its own special advantages, so that what is lacking in one, can be won in the other." But he immediately added: "a wise Inhabitant, Owner and Economist knows how to use this to his own advantage, so that he and no one else gains that which is lacking."[16]

At other times, Linnaeus miniaturized such comparative advantage into a question of regions within countries. In his Dalarna *resa* of 1732, he aimed to "see how each province has its advantage; see how it can be improved . . . see how one province can be helped by the practice of another."[17] His Uppsala colleague Anders Berch expressed the same thought in a letter to the Academy of Science in 1759. If naturalists directed the domestic economy, he wrote, they could "work out a general method, according to which political values [*politiska värderingar*] could be employed, not only in farming but also in mining, manufactures, trade." In this way, he continued, "each county could have its own main occupation, one grow grains, another keeps cows, butter and cheese, leather, a third takes care of sheep, a fourth rears horses: another grows linen and hemp, another fishes, another works wood, builds barrels, burns tar, saws [timber], and so on."[18]

Linnaeus' cameralism also colored his understanding of tribal people and non-European societies. Generally, he invoked the customs and habits of foreign cultures, as we moderns do, to widen and expand his own culture's sense of the natural and the proper. "All four-footed animals can be eaten," he asserted, "as the Chinese prove with their own Example. The most cruel snakes are Americans' food."[19] As we saw earlier, Linnaeus also described Scandinavia's indigenous nomads as "happy Lapps" or ideal cameralists, whom other Europeans would do well to emulate. Linnaeus' students, too, attempted to pinpoint an ideal cameralist state to a real place and time. Most often, they chose China and Japan as their examples. When Carl Peter Thunberg returned from Japan in 1779, he gave a speech at the Swedish Royal Academy of Science. Thunberg, who was the only European naturalist to have investigated the country since the late seventeenth century, fashioned his report as a panegyric to "a foreign most respected Nation."[20] At the same time, he prefaced the account of what he had learned during his six months in Japan by condemning the practice of traveling. His speech makes clear that he disapproved of all contacts, and especially trade links, between nations.

In 1775 Thunberg had taught the Shogun's physicians and Dutch translators European medicine in his hotel rooms in Jedo (present-day Tokyo) and in Nagasaki. These hotel sessions marked a turning point for Occidental medicine in Japan: they were elaborated by Yoshio Nagaakira in *Komohijiki* (Secrets about the Red-Haired Ones), *Komoryukoyakukata* (On Western Ointments), and *Thunberg koden* (The Oral Traditions Surrounding Thunberg).[21] Thunberg had seen how painstakingly these learned men wrote Dutch, in watercolor and with paintbrushes. He knew how keen they were to get hold of European books, and how they had learnt by heart the few seventeenth-century medical texts they did possess. He had been overwhelmed by their questions, delivered in rudimentary Dutch. (Some translators spoke no Dutch at all. Others had generated Dutch from old books, and spoke a home-made version only tenuously related to the spoken language of Holland.)

Yet Thunberg by no means lamented the isolation in which his Japanese friends found themselves. Rather, he told his listeners that he brought no economical or ethnographical objects to demonstrate, because the Japanese had wisely forbidden such exports. He strewed on the table a few small coins instead, as his only souvenirs and mementos. As Thunberg explained to his audience, the Japanese had long ago expelled the "immeasurably greedy" Portuguese. After they "happily had rooted out, wholly and completely with its roots, this countrywide malignant canker, a wise, sensible and clever Government" forbade Japanese to travel abroad, and Chinese and Dutch merchants to travel within the country.[22]

Thunberg admitted that the Japanese traded copper and raw camphor for raw materials such as sugar, tropical hardwoods, ivory, tin, lead, and tortoise shells. He passed in silence over manufactured imports (which indicate a technological lag), such as eyeglasses, mirrors, pocket watches, and medicines.[23] Instead, he stressed that apart from a few imported trinkets, the Japanese were economically self-sufficient, and thus politically independent. "No Nation in the Indies more guards its freedom than the Japanese, and none is more free from the Europeans' violence, crookery, and force that is so commonly practiced in the Indies."[24]

Thunberg also admired Japanese medicine and politics. He claimed that the Japanese drank no alcohol, used tobacco only sparingly, and hardly knew even the term for coffee. They cared nothing for fashions. As he saw it, all Japanese dressed in a "National dress, with consists in one or several foot-length Nightgowns."[25] This would have interested his Swedish audience in 1779. To improve the Swedish trade balance, and to discourage vanity and social differentiation by rank and degree, Gustav III had legislated a "national costume" made from all-domestic materials. The same idea moved Linnaeus to suggest that Uppsala students wear uniforms, "be they Counts, Barons, Nobility or sons of Landless Peasants."[26] Japan provided a model. Their "Nightgown," Thunberg improbably claimed, "is so commonly worn and so alike for the entire nation, that there are no differences, even from the Emperor, unto the lowliest Fisherman."[27]

Another of the Linnaeans' favored example of a cameralist nation was China. Consider, for example, a 1762 verse lecture, written by a student of Linnaeus, "on the necessity and utility of plantations." As this student saw it,

> The Chinese urges us to follow his example,
> His industry and household sense can not be praised too highly.
> From travel books one sees, with great amazement, how they grow tea
>   and rice and spices. . . .
> And tempt thousands of trading ships to their harbors. . . .
> I can predict with assurance, that it will be less a disgrace
> To learn how to behave from the Chinese and the Heathen,
> Than to pay fines to him, as we do now,
> And give him gold, for clay and toys.[28]

Despite the high moral tone of this Linnaean poem, its message is not merely one of renunciation (of a misplaced pride in one's own customs and ethnicity). It also offers a positive model of a unified economic order, where all hands are kept busy for the common weal. As we saw earlier, Europe had few manufactured goods to offer in return for Asian "clay and toys," or porcelains, spices, silks, and teas. It was to reverse this trade pattern that the Linnaeans held up the ideal Chinese, gleaned from travel accounts.

Historians have interpreted eighteenth-century European contacts with the non-European world as a prelude to the high imperialism of the late nineteenth century. The future of global relations, however, was not thus transparent and preordained to Europeans of the time. In the mid-eighteenth century, Europeans considered the Chinese their technological and economical equals. As the French historian Fernand Braudel has noted, "the gap between the West and the other continents appeared *late in time*."[29]

(Indeed it may be only a historical parenthesis, now vanishing.) As Linnaeans saw it, Europe should cut its links to Asia by modeling itself on Asia.

A student annotation from one of Linnaeus' lectures summarizes that political goal (at least as the student heard it): countries should "investigate what they themselves have and don't need to rely on the foreigner for . . . because this is the ground for [a good] economy."[30] This view also colored Linnaeus' visit of 1746 to the Alingsås textile factory, an enterprise founded with much fanfare in 1724. After twenty years of state subsidies and trade barriers, this family-owned garment industry loomed, cut, and sewed only a few pieces of badly made garments, unable to compete with either smuggled foreign goods or homespun peasant wares. Yet Linnaeus, who was a friend of the owner (a fellow member of the Swedish Academy of Science), saw no problems with the losing enterprise. He gloried in the fact that "our own countrymen"—or "Swedish hands in Sweden"—now produced cloth as good "as ever other nations abroad."[31]

In a speech delivered during Uppsala University's doctoral commencement of 1759, Linnaeus demonstrated how he defended this myopic productionism. After "greet[ing] all family fathers and inhabitants of this academy and city, both from the higher and lower estates," he deduced "the birth of that science, which is called Economics," from his natural theology—his fervent, if troubled, faith that nature still existed in a state of prelapsarian grace, and for the benefit of man.

As he explained to Uppsala's "family fathers," the science of economics derived from the truths of theology. "God has now given man for his needs and comforts everything under the Sun. . . . He can only sustain himself and thrive from their use, if he should take as his goal to use each one according to its correct purpose." The professor concluded triumphantly, if clumsily: "Therefore nothing else could have come about among mankind apart from the birth of that science, which is called Economics."[32]

To Linnaeus, then, "economics" did not mean the study of how most efficiently to allocate scarce resources given infinite demand. He viewed it instead as the discipline of how to husband the natural world. As he put it in 1740: "The science that teaches us to use *Naturalia* through the Elements (4) for our use is called Economics."[33] To Linnaeus, "economics" was a conglomerate of applied forms of natural knowledge. It was a technology, subdividable even into "mineral economics," "vegetable economics," and "animal economics."[34]

At the same time, Linnaeus did not import into his national economics the concepts that he employed in his analysis of the divine economy. While in a general sense his human economics is predicated on the divine economy (and especially his conceptions of natural plenitude and prelapsarian nature), we do not find here the notions of equilibria we encountered in the previous chapter, such as checks-and-balances and feedback loops. He modeled the intermediate-sized unit of analysis, the economy of the nation, on mechanistic notions of force, even as he modeled global nature and individual households (and also the body and its health) around notions of equilibria.

Linnaeus' national economics is a machine. Nature itself existed in a state of harmony. But he denied a similar *harmonia praestabilita* in the economic realm. "Good economics" was not a matter of an initial calibration of balance. Rather, he imagined entirely different economic mechanisms and processes in the human realm. Its governing metaphor, as we may put it, was that of pulleys and levies, and its inherent tendency, again in our language, was reversion to entropy.

Linnaeus' economics was also an optimistic—even a Candidean—enterprise: "The most savage wilderness, where hardly a sparrow can feed itself, can through good economics become the most wonderful land."[35] Yet at every turn, his imaginary economy, his "most wonderful land," is seen to depend on the ongoing intercession of the naturalist civil servant. This also means that Linnaeus himself stands outside that more general reception of his thought when Adam Smith, upon reading the great botanist, imports the cybernetic concepts governing Linnaeus' natural theology into his human economics of the "invisible hand," and Charles Darwin in turn re-imports Linnaeo-Smithian conceptions of the economy and its self-regulatory features into the realm of nature.[36]

Linnaeus was under the sway of a fashion for economics that swept Sweden in the earlier part of the Era of Freedom (1718–1772). His Candidean cameralism, his views of how economics and the economy were related to the natural sciences, were widely shared by his countrymen. In the Great Northern Wars with Russia (1700–1718), Sweden had lost her Baltic colonies—Estonia, Stettin, Bremen, Verdun, Livonia, Ingria, and most of Pomerania and Karelia. Only Finland, Wismar, and Swedish Pomerania were left. Contemplating this defeat, the Swedish state elites determined to continue the imperial policies of their seventeenth-century forefathers by other means. Increased productivity would replace territorial conquest. Aided by naturalists, Sweden's nobility thus hoped to exchange the sword for science, a Baltic empire for a Swedish nation, and military victories for manufacturing ingenuity.[37]

One of Linnaeus' colleagues at Uppsala University, Anders Berch, typically made the need for improving manufactures and agriculture the center of his teaching.[38] The first professor of cameralism in Sweden, Berch modeled his task on the first chairs in cameralism at Halle (1727), Frankfurt-am-Oder (1729), and Rinteln in Hessen (1730). In his teaching he merged law, physics, and natural history, in order that they be combined and applied to economics.[39] As Berch saw that task, it was a practical and concrete one. (His students complained bitterly about their "Dung-Exams.")[40] His 1747 textbook in economics—which upheld the dogma of seventeenth-century German cameralism—was used at Swedish universities until 1829. For over fifty years, it remained the only introduction to economics written in Swedish. Generally, too, the works of economics that were translated were German cameralist tracts. Adam Smith's *Wealth of Nations* of 1776 was first published in Swedish in 1800, and then only in the form of a translation of a short German summary.[41]

Another acquaintance of Linnaeus', the engineer and mathematician Christopher Polhem, argued upon the founding of the Swedish Academy of Science in 1739 that "the economy is the key goal" of the new institution, and that the academicians ought to work "as tireless servant maids in all areas that can help improve the economy." The Academy's secretary, the statistician Pehr Wargentin, more grandly claimed that its founding had inaugurated a new era, in which economics was "like a sea into which all rivers ought to flow."[42]

In his first sustained formulation of his view of economics, published in the first volume of the Academy's Acts in 1740, Linnaeus similarly stated: "No science in the world is more elevated, more necessary, and more useful than Economics, since all people's material well-being is based on it . . . thus, also the means of Physics and Natural sciences, without which no Economics can survive."[43] In a sermon of 1763, he elaborated: "nature's economy shall be the base for our own, for it is immutable, but ours is secondary. An economist without knowledge of nature is therefore like a physicist without knowledge of mathematics."[44]

Unsurprisingly, Linnaeus' students came to consider a knowledge of economics a useful credential. It was typical that, when Daniel Solander was about to embark on his foreign journeys in 1759, he asked Linnaeus for a "certificate about his progress in those parts of Natural History which lay the grounds for economics."[45] Concerning another student, Eric Gustaf Lidbeck, Linnaeus approvingly noted that Lidbeck understood his task to be to "apply natural history to economics."[46] Yet another Linnaean student, Pehr Kalm, acknowledged to his teacher "that Natural History is the base for all Economics, Commerce, Manufacture . . . because to want to progress far in Economics without mature or sufficient insight in Natural History is to want to act a dancing master with only one leg."[47]

In the Uppsala University lectures Linnaeus gave after 1741, he underscored that "our own economy is nothing else but knowledge about nature adapted for man's needs."[48] In turn, this reading of economics was linked to the ambitions of the state and its tax-collecting organs. For example, the estates' committee on commerce and trade financed Linnaeus' 1741 voyage to look for herbs in the Baltic archipelago, so that the estates could continue to safeguard a healthy populace, while "restrict[ing] apothecaries' freedom from custom duties."[49]

Perhaps the farthest-reaching political reform suggested in Sweden during the Era of Freedom was a 1738 proposal to the estates made by the mathematician Anders Gabriel Duhre. He envisioned an Economic Society, or state department, that would be run by naturalists and would own all means of production—including manufacture, agriculture, fisheries, international trade, and shops.[50] Rather vaguely, Duhre suggested that in order to prevent bribery and corruption, the scientist managers should not be allowed to become "too" rich.

Less drastic suggestions for reforms that were still formulated upon this conception of state agencies controlling means of production were common in the period.[51] They often originated from supporters of the mercantilist, anti-Russian, and lesser nobility Hat party (*hattarna*). At the same time that modern party politics were emerging in Sweden in the 1730s, the participants in this novel experiment debated in an innovative spirit how to theorize the limits for state intervention. (One sign of this exploration is that Sweden was the first country in Europe to abolish censorship laws.)

Linnaeus never committed himself publicly to either the Cap or the Hat party. As a royalist and old-fashioned moralist, he was suspicious of Sweden's new-fangled party politics and the estates' near-absolutist powers. Such politics contrasted starkly with the royal absolutism under which he had grown up, and in celebration of which he may have received his Christian name, after the Carolingian kings. Nonetheless, he moved in Hat circles, and his public career closely coincided with the Hats' parliamentary rule (1739–1765).

During 1739, when the Hats ousted Count Arvid Horn's cabinet, Linnaeus lived in the Stockholm palace of a leading Hat and future chancellor, Count Carl Gustaf Tessin. As he later remembered, "all the Hats called Linnaeus their chief doctor (jokingly)." That year, his practice "grew incredibly." As "Cleon" and "Seminte," Linnaeus and his young wife also joined a pastoral order in Stockholm which had many Hat members.[52] In the same spirit, he supported industrial ventures like Alingsås. These manufactures were favorite Hat projects (even though some had been started as early as the 1720s). Linnaeus deviated from the Hat party line only in that he was also concerned about agriculture (which employed about seventy-five per-cent of the Swedish work force). Unlike many Hats, he did not worry only about manufacturing (which employed less than a tenth the number of people).[53]

In 1739 Linnaeus was also one of the six founders of the Swedish Acad-emy of Science. The Academy was widely assumed to be a Hat institution. Most early members were Hats. Obvious candidates who belonged to the ri-val Cap party were not elected. One prospective member, an Uppsala profes-sor of mathematics, suspected that its catch phrase, "honest Swedish men," was a code for "Hat supporters," and therefore declined to join.[54]

At its outset, the Swedish Academy of Science narrowly valued utility over curiosity. It favored only those applied sciences which could be har-nessed in a state-building effort. "Here only those sciences are dealt with which serve the Fatherland's development."[55] This patriotic cause also in-volved discarding ranks and degrees. The Academy emphasized the equality of a shared Swedish nationality among its members (while Sweden's territo-ries encompassed several nationalities, most importantly Finns, the political nation during the Era of Freedom was largely ethnically Swedish). As one founder emphasized, at the Academy's meetings "a Chancellor will not con-sider himself too good to sit down next to an Artisan."[56]

This radicalism had precedents in the political movements of the 1730s, which challenged the Swedish state to provide a social order in which status would derive from state positions, not from family lineages. In that decade, the clergy, the parsons, the burghers, and the lower nobility (*ämbets-mannaadeln*), a permeable class of civil servants, seized the political powers previously held by the high nobility and the Wasa kings. But the academi-cians were even more radical levelers. Because of their egalitarian point of view, Linnaeus could become their first president, even though in 1739 he was not yet a professor and was not yet ennobled. For the founders decided to award this honor by lottery, "since the Gentlemen had determined that no positions in this Academy should be distributed according to the dignity of position and profession."[57]

The founders of the Swedish Academy of Science also planned to publish their Acts (f. 1739–) "only on economic and practical matters, and this in the mother tongue."[58] Partly thanks to their new emphasis on the vernacular as a language of science, the Swedish language rapidly modernized during the 1740s into early new Swedish (*yngre nysvenska*). At the same time, Swedish printing was shifting from the Gothic to the Antiqua typeface. The type in which the Acts of the Academy of Science should be set was heatedly debated among the founders. Reflecting a typical cameralist worry over trade balances, as well as the prickly self-esteem of a small nation, the supporters of Antiqua type argued that because of Gothic script, foreigners "hold us to be an ignorant people, rarely read our books, hold our language in contempt, and thus ensure that we yearly have an incredibly negative balance [of payments] in the book trade with foreigners." The Gothicists responded, however, that "a large part of our women and ordinary farmers"[59] only read Gothic script—a valid objection, given the Academy's popularizing goals.

In the beginning, the founders of the Swedish Academy of Science called it the "Economic Society of Science." In their first protocoled meeting, chaired by Linnaeus, they changed the name to the grander-sounding "Academy." But they did so stressing that "for goodness' sake, no one will be admitted as a Member who does not love useful sciences and also does not have some insight into some part of them." Having settled on the name, "they also agreed that in order for the Public to understand more easily this institution's and its chosen name's *actual* intention, the title of its to-be-published Acts shall be as follows: 'The Academy of Sciences in Stockholm's Acts, containing new remarks, inventions, discoveries, and experiments, which will serve the growth and development of useful Sciences, Economy, Trade, Manufactures, and several publicly necessary Arts and Artisanal trades.'"[60]

After his call to a chair at Uppsala University in 1741, Linnaeus began to plan how to reform Baltic universities, too. In 1756, his student Pehr Kalm wrote to him from Philadelphia to describe the founding of the University of Pennsylvania: "Nat Hist and useful sciences are hardly mentioned; Latin, Greek, Logic, Rhetoric, etc., get first place, and those professors the highest salaries: never has the English nation embarrassed itself as much as in this . . . I think your stomach would ache, Sir, if you read" the university bylaws.[61]

It might indeed have ached. In 1746, Linnaeus had helped to compose a "pro memoria," submitted to the diet of the national estates by the chancellor of justice, which urged that all university students be compelled to study natural history—including the care of Spanish sheep and silk worms.[62] Theology students, he wrote, must be required to take a degree in medicine before they were admitted into the Church. "This whole science the students could easily learn . . . in eight days at the most."[63]

Two years earlier, Linnaeus had pleaded with the estates that university degrees only be granted after students passed exams in botany, "and especially, all its uses in *oeconomicis*." Taking the fir-tree (*tall*) as an example, he outlined the kinds of things a qualified graduate should know about it: how to harvest resin, how to produce rosin, pitch, tar, charcoal, firewood, and timber, how to bake bark bread, and how to use saps and shoots to cure scurvy. He closed his massive list by an off-hand remark. "And in the like manner with all other plants."[64]

For Linnaeus regarded the clergy as crucial mediators of his science. "The Gentlemen Graduates become most all of them Parsons, spread over the entire country, mostly in the Countryside. . . . The common Man's inclination and money don't allow him to do experiments; but [he] copies everything that he sees in his Church that his Parson succeeds with."[65]

When Anders Berch arrived at Uppsala University in 1741 to take up his position as Sweden's first professor in cameralism, he founded an "economic-mechanical theater"—at once a library, a collection of production samples, and an array of models of agricultural and manufacturing tools.[66] It complemented Linnaeus' collections of *naturalia*. The natural and the artificial were thus both represented at the little university, with its encyclopedic effort to catalogue and display all that people might grow or manufacture.

On Linnaeus' instigation, additional professorships were endowed in what he termed "practical economics, based on natural science."[67] Except for Greifswald University, every Swedish university received such a chair: Åbo University in 1747, Uppsala University in 1759, and Lund University in 1760. In their teaching, these chairs typically combined cameralist theory with the technologies of mining, manufacture, and agriculture. As Linnaeus defined the duties of the holders of these chairs, they should "apply Natural history to private economics" by giving courses "in the mother tongue, . . . [in the] first year Earth and Minerals, care of Fields etc., second year Plants and their uses, Plantations, Dye Plants, Hedges, Forests, etc., third year the Animals: hunting, bird catching, fishing, silkworms, etc., and thus within three years all parts of economics."[68]

Already in 1758, a leading German cameralist claimed that Sweden taught economics better than the polities of his homeland.[69] In 1780, Pehr Kalm's funeral orator noted that the Era of Freedom was "the period, when Sweden's ancient pleasure in wars and battles turned instead . . . to peaceful achievements." "Economics was encouraged, and Sweden had the honor of being the first to transform it into a proper science, and graft it onto the Academic disciplines."[70] Yet the Linnaeans' rage for utility contained a fair dose of Romanticism. The Uppsala professor in "practical economics" was to live on an experimental farm, and his first three years were to be spent on voyages of discovery.[71] Because of the contingencies of eighteenth-century academic patronage, however, little came of this ambitious project. The Uppsala chair, for instance, was donated by an acquaintance of Linnaeus' and the owner of an iron works, Eric Ericsson Borgström. According to academic custom, Borgström could appoint its first holder. He settled on an obscure man, a certain Johan Andersson Låstbom, after having restricted the search to candidates born in Värmland, his own forested and sparsely settled home province.

Låstbom soon resigned and became a parson (1771), then a professor of the Uppsala theological faculty, and finally a dean of Uppsala cathedral (1790). Linnaeus had greater influence, however, on candidates for the Åbo chair of 1747. This position was given to one of his favorite students, Pehr Kalm. The second short-listed candidate was Pehr Adrian Gadd, also one of Linnaeus' followers. Like Kalm, Gadd specialized in botanic acclimatization experiments. Similarly, in 1760, the first holder of the Lund chair was Eric Gustaf Lidbeck, also one of Linnaeus' pupils and a keen experimenter with floral transplants.

In the field of economic science Linnaeus always favored those of his students who specialized in transmutationist botany, a science that assumed that nature was so malleable that by means of floral transplants naturalists could assure independent yet complete state economies. For he believed that in order to accommodate the political fact that nations prosper best in a state of self-sufficiency, God had so created the natural world that each principality duplicated in miniature the world economy. Nature provided all the ingredients necessary for a complex and complete economy within each geographic area constituting an independent commonwealth.

Linnaeus was also profoundly troubled by human suffering, by "little children's whimpering, suffering and death agonies" during famines, say. This is also why he coupled nature (*natura*) and nation (*patria* or *Faderland*). By using his natural knowledge to alleviate human suffering in nature, and by extension, in the cameralist state, he hoped to provide a material theodicy.

Through this intertwined understanding of nature and nation, Linnaeus questioned not only the mercantile imperial impulse, but also, in effect, what economists since Adam Smith have regarded as the engine of economic growth: international trade and its concomitant global specialization of labor (which we ground in nations' observably diverse natural and human resources).[72] Linnaeus' botanic transmutation provided an alternate reading of international economics. He suggested, for example, that growing tea in Sweden was equivalent to a war victory. "Imagine then what great provinces are not added through this to our land."[73]

Linnaeus held to this understanding of economics until his death. When as an old man he summarized his achievements, he only remembered his economic work. In 1775, on his election to the Royal Patriotic Society, he composed a "merit list": "to apply nature to economics and vice versa." Here he listed his travel journals and his cameralist pamphlets on topics such as "medicinal herbs that grow wild and that could be grown within the nation," "native plants that can be used for dye factories," and "plants that serve as food in times of famine." At the end of this list, in that shorthand that senility brings, Linnaeus appended his last words on the achievements of his science: "first produced rhubarb and 600." To fill in the blanks between memory, thought, and hand: "I first procured rhubarb in Sweden and 600 other plants."[74]

Yet Linnaeus' native ideas of "improvements," his notion of the technological, economic, and social benefits to be captured within his localized realm, or his *fädernesland,* were narrowly circumscribed. Obviously, he could not imagine chemical and electrical industries, precision engineering, or inanimate and nonrenewable energy sources. He had few notions of mechanized manufacturing of any kind, except for some mining machinery. He did not understand Newtonian sciences such as astronomy, mechanics, and mathematics. Mathematical analyses of inductions based on systematic experiments were foreign to him. Instead, he turned numbers themselves into mystical principles. "Nature is balanced between opposites and always divides itself into quintuples." At other times he suggested that nature is organized around the number seven, or twelve. Invoking the rhetoric of a new and empirical science, he noted about his numerology: "with examples all this was proved. What can be more powerful?"[75]

Linnaeus' antiquated natural philosophy blended biblical creationism and Empedocles' and Aristotle's two thousand-year-old cosmology, with its four elements of water, earth, fire, and air.[76] It was a philosophy largely in place by the mid-1730s, and one Linnaeus would live to see outmoded. By the 1770s, students mocked the old professor, the Order of the Polar Star dangling from his soiled coat, repeating antiquity's zoological commonplaces about such things as swallows wintering at the bottoms of lakes.[77]

Even Linnaeus' living spaces resembled a Renaissance *Wunderkammer*. As he saw it, the dwelling reflected nature's harmony, which in turn was analogous to the orderings of his study. "The earth is then nothing else but a museum of the all-wise Creator's masterpieces, divided into three chambers."[78] From his student days on, he arranged around himself a microcosm of that world museum. "You ought to have seen, my reader, his museum, which was available to all his auditors, and you would have been overtaken by admiration for, a sense of well-being in, yes love for this home of his."[79]

In Linnaeus' house parrots and monkeys played among stuffed animals, potted plants, insect specimens, mineral samples, scientific instruments, and herbaria sheets. The walls of his rooms disappeared behind tangled branches—some thirty species of songbirds nested in them. Using fish-glue, Linnaeus pasted botanic prints as wallpaper. He also hung on the walls framed portrait engravings of botanists, sheets of paper with handwritten botanic annotations, and pressed plants (they looked like silhouette portraits). Shells and conches dangled from iron nails. Next to family portraits and plaster medallions of royalty, he arranged likenesses of guenon monkeys, a sketch of his tame raccoon, a drawing of a whale captured off Norway in 1719, and a porcelain and plaster double medallion of Solander and Banks, marked Wedgwood & Bentley.[80]

Over doorways Linnaeus pencilled Latin mottoes.[81] And on top of cabinets, he balanced pieces of china decorated with his own heraldic flower, *Linnaea borealis*. He added Chinese shell arrangements and Spanish cork statuettes of a type sold to sailors, depicting Africans covered by artfully arranged mussel-shells.[82] Over the sanded, broad-planked floors, he strewed his botanic manuscripts, which blinded nightingales splattered with droppings while raccoons played and clawed among them. He clad the ceilings in birdskins and hung his Lapp costume on the wall "together with other curiosities."[83]

In this emporium of art and organic nature, the materials and methods of what would come to be termed the 'hard' sciences found no place. Nor did Linnaeus interest himself in the instruments that excited so many in his generation, such as diving-bells, steam-engines, air pumps, telescopes or even, though he used them, microscopes.

When Linnaeus lectured in botany to the Swedish House of Nobility in 1739, he promised the listener "more profit in these [lectures] than [if] he learnt all logic, metaphysics, History, Poetry, rhetoric, Greek, Hebrew in the world."[84] He did not, however, compare his efforts to earlier lectures on natural philosophy given in the House, and especially not to those on "the new science" which Mårten Triewald had delivered there in 1728 and 1729. In these popular talks, Triewald had explained Newton's mechanics. He showed how it informed the machines he was then constructing: Sweden's first steam engine, and Sweden's first diving-bell. Triewald had also demonstrated his instrument collection, including an air pump, and performed chemical experiments.[85]

When Linnaeus spoke of economics as a conglomerate and all-encompassing technology, his vision of the potential of this technology was confined to the betterment of flora and fauna. And when he lobbied for educational reforms, he was concerned only with his amalgamation of economics and natural history. In many ways he was a typical Enlightenment improver. But he ignored the power of the physical sciences to improve society, never reflecting, for example, on the progress made during his life-time in ferrous metallurgy and hydrodynamics.

In the world of Linnaeus forces were animate (such as horsepower and human muscle) or renewable (such as wind and water). His shoes were stuffed with grass, his pillow filled with hay, and his clothes were woven from sheep wool and dyed with herbs. He rode horses, wrote with goose feathers, and read by the light of ox-tallow candles.

In turn, his understanding of economic "improvement" was confined to a qualitative elaboration of this living world which he inhabited. He wanted to perfect, not to break, what he saw as a God-ordained link between nature and man. In his projected future, shoes would be stuffed with cotton grass, pillows filled with eiderdown, and cloth woven from buffalo wool and dyed with tropical insects. He hoped to ride elks, write with swan feathers, and read by the light of seal-fat lamps.

Linnaeus simply could not envision economic growth. At times even his philosophizing about projects of improvement appears as a form of elegiac contemplation. For example, in his first cameralist program-statement, "Thoughts about the Foundation of the Economy" (1740), he began by pointing to the enormous potential of his knowledge of nature for generating wealth. Towards the end of the treatise, however, he retreated to the modest hope that he might alleviate Scandinavia's recurrent famines and epidemics by teaching his theology students about bark breads and herbal medicines.

If the rural clergy knew some botany, "the Farmer could be taught which [wild plants] can serve as bread during scarcity times; also he could more easily find during times of illness House Medicines growing by him." Linnaeus closed the essay yet more modestly. "But I wish for too much; for however small this matter may seem, there still does not exist a Polity in the world which has enjoyed this benefit"[86] of saving its people from plague and starvation.

CHAPTER 6

"Should Coconuts Chance to Come into My Hands": Acclimatization Experiments

Apart from his travels in the Scandinavian provinces, Carl Linnaeus only once embarked on an extended journey, at the time of his youthful sojourn in Holland, France, and England between 1735 and 1738. He passed up offers to explore the Cape, Pennsylvania, and Surinam. Starting in 1741, when he became a professor at Uppsala University, he chose instead to sponsor long-distance travels undertaken by his pupils—whom he called his "apostles" or "disciples."[1]

Between 1745 and 1792, nineteen first-generation students of Linnaeus' left on far-flung voyages of discovery that he helped arrange. With typical hyperbole, Linnaeus spoke of the voyage of discovery as a Swedish invention that was later copied by Joseph Banks (Australia and the Pacific), Johann Georg Gmelin (Siberia), Michel Adanson (Senegal), and José Celestino Mutis (South America).[2] Yet Linnaeus had reasons for boasting. Daniel Solander was Joseph Bank's botanist on Captain Cook's first circumnavigation of the globe (1768–1771). Anders Sparrman was the botanist for Johann Reinhold and Georg Forster on Captain Cook's second circumnavigation (1772–1775). Carl Peter Thunberg, working as a ship's surgeon in the Dutch East India Company, was the first Western naturalist in a century to visit and study Japan (1770–1779). Pehr Kalm, financed by the Swedish government, criss-crossed northeast America (1748–1751). Pehr Löfling, employed by the Spanish crown to revive natural history in Madrid, explored parts of Spain and Spanish South America (1751–1756). Pehr Forsskål took part in a Danish royal expedition through the Ottoman Empire and the Arabian peninsula (1761–1763). Johan Petter Falck, as part of the Russian imperial Orenburg expedition, explored the Caucasus, Kazan, and Western Siberia (1768–1774). Other, now obscure, Linnaean voyagers took the well-worn Guangzhou route, abandoned their science during their treks, returned insane or mortally ill, or died early on in their travels.

The Linnaean voyagers belonged to one and the same school of knowledge and were involved in what they understood as a single enterprise.[3] Yet historians have discussed their shared purpose only vaguely, as an expression of an age of heroism[4] or as an example of "a global data collection."[5] As I will argue, however, their travels were part of their larger strategy to create a miniature mercantile empire within a European state.

In 1746, appealing to the Swedish Academy of Science to fund Kalm's North American journey, Linnaeus explained how explorers fostered strategies of national improvement based on ecological diversification rather than on territorial expansion. "If Oaks did not grow in Sweden, and some mortal wanted to get Oaks into [the country], and they then grew here as they do today, wouldn't he serve the country more than if with the sacrifice of many thousands of people he had added a Province to Sweden?"[6]

Using arguments of this kind, Linnaeus solicited money or other aid (especially free passage) from various public and semi-public institutions, such as the Swedish Levant Company; the Greenland Company; the East India Company; the Bureau of Manufactures; the universities of Uppsala, Åbo, and Lund; the Academy of Science; the Uppsala Science Society; the diet of the national estates; provincial governments; and the cabinet and the court. He also courted private patrons. Some belonged to the high nobility, such as the Swedish chancellor Carl Gustaf Tessin, the entomologist Charles de Geer, and Sweden's reforming landlord *par excellence,* Sten Carl Bielke. Others were of less exalted heritage, such as the Surinam planter Carl Gustaf Dahlbeck, the Swedish court physician Abraham Bäck, and the East India Company director Magnus Lagerström.

Linnaeus chose his travelers carefully, too. Candidates were easy to find, for voyagers were respected and rewarded in mid-eighteenth-century Sweden. The Linnaean chaplains who had served on Swedish East India Company ships were preferred for parsonages, and explorer-economists were favored for academic posts. As Linnaeus wrote in 1750 to Fredrik Hasselquist, his student and a Levant explorer, as the young man lay alone and dying in Smyrna: "Our young graduates from medical school jump when they hear your letters, they cry and shout: 'Help us abroad so we too can gather laurels.' Löfling sheds tears every day because he can't leave for abroad."[7]

These students took their cue from their teacher (and the "crying" Löfling soon left for Spanish South America, where he promptly perished). In 1754, when Linnaeus heard that a Dutch merchant marine officer had offered to pay for a student's voyage to Surinam and had even invited him to stay at his sugar plantation, he displayed his usual mix of exhilaration and anxiety: "Dear, tell me, is it serious . . . is it sure; shall I believe it is serious? May I select a student? When should he be ready? Tell me, answer me."[8]

Tabulating the traveling naturalist's ideal attributes, Linnaeus registered first "that he be a native *Swede*, so that foreigners can't take what others have paid for."[9] He cast his students as his hungrier, leaner self. They must be young, penniless bachelors, sleeping as well "on the hardest bench as on the softest bed, but to find a little plant or moss the longest road wouldn't be too long."[10] It was in this spirit that he urged the Academy of Science to use Pehr Kalm and use him now: "Now is the time, another time he will be heavy-footed, lazy, and comfortable, and too fat to run like a hunting dog in the forests."[11]

The professor also encouraged his students to prepare for their longer voyages by shorter domestic trips.[12] These briefer voyages, he held, were valuable in themselves, too. "Good God! how many, ignorant of their own country, run eagerly into foreign regions."[13] Indeed, he first lectured at Uppsala University in 1741 "on the utility of scientific journeys within the fatherland," and he defined that "utility" as the domestication of indigenous wild flora into cash crops that would displace imports.[14]

Linnaeus undertook five such regional explorations, through Lapland (1732), Dalarna (1734), Öland and Gotland (1741), Västergötland (1746), and Skåne (1749). His 1732 Lapland report, delivered to the Uppsala Science Society, established the Linnaean sequence of reportage: first minerals, then plants, animals, and local technologies, and finally ethnography. The structures of the 1732 Lapland voyage (spelled out in Linnaeus' 1741 inaugural address in Uppsala, and in the "laws and statutes" of the Dalarna and the Öland and Gotland explorations) governed later Linnaean travels. At their center lay an earnest validation of work. "The voyage should not be frittered away with gossip, chats, songs, fairy tales, jokes, playing, and vanities."[15]

Linnaeus' students imitated their teacher by giving primacy to "oeconomy" in their domestic journeys and observations. The China traveler Pehr Osbeck presented his Atlantic travels of 1749 in an "oeconomical speech" entitled "The Necessity and Utility of Natural History, Particularly near Beaches." (Coastal Swedes, he argued, should eat seals.)[16] Pehr Kalm drew on his economic journeys through western Sweden (1742), the mid-Baltic archipelago (1743), Finland, Ingria, Carelia, northwest Russia (1744), and Västergötland (1745) in planning his doctoral dissertation on "which of our

domestic plants can be used instead of bread, and for porridge and gruel."
Kalm abandoned this project when he discovered he was unable to finish it
in three days. Then he turned to another "oeconomic" subject, also inspired
by his "own observations," namely: "how a Farmer ought to follow the example of the Chinese . . . and [use] the smallest pieces of land on his farm."[17]

Kalm planned to devote eight precious pages to show how the slash-and-
burn corn farmers scattered across Scandinavia's vast taiga ought to emulate
the rice growers of the crowded Yangtze plains. This, too, he failed to write.
Nonetheless, the Swedish cabinet considered Kalm's work sufficiently important to appoint him professor in economics and natural history at Åbo
University in 1747, at age thirty-one, and without a doctoral degree.

When his students left for abroad, Linnaeus issued them with "orders,"
"instructions," and "memoranda."[18] These instructions are reminiscent of
earlier efforts in the genre, such as Francis Bacon's "On Travel" of 1625.
They were not intended, however, to inspire all travelers. They were indi-
vidualized commandments, inspired by the instructions Linnaeus himself
had received as he set out on his provincial journeys.

Before Linnaeus left on his Öland and Gotland journey of 1741, the Manufacturing Bureau requested that he bring back porcelain clay, dye grasses,
and medicinal herbs. "Hitherto it has been usual to import from abroad."[19]
Before he left on his Skåne journey of 1749, the Estates' Commerce Committee instructed him to find gypsum, "which now yearly costs the Realm
almost One barrel of Gold, and also flintstones of the better kind, which pull
out [of the country] c. 12,000 *daler* copper yearly."[20]

Linnaeus' "memoranda" to his students similarly centered on economic
matters. To express it in modern terms, he requested life-forms, material
samples, and production technologies. This was true also for those students
who were on foreign payrolls, such as Daniel Solander, Pehr Löfling, and
Pehr Forsskål. Linnaeus even instructed them to smuggle home the specimens and know-how they came across when in the employ of other nations.[21]

How Linnaeus ranked utility before curiosity in these "memoranda" is
nicely demonstrated in how he applied the term "use" in dispatching Pehr
Osbeck to Guangzhou in 1750. "No instruction is needed for him who understands as much as you, Sir; everything is [to be] observed that isn't [al-
ready] used and found by us."[22]

At other times, Linnaeus' "instructions" were more detailed. In 1745, he
ordered Christopher Tärnström to bring back from China "a Tea bush in a

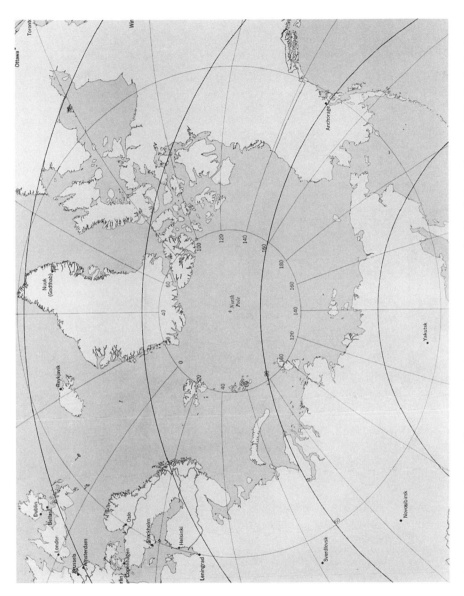

Figure 3. Azimuthal equidistant projection map centered on the North Pole. Source: U.S. Geological Survey, 1990. Courtesy of the Harvard Map Collection, Harvard University.

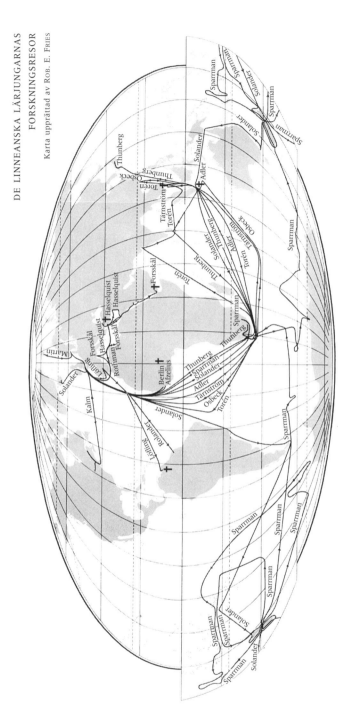

DE LINNEANSKA LÄRJUNGARNAS
FORSKNINGSRESOR

Karta upprättad av ROB. E. FRIES

Figure 4. The voyages of Linnaeus' students, drawn by Gustav Henrik Brusewitz in the second half of the nineteenth century and reproduced by Robert E. Fries. Source: *SLÅ* (1950–51). Courtesy of the Linnémuseet, Uppsala.

pot, or at least seeds," "an untilled piece of the original porcelain soil" (as a sample to compare with Baltic clay deposits), "seeds of the Chinese Mulberry tree," palm fruits, lily bulbs, aloe, myrrh, rubber, and "live Goldfish for Her R Highness."[23] The Academy of Science added a request for a description of Chinese technologies for zinc alloy smelting.[24]

In like fashion the Academy of Science asked Pehr Kalm, as it planned his American journey, to find a North American "kind of Rice, tastes rather pleasant, and is used by one Nation, called Ilinois, as food instead of grains, grows without cultivation there in all marshes and lakes." It also demanded "two species of mulberry trees that grow on hills among birches, firs, and spruces, which we here in the North in the future could have for silk cultivation just as well as in Europe's southern countries." The Academy's final request was for "wild oxen, big-bodied but short-legged; they have a rather fine and long shaggy hair, useful for spinning, but it doesn't mill. The Colonies there have found ways to tame this kind of creature, employ them for everyday haulage and use their milk as food."[25]

Linnaeus' instructions to Kalm echoed those requests, which he had suggested to the Academy in the first place. He, too, stressed mulberry trees. "How much money yearly leaves for silk everyone knows." He also asked for American oaks, medicinal herbs, sassafras, walnuts, chestnuts, hay grasses, and magnolia. As the academicians had done, he mentioned that American "cattle with long wool or shaggy hair suitable for spinning and weaving."[26]

Thus Linnaeus asked Kalm to learn from indigenous Americans about the habitat and harvesting of a wild crop, and from European settlers there how to handle an indigenous mammal. Only at the end of his "memorandum" did he suggest, in a single line, that Kalm make "observations on Birds and Fishes, on Snakes and Insects, on Plants and Trees, on Stones and Minerals." Even at the stage of projection, the Linnaean voyage of discovery was a meeting of cultures and not a contemplation of nature.[27]

That Linnaeus favored economics is exemplified in how he planned Kalm's travels, too. He rejected Kalm's bid for Iceland as a destination, declaring that it was too similar to Lapland. Subtly signaling his doubts about his teacher's theories of acclimatization (discussed below), Kalm himself rejected the equatorial tropics as "more curious than useful for our cold Sweden."[28] Linnaeus also rejected Kalm's second suggestion, southern Africa, with its indisputably rich and unknown flora, but (as he saw it) ignorant and lazy local people, who would not usefully mediate its economic and medical uses. And he was lukewarm toward Kalm's third idea, the Holy Land—although he did admit that it would be easy to raise the money for that trip from Uppsala's theological faculty. Kalm's fourth alternative, North America, he felt had merits. But it paled next to Kalm's fifth suggestion, China.

As Kalm wrote to his patron, Sten Carl Bielke, Linnaeus "became so happy that he simply danced and jumped about in the room, and I think that he must have said ten times that this journey would be far better than the one to the Cape or to the northern part of America." What gladdened Linnaeus as he danced in his chambers was that Kalm planned to enter China by the northern land route, "with the Caravan that every second or third year leaves from Russia to China." Thus, he enthused, "there would be an opportunity to bring from there to Europe seeds of tea which otherwise are impossible to bring." As Kalm wrote his patron, Linnaeus "will teach me the method as surely as $2 \times 2 = 4$ to bring them undamaged here to Sweden, and in such a condition that they later, when seeded, will grow."[29]

Linnaeus "danced and jumped about" in a celebration of sorts. Peering at nature through a physico-theological looking glass, he and students regarded nature as a benign—indeed, prelapsarian—-superorganism, created by God for humankind's use. The only challenge, as they saw it, was to detail nature's history, or what we moderns might term its geography, anthropology, geology, zoology, and botany. This is why Linnaeus could so confidently exhort his students, as they ventured outside Europe, to bring home natural products, human technologies, and even models of society. Some contemporary proponents of international trade, such as the clergyman Anders Chydenius (a brilliant economist), argued that the uneven scattering of natural resources signified God's desire for transborder commerce. But the Linnaeans were sure that the Lord had granted them the potential satisfaction of all their needs and desires within the borders of their own frosty nation.

Since the Baltic climate was the key limiting factor in Linnaeus' efforts to promote national self-sufficiency, his feeling about it was understandably manifold and ambivalent.[30] Following the Gothicist tradition that he had encountered as a tutor in the home of Olof Rudbeck the Younger, Linnaeus at times represented the Scandinavian winter as a kind of stately pleasure palace.[31] As he wrote in a dissertation of 1747, "we can run and ride chariots on our Crystal-ices over all lakes and swamps." The "chalk-white snow shimmers against the daylight, as if it were strewn with the clearest Diamond powder." The aurora borealis "plays each winter night with thousands of kinds of Comedies, the further north, the more."[32]

Rudbeck the Elder's *Atlantica* (1679–1702) and Rudbeck the Younger's *Nora Samolad* (1701) both had asserted that, contrary to popular belief, Lapland's climate was quite mild. Linnaeus more cleverly celebrated the supposed advantages of its Arctic cold.[33] He expressed his wonder at the beauty of the Nordic winter in more spontaneous and less rhetorical ways, too, describing its cold spells in a homey, paratactic prose largely free of figures. "Strong cold. The smoke rises high. Animals become nimble, alert, and greedy. In the air a fine silvery powder hovers. On the birch's branches and twigs little leaves made of crystals settle. Hair and beard are powdered white. Iron sticks to the finger. Horse shit spatters. At night there are bellows and wails from [ice shifting on] the lake. On the windowpanes a new world is magically created."[34]

Linnaeus' *Calendarium flora* of 1757, which presents itself as a farmer's almanac and tracks the mid-Baltic growing season, also shows his sensitivity to local climates. It begins the year in a highly unorthodox manner, at the spring equinox, and it punctuates time by natural events rather than by Christian holidays. "Month II" begins when "the snow first starts melting." "Month IX" continues until "the Swallow drowns itself." Spring has arrived when "the butter comes loose from its tub," "the house corners crack and bang in the night," "swans fly by," and "underground cellars are filled with water." The dramatic end to Linnaeus' floral year suggests that its author grasped that cotton, mulberries, nutmeg, and tea might not thrive in his northern abode: "The rivers freeze over with / thick ice. / The earth is covered with snow. / Snow and soil mingle as they freeze. / The snow grow harder. / It thaws. / The graves are filled with water. / The winter comes with the ice."[35]

Linnaeus understood the rigors of the Nordic winter. He also knew that plants have native habitats. He recognized the existence of distinct and mutually dependent communities of life forms. He knew, too, that these are linked to differing geographic and climatic conditions.[36] If anything, he overemphasized the role of temperatures, and especially of minimum winter temperatures, over other factors of which he was also aware, such as light, salinity, rainfall, and wind conditions. Indeed, the question of climate (defined in this narrow way) preoccupied him. He sensed that upon it turned the success of his cameralism.

Perhaps this is why Linnaeus developed several different hypotheses about plants' relation to climate. To use modern terms, he variously advocated a latitudinal, a global, and an alpine acclimatization theory. At times,

Linnaeus predicated plant geography on two variables: latitude and soil. As he explained to the Academy of Science in 1746, plants "thrive and grow without difficulty at the same height wherever they are planted, if only the soil is similar; just so the plants from the East Indies grow easily at the same height in the West Indies." Illustrating how, in his own mind, he meshed natural history with economics, he then in practice (and practice alone counted to Linnaeus) reduced his two variables to one—latitude. "The soil types are the same mostly over the whole world, so that one has never had to put aside any plantations for their sake."[37]

Here Linnaeus assumed that plants had an inherent temperature tolerance range, in the simplified sense of growing within a latitudinal band. And since the earth exhibited differing plant communities at equal latitudes, botanic exploration of unknown regions remained profitable, in a narrow cameralist sense. Already in 1740, in the *Acts* of the Swedish Academy of Science, Linnaeus listed fruitful transplants such as potatoes and tea. "Who would have thought that Rhubarb . . . would have grown in any other Climates but the Orient, until Botanical experiments showed, that she grows as heartily and strongly in Holland, as there? Who would have thought that Tobacco, that first grew in Florida, would sprout by us in the North?"[38]

Noting that "in north america completely different plants grow than around us,"[39] Linnaeus asked Pehr Kalm to search there for hay grasses "that tolerate our climate" and "many sorts of trees useful in economy, Manufactures, medicine; that could grow here."[40] On his North American travels of 1747–1751, Kalm dutifully pondered latitudes and winter temperatures. In 1748 he reported from Philadelphia that although "New Sweden" (as Kalm and other eighteenth-century Swedes called Pennsylvania, asserting their moral ownership of Sweden's erstwhile Delaware colonies of 1638–1655) lay south of Scandinavia, its winters were just as cold. "It follows therefore, clearly, that those trees and plants which thrive here and tolerate the winters will also do the same for us in Sweden."[41]

In his botanical lectures, Linnaeus divided the world into five climate zones: Australian (Ethiopia to southern Africa); Oriental (Siberia to Syria); Occidental (Canada to Virginia, and also China and Japan); Mediterranean; and Boreal (Lapland to Paris).[42] These climate categories spanned vast areas: indeed, the Oriental zone stretched from the Arctic Sea to the Middle East. They were larger frames within which bio-geographical regions fit. As such, they promised profitable botanic transplants.

At other times during the 1740s Linnaeus further widened the scope of his economic science. He radically hypothesized that all plants were globally adaptable. Even crops such as "Indian fruits" might grow in Scandinavia, "at least after a few Centuries"[43] or "within a Century."[44] He celebrated his homeland's "pleasurable summertime, when the sun hardly leaves us during the nights, but gives a steady day, which causes the birds to abandon the Southern countries and sail to us yearly, here to build and live and breed their young."[45]

Might not also tropical plants, like birds, "sail to us" and our midsummer sun? In his high optimistic mode, Linnaeus believed they could. He even stated that "there is no plant that cannot be cultivated also by us."[46] To quote a verse description of his Uppsala lectures, written by one of his students, he "teaches with good reasons . . . where [plants] have their home, how they are tempted / to move their nest to another place."[47]

Reaching out to the widest possible audience, Linnaeus asserted in the Swedish farmers' almanac for 1746 that plants could migrate from Mediterranean to Boreal zones. "Most all plants that grow in Italy never freeze to death in Skåne, and rarely here in Uppland. Thus I see very clearly that Tea could thrive here."[48] In 1750 his student Fredrik Hasselquist similarly reported from Smyrna on plants that inhabited both Oriental and Boreal climates, or "which I find to be both Anatolia's and Sweden's guests."[49] The same year, Pehr Kalm observed from Philadelphia that "certain trees stop at a certain degree toward the north, and don't go farther by themselves, although if they are moved by human hand farther north they are not damaged by the cold."[50]

One of Kalm's examples was the mulberry, the material basis for the silk industry. Linnaeus also made lists of "Foreign Trees (*Exoticae*) that have been naturalized by us, and are beginning to tolerate the Climate well."[51] His examples included horse chestnut, Swiss maple, and cherry. It shows that many now unremarkable cultivars were noteworthy in Scandinavia in the mid-eighteenth century. His students' inventories of "*Exoticae* that tolerate Uppsala's Climate" also include such modern commonplaces as red currants, larches, and honeysuckle.[52]

Linnaeus and his students spoke of their efforts to "fool," "tempt," "teach," or "tame" tropical plants to grow in Arctic lands. They claimed that plants could "get used to the Swedish Climate," "move their nest to another place," "tolerate our winters," or live "in our Climate."[53] To put it in modern terms, they believed that any given plant species exposed to colder temperature ranges changed into a hardier variety. Linnaeus even envisioned that these new varieties would turn into self-seeding feral tramps, vigorously colonizing their novel home.

From extant sources it is unclear, however, how Linnaeus and his students conceptualized the mechanisms of this botanical acclimatization. In other contexts, Linnaeus had argued that at least some present-day plant species were the result of earlier hybridizations. But he also suspected that flora was fixed and immutable in the present. In later life, he wavered on this question of botanic hybrids, at times even positing them as the key principle of natural diversity.

Yet Linnaeus never named hybridization as a mechanism of climate acclimatization. As we saw earlier, he understood all life-forms to be what we call symbionts, in that he cast the relations between predator and prey, parasite and host organism, as part of a larger natural teleology. Indeed, Linnaeus considered the earth itself (which he conceptualized as a self-regulating superorganism) as the appropriate explanatory level for his natural philosophy. Within this larger, framing natural philosophy, Linnaeus may have understood the physiological mechanisms of his hypothesized adaptationism in semi-Aristotelian, teleological terms, as a shift from potentiality to actuality. Yet his writings never document why, to put it in modern terms, he assumed that plants were eurytropic—that is, inherently able to adapt to variant environments.

Again from the viewpoint of the present, it seems that Linnaeus did not empirically test his adaptation hypothesis (however constructed and arrived at). He was not a selective breeder. While he occasionally listed the contents of his garden, he did not record plant survival rates over time. "Plant experiments" is too grand a term to use for his cultivation of exotic plants. Indeed, Linnaeus seems to have set against "experiment" the higher order of "experience" as a more immediate vehicle for knowledge.[54]

As Linnaeus saw it, the problem was not one of testing the hypothesis of transmutation. Rather, it was to explain why, given the natural truth of this hypothesis, there were great numbers of what we call variant endemic ecosystems. He addressed this question most interestingly in his dissertation *Coloniae plantarum* of 1768. Here he discussed the fact that plants have many natural mechanisms for dispersal. He noted that they can be carried by air, water, animals, and humans; and by means such as trade-route corridors, weed-infested grain imports, and cultivar feralization. He also argued that physical, or abiotic, barriers (such as oceans, deserts, and mountains) sufficiently explained the earth's observable diversity of plant communities—or what in his eyes was a "lack of seeds," a relative floral poverty of each geographic area of the earth.[55]

Putting aside the physiological mechanisms for floral transmutationism, how, as a matter of practice, did Linnaeus and his circle propose to encourage the botanic acclimatization they imagined? Describing how he grew six almond trees, Mårten Triewald argued in 1740 *Acts* of the Academy of Science that the whole secret was to plant in the spring, and not in the fall.[56] Pehr Kalm wrote Linnaeus from London in 1748 to outline a more complicated measure to "fool" plants northward. If English saplings were moved directly to Uppsala, they "freeze during cold winters because they have made too long a jump north in one go." If they were first moved to southern Sweden, "one could fool them, little by little, into getting used to the Swedish Climate; from there they could be moved to Uppsala, and so farther north."[57]

This gradual exposure to colder temperatures, or to longer periods spent in cold temperatures, was a favored Linnaean means of acclimatization. Linnaeus sought to establish an acclimatization garden in a sheltered plot in Ystad in southernmost Skåne, to grow herbal medicines, food crops, and "Lilies and Bulbs of Tulips, Hyacinths".[58] Farther north, in Uppsala botanic garden, commercially valuable cultivars such as cacao, rice, coffee, sugar cane, ginger, pistachios, olive trees, and sago palms were hardened against the cold. Even as they withered in the northern air, their tormentor cried out for more victims. "Should Coconuts chance to come into my hands, it would be as if fried Birds of Paradise flew into my throat when I opened my mouth."[59]

Linnaeus ascribed the failure of cold-sensitive seedlings to grow in mid-Baltic regions to chance or circumstance, or to extraordinary weaknesses resident in the individual specimen rather than in the species as a whole.[60] At any given moment, too, Linnaeus' botanic garden flourished with many tropical specimens, even if most were about to perish. Therefore, in a sort of circular reasoning that ignored the turnover rate of particular specimens, the garden as a whole was presented as living proof that "sensitive foreigners" survived in the "cruel cold" of Scandinavia. As late as 1778, at Linnaeus' memorial service at the Academy of Science, it was claimed that his cultivation of bananas and teas proved how "seeds and plants, originating from under another sky . . . thrived in our cold North under the care of Linnaeus."[61]

On a theoretical level, however, the Linnaean theory of floral acclimatization had already been severely criticized in Pehr Löfling's dissertation, *Gemma arborum,* of 1749. Linnaeus wrote almost all of the 186 dissertations he supervised. This text was unusual in being written by the student.[62] Along with Pehr Forsskål, Löfling was Linnaeus' most brilliant disciple. Perhaps he chose the topic because he thought his teacher's transmutationism was too facile. More likely, he wished to address the problem Linnaeus was interested in just then: the timespan between bud formation and flowering in plants. (Given Linnaeus' style of thinking and his lack of interest in plant anatomy, it is unlikely that he had asked Löfling to investigate the physiological mechanisms of his acclimatization doctrine.) Be that as it may, investigation of those mechanisms became one of the central themes of the thesis. Löfling argued that tropical trees and bushes do not have buds; that buds are necessary to survive European winters; and that, therefore, tropical trees and bushes could never habituate themselves to Scandinavia. To support his hypothesis, he "enumerates several examples" of frost-bitten, ravaged plants "that he has seen in the Academic Garden" of Uppsala.[63]

Immediately upon the presentation of *Gemma arborum,* Linnaeus announced in a Stockholm magazine that Löfling had written a "tragic thesis." "Since the time we saw that so many trees from southern countries over time settled in our Gardens and tolerated the Climate, we have flattered ourselves that the Indian trees themselves could, over time, at least after a few Centuries, be moved to us, so that the inhabitants of the North could enjoy Indian fruit." Now such hopes must be discarded. Löfling was "the first who has discovered that all trees within the [equatorial] line have absolutely no buds that can protect them from cold; therefore they can never be taught to tolerate our winters."[64]

The same year, Linnaeus naïvely rehearsed his student's argument in a melancholy letter to the court physician Abraham Bäck. "All buds are only created to protect from the cold the little leaves and flowers that will grow next year. But think, an astonishing thing, that [on] all trees under the line . . . the little leaves are born completely naked like little dogs; therefore it will never be possible for them to get used to our climate."[65]

As Linnaeus summed it up, the *Gemma arborum* "supports our assumptions very little."[66] He meant to say that Löfling's "tragic thesis" tore apart the theory of botanic acclimatization that underwrote the Linnaeo-camralist project of import substitution. After this sad discovery, he added, he had decided to retire from science. "All obstacles can be conquered [only] by being accepted. I will sail to harbor; and never will anyone hear even a piffle from me, once I have finished the Skåne journey."[67]

Linnaeus did not "sail to harbor," or cease his inquiries into nature. After 1749, he began to ask instead what other factors determine the geographic distribution of plant species, and hence account for what we term ecosystems. At first he considered the fluctuations of winter temperatures. In a letter to the Swedish king, written in November 1749, he discussed a tree the national estates had asked him to acclimatize in Sweden the year before, namely walnut.[68] "The most costly plantations are devastated by harsh winters," he lamented, "roughly every 20th year, according to what the common folk say."[69] In 1754, he published a plant geography, *Stationes plantarum*, in which he distinguished among different habitats such as saline, wet, high altitude, and shady.

The idea of Linnaean acclimatization sometimes lapsed into parody. A year after Sweden's famine of 1756, a craze for botanical alchemy swept the country. A fortifications officer reported that he had transmuted black oats into rye. In a pamphlet of 1757, *Observations and Experiments of a Wonderful Transformation of Grain from a Worse to a Better Kind*, he described how he had seeded a field in early spring with black oats. When it formed ears, he cut it at the root. After leaving the field untilled for a year, he harvested a fine rye crop. The officer calculated that by thus cutting down on South Baltic grain imports, every year Sweden would save "24 barrels of gold."

In Linnaeus' response of 1758, *De transmutatione frumentorum*, he asserted species stability on a theoretical level. He patiently explained that at harvest grains fall to the ground, so that a former rye field will have aftergrowth the next year. He also set his student Pehr Forsskål to vernacularize his Latin theory in one of the capital's weeklies, *Stockholms Wekoblad*.[70] The transmutationists persevered for a decade after the famine. Linnaeus steadfastly mocked them. As he put it in a speech he gave to the Swedish royal family in 1759, "Without sciences our economy would be run by charlatans. . . . Pussy-willow would bring forth Cotton/ Oats would be transformed into Rye."[71]

In the winter of 1760, Linnaeus' student Daniel Solander wrote his teacher of "my victory over the Grain-transmutators at Ramlösa spa. A Baron Gyllenstjerna was especially their Patron, and this from his own experience; the topic was discussed daily and everyone asked Professor Lidbeck for advice."[72] Another student of Linnaeus, Erik Gustaf Lidbeck, who was professor of natural history and economics at Lund University, called for "more accurate experiments," or so Solander snorted. "I on the other hand claimed it was impossible."

The baron "sent [a servant] home to his estate [to fetch] these topics of debate for Parliament, only so that he could convince all Spa visitors about the transformation, and publicly turn me into a heretic. The sheaves of rye arrived, and everyone expected to amuse themselves by my surprise . . . everyone found that I was right: yes Baron Gyllenstjerna himself found out his proofs were false . . . in this manner I resolved what would have been a Parliamentary *affaire*."[73]

In the 1750s, Linnaeus began developing his second acclimatization theory. Here he turned to Scandinavia's "totally useless" Arctic region: "Our Lapp fjeld mountains make up one third of Sweden, where no Grains can grow, therefore they are totally useless; but in other countries such highlands are made useful through plantations of several plants. . . . If we shall get any use out of our highlands, we must get the kind of plants that grow there, and that we can use in the Oeconomy and medicine." Implicitly chastising himself for his earlier optimism, he added: "We can't take just anything we find and happen to have on hand, but [must] get those which in other places grow in similar highlands."[74]

Linnaeus now hypothesized an Alpine climate zone that "includes all mountains" and that existed next to his Australian, Oriental, Occidental, Mediterranean, and Boreal climate zones.[75] In turn, he deduced from the (hypothetical) existence of this climate zone that alpine plants could be transplanted from mountain chain to mountain chain across the globe. Equating altitude with latitude, he especially conflated the northern and the upper-altitudinal tree lines.

Linnaeus had presented an early version of this theory in 1744, as he attempted to reconcile Scripture and science in a dissertation entitled *Oratio de telluris habitabilis incremento*. There he had described Eden as a lone, mountainous isle in a watery world, containing all present-day flora and fauna in climate zones defined by altitude.[76] It was thus not surprising when he asserted, in 1754, that at high altitudes all mountains "become bald [treeless], get the same soil, same snow, same plants: this is the foundation for the improvement of the Economy in our Lapp Highlands."[77]

Elaborating his alpine hypothesis in another dissertation, the *Flora alpina* of 1756, Linnaeus attributed the Arctic's present-day floral poverty to its physical barriers to plant migration. "It is because of a lack of seeds that not all [alpine plants] are found everywhere. Therefore it cannot fail that all kinds of foreign alpine plants could thrive in our highlands, quite as well as in their real Fatherland."[78] Linnaeus calculated that out of the four hundred alpine plants growing worldwide, Lapland housed one hundred. Alpine explorers should travel abroad to collect the remaining three hundred, and to study their uses in "Farming, Medicine, and Factories." If "the Public" undertook to "plant a little Garden up there [in Lapland], and to entrust it to the care of a skilled person, one who understands how to tend these foreign plants,"[79] the floral newcomers would transmogrify into self-seeding ferals and invade the tundra.

Linnaeus' alpine acclimatizationism of the 1750s was narrower in scope than his global acclimatization notion of the 1740s. But both hypotheses served the same political ideal, that of a nation that "doesn't have to rely on foreigners."[80] Both projected onto a not-as-yet self-sufficient homeland the more favorable conditions found abroad, in order to eradicate the differences in choices available to different nations in responding to a modernizing global economy. Within their frame, nature and nation coalesced into a manageable microcosm.

Earlier, and working under the assumption that plants could be transplanted across the globe, Linnaeus and his students had anticipated Baltic plantations of naturalized warm-weather cultivars.[81] As an idea, Linnaean acclimatizationism infiltrated the highest levels of Swedish society. In a 1751 "Mirror" (book on the royal person), the royal tutor Carl Gustaf Tessin admonished his charge, the future Gustav III, to admire civil servants and licensed monopolists such as Johan Alströmer, owner of the Alingsås manufactures, and Colin Campbell and Claes Grill, directors of the Swedish East India Company, as well as their learned intermediary, Linnaeus. These men (from our perspective typical rent-seekers) "tidied up, and made useful the smallest plants and insects." Natural knowledge, the princeling was instructed, allowed the state to turn Chinese manufactures into Scandinavian products. Tessin specially praised Pehr Kalm, "who in American wild deserts established new Colonies of Trees, Plants, Spices, and Fruits; introduced them into the Country; and is now taming them to our Climate."[82]

On returning from America, Kalm had advertised the many American plants "whose seeds he had brought home, and gave free of charge to lovers" of science. Yet as it turned out, the professor at Åbo University was for many years "forced, in a harsh climate, and with heavy costs, to tend his strangers in his own garden." He especially resented having to pay for his day-laborers' "morning draft of aquavit."

"Many people believe," Kalm angrily wrote at one point, that growing exotics is only a matter of "tossing seeds into the soil, then they grow like mushrooms." Late in life, he condensed his life's labor in a pamphlet which, interestingly, reversed Tessin's argument by addressing "the effects that the cultivation of a Country has on its Climate."[83] As Kalm now saw it, there would be no "mildening of these cold winters and the frequent summer frost nights until we change this custom, so common in this country, but most destructive, to practice swidden farming," or burn wildwoods to grow rye in the resulting ash fields.[84]

Earlier, the Linnaeans had fed with rumored achievements their collective hopes of "taming" exotic cultivars. For decades around mid-century they circulated such success stories in their letters. Lacking (in our eyes) procedures for verification, a student would write from some foreign location that he had heard that in some as-yet-unvisited place nearby, plants had adapted to new climates. One of Linnaeus' students reported from Moscow that "the Coffee-Trees are quite successful in Petersburg; they have grown to a height of a couple of arms: at the Imperial Court there is hardly drunk any other coffee but that which is grown in Petersburg." Another student declared from Jena that "a year ago a Tea tree in Berlin and a large Coffee tree in Karlsruhe and Danzig flowered." A third student wrote from London how he had "heard that outside Cambridge a saffron plantation is established that has succeeded quite well."[85]

The Linnaeans hoped to grow herbal medicines, particularly from China, at home. In 1741, the estates instructed the Academy of Science to ensure domestic supplies of simples, so that "all import of such herbs can be forbidden altogether, and our own instead exported to the Foreigner."[86] Linnaeus, too, entreated apothecaries to plant foreign simples. Opium they could "grow at home as well as in Turkey, I should think. An area as big as my plot of land in Uppsala could, in two years time, yield 3,000 Swedish copper coins."[87] He also rejoiced that Sweden produced its own leeches, and considered his discovery of indigenous Spanish flies as important as his cultivation of rhubarb.[88] In 1757, he also supplied a Finnish army orphanage with an Asian rhubarb seed.[89] It "not only grows heartily, regardless of our cold winters, but also so multiplies that one no longer needs to buy it from the Chinese, Turks, and Russians."[90]

"Tartarian Rhubarb I have made common in my fatherland," Linnaeus bragged in a merit list, "and the costly Chinese Rhubarb, which is consumed so much in the drugstores and never has been in European gardens, I just received; it tolerates our climate and ought to be planted with utmost care as a treasure and a domestic product."[91] In the same breath, he advocated

Swedish tea plantations and quinine groves. Malaria was common in Scandinavia in the period. In low-lying, swampy Uppsala, it was so rampant that it was called "Uppsala fever." "If someone could get tea or the China bark alive and growing to Europe or Sweden, would he not be worthy of a reward . . . as these two things draw so much money [out of Sweden] and surely can grow here, at least in Skåne."[92]

Linnaeus and his students hoped for new naturalized food crops as well. In 1748 Kalm promised maize, chestnuts, ginseng, walnuts, persimmons, cranberries, and sugar maples. (He also recommended leatherwood, "to spank children.") Perhaps, added Fredrik Hasselquist after traveling through the Holy Land in 1750, almonds and olives could grow in Sweden.[93] Two years later, in 1752, another student promised Linnaeus "a form of potatoes Mr. Osbeck brought with him that are curious, and are called in Chinese *Yams.* . . . I have planted a piece in a pot; if it succeeds, I shall humbly report. If this plant wants to grow here, it promises advantages to the nation."[94]

Linnaeus published lists of "those kitchen garden plants which one can, and those which one really does, plant in Swedish soil."[95] He advised on the planting of "Egyptian rye" and "Sicilian wheat."[96] And in 1763, he became involved in Scandinavian rice plantations, after hearing from a navy officer who was "becoming more and more convinced that our land ought to be as profitable as the English."[97] "I have been told," the officer wrote, "that the English plant rice in North America in the same Climate as ours. I have in finland a piece of land [which should yield] 24 barrels [of rye] which is flooded by salty seawater every spring." Earlier he had planned to drain the land and plant potatoes, "like the Dutch do," but now "I think it would be a shame if we couldn't do what the English do."[98]

In a glowing review of his own work that he published anonymously (as was his custom) in a Stockholm newspaper in 1752, Linnaeus described Scandinavia's indigenous flora with optimism, too. "Yes here are some that can be used instead of Capers, Groceries, Coffee, Lemon juice, Sugar, Tea, and Saffron."[99] During those years he devoted much energy to identifying native substitutes for imports. (This earned him fresh fame in Sweden during the First World War, when a "Linné-Kaffefabrik" was opened in Falun. Kathreiners Malzkaffee-Fabriken GmbH of Munich immediately sued: the drawing of Linnaeus gracing the Falun tins looked suspiciously like the picture of a Catholic priest named Kneipp, which had long guaranteed the Munich *Malzkaffee*).[100]

Medicines continued to be a worry. As Linnaeus wrote to a friend in 1757:

"It is horrifying to reflect that a poor man here at home can hardly be given a purgative without it being prescribed from the East or West Indies, when after all we have just as good purgatives in the Fatherland as ever from outside."[101] Seven years later, in 1766, Linnaeus penned a treatise, *Purgantia indigena,* listing twenty-four native options.[102] He argued that Sweden contained within its borders many other, as yet undiscovered simples, "which a generous nature has prepared even for our poor fellow countrymen, and whose potential, today still unknown, a future time will realize to cure thousands of diseases."[103] Taking up a request by the national estates, in *Plantae officinales* (1753) he attempted to limit the apothecaries' stock of herbal medicines to indigenous flora.[104]

Linnaeus especially searched for a native substitute for coffee, which he considered a French fashion and therefore a moral and medical hazard.[105] In *Potus coffeae* (1761), a tract devoted to coffee and its dangers, he drew on a nativist ideology that saw the exotic as infectious. "There are still living the most trustworthy old people, who assure us that [coffee] was brought into [Sweden] by travelers returning from France, and infecting our people [*folk*] with this, as with other foreign customs."[106] Linnaeus himself was "infected" in this fashion. He was a compulsive coffee drinker, downing cup after cup as he worked, and at family coffee parties. Complicating matters, he regarded coffee as a set of social customs as much as a potable liquid. Drawing on his own experience, he warned that the habit of coffee drinking required a capital outlay of one thousand Swedish silver thalers. In *Potus coffeae* he ominously listed the objects of conspicuous consumption demanded by this custom, such as a silver coffee pot, Chinese porcelain cups, a coffee-bean grinder made of steel, trays, linen cloths, and "a round table, painted and lacquered."[107] He himself had had to order his porcelain twice from China, with great trouble and expense. The first set had arrived largely smashed, and with the *Linnaea borealis* decorating it rendered in a peculiar dark red.[108]

Ideally, Linnaeus felt, Scandinavians should abandon such immorally wasteful forms of sociability. Short of that, because (as he sighed in *Potus coffeae*) "it is obvious that [coffee] can never successfully be planted and multiplied by us, since its homeland lies on the equator," they should drink a brew of boiled water mixed with burnt "peas, beechnuts, almonds, beans, maize, wheat, or toasted bread."[109]

The estates intervened more concretely. They repeatedly banned coffee, along with other luxuries such as chocolate, clocks, wigs, silk stockings, tobacco, and horse-drawn carriages. In 1766, after one such ban on coffee—they always proved ephemeral—the brother of Johan Peter Falck wrote a sad letter to St. Petersburg, where Falck tended the botanic garden of the Scientific Academy. "I mourn only for this wonderful coffee . . . I can just imagine how you will be drinking your cup of Coffee in peace and quiet and how I will be pottering about here in Sweden my mouth watering."[110]

While Linnaeus wished to substitute wild flora for foreign luxuries like "Capers, Groceries, Coffee, Lemon juice, Sugar,"[111] his overriding guiding impulse was more simple. He wanted to feed the peasants. Famines still decimated the Scandinavian countryside. As Linnaeus lamented in 1773: "Never have more people died in Sweden than this year, first from hunger, then malignant fevers, towards the autumn from dysentery."[112] Nearly every decade throughout his 1741–1778 tenure at Uppsala, the university hamlet was overrun by emaciated beggars.[113] (Sweden's last famine came in 1867, and struck among the Lapland homesteaders.) The year 1756 was particularly dreadful. Linnaeus grieved for the "many thousands" who, as they lay dying, must "hear their little children's whimpering, suffering, and death agonies."[114]

Linnaeus was not posturing when he wrote these lines. He generously handed out food and small coins to the hungry. His Uppsala colleagues noted—with obvious surprise—that he was unable to walk by a starving mother and child without some act of charity.[115] Yet, because of his natural theology, Linnaeus was puzzled as to *why* the peasantry starved. Perhaps they were ignorant of his own amalgamated natural history and economics? "It seems strange to me that people here at home, at least in the summer and when native plants are available, should ever die of hunger if they knew and could select between healthy and dangerous" plants.[116]

In a sermon of 1763, Linnaeus pronounced more emphatically: "He who knows his plants shall never need to perish during the crop failures in our country."[117] As rye fields froze into blackened straw and turnips rotted on the root, Linnaeus advised the famine victims to eat asparagus, cherry-tree resin, and house leeks, which "grow in masses on the roofs in the southern provinces." He proposed tulip bulbs "cooked with butter and pepper," black currants mashed with sugar ("anything but powdered sugar will *not* do"), and wild strawberries soaked in wine and sugar.[118]

Linnaeus also (and more realistically) told the starving to eat fir bark, nettles, acorns, Iceland moss, seaweed, burdock, polypody, bog myrtle, and thistles. He advised adults to imitate their scurvied children who, scrambling for greenery at the end of the long Scandinavian winter, "rub forth the sap from this tree," the Norway spruce.[119]

It was in this context of a crusade against both periodic famines and endemic hunger that in 1775 the Swedish cabinet requested from the Academy of Science a plan for the production of grain-free aquavit. Most of the naturalists who were asked for an alternative source suggested potatoes. Linnaeus was too senile to respond. But his former student Pehr Kalm answered in a more imaginative, which is to say, orthodox, Linnaean manner. He recommended that aquavit be distilled from "cabbage leaves," alder buckthorn, burdock, "fresh shoots of Norway spruce," and "unripe apples, half frozen."[120]

Earlier, Linnaeus had endorsed marihuana, "which has the same effect as aquavit, chasing away melancholy" and making you "happy and funny."[121] The Linnaeans also contemplated other, less colorful food substitutes and economizing measures. Linnaeus' 1749 *Flora oeconomica* lists items of Baltic wild flora that kill lice, mosquitoes, and moles; fatten pigs, cage-birds, and beavers; and spice aquavit, beer, and wine. He recommended swamp dogwood, sloe, and bearberries for tobacco; wild lucerne for imported clover; pulverized moss for wig powder; birch sap for cane sugar; and salep for coffee.[122]

In animal husbandry he proposed to use moose for horses, and black grouse for hens. Of the grouse, Linnaeus wrote: "One teaches him to eat grains like the hens, but he ought to be granted an occasional juniper berry."[123] Another substitute meat were guinea pigs, "slaughtered, shaved, and fried."[124]

Linnaeus' most popular physico-theological tract, reprinted three times in Swedish and translated into Dutch, Latin, German, and English, focused on native insects. "There are surely some that are sweeter than honey, tougher than silk, redder than Cochineal."[125] Continuing his father's efforts, Linnaeus the Younger wrote a family friend in 1778: "I am trying to tame native animals, to begin with I now have a moose that is as tame as a dog."[126]

Such substitutes appealed to the Linnaeans' Gothicist moralism, their valorization of a "northern" rusticity. They also functioned as a back-up option to their botanical adaptationist schemes. In 1764, for example, one of Linnaeus' students, now a professor of natural history in Stockholm, wrote to the Academy of Science to recommend that Swedish bean sauce be substituted for Chinese soy sauce. "Since Soy-sauce has become an attractive luxury that pulls specie out of the Realm, it would be good if the Soy-bean could be tamed to our Climate, so that it could produce ripe seeds here and multiply on cold soil, but it is uncertain if such a thing can be done." Yet, the professor continued, "I have every reason to believe that Soy can be made from our ordinary Turkish Beans, since the taste of Soy-beans is close enough to theirs." The Academy of Science asked Linnaeus what he thought of his erstwhile student's suggestion. He enthusiastically endorsed making "Soy juice from our Turkish beans," adding that he had other acquaintances who had hit on the same idea.[127]

In another area, that of clothes and their production, the Swedish state elites have traditionally encouraged the transfer of foreign textile technologies to domestic grounds. Until recent times, they have protected domestic textile industries from international competition. In the Linnaean era, the production of wool, silk, cotton, and linen dominated the manufactures. The most important of these textile ventures were, as mentioned earlier, the Alingsås manufactures.[128] Owned by the Alströmer family, they were founded in 1724 with "no less of a goal than to make the Fatherland, in relation to its needs, independent of the Foreigner."[129]

At fantastic public expense, workers raised imported merino sheep, produced unfashionable hats, and wove and dyed woollen cloths. (A common complaint was that the Alingsås dyes bled even before the cloth was washed.) Mostly, the products found no outlet and were simply stored in warehouses. In the meanwhile the production was financed by state cash grants. The well-connected owners ran the operation as a small police state, meshing their powers over the means of production with all other police powers.

In this textile project, silk was accorded special importance. Linnaeus' circle believed that about three-quarters of Sweden's export earnings were frittered away on imports of silk.[130] "The otherwise considerable profit of our iron trade does not even nearly answer," Linnaeus moaned, to the cost of "this finery."[131] To solve this (quite imaginary) problem, he advocated domestic mulberry plantations, to create a botanical base for a home-based silk industry.

At Uppsala, Linnaeus noted, the mulberry tree "is kept over the winter in Hothouses, but in Skåne Gardens it grows under the open Sky, and thrives so that it also survived the hard winters of 1739 and 1740."[132] Later Linnaeus admitted that even the black mulberries planted in Sweden's southernmost town, Ystad, had frozen between 1739 and 1741. One of the tasks set for Pehr Kalm, therefore, was to bring home hardier New World mulberries. In 1751, Kalm imported red mulberries from Canada, and planted them around Åbo, Finland, in the hope that they might fare better.[133] White mulberries, too, were tried in Sweden.[134] In 1749, on Linnaeus' recommendation, the Swedish Chamber of Commerce dispatched one of his students as an industrial spy specializing in silk-making. He returned from his European *Rundreise* five years later, in 1754. Setting up a state-financed venture, he planted some 25,000 white mulberry saplings at Ladugårds-landet, on the fringes of the city of Stockholm.

By 1762, nearly nine hundred silk looms were busy in Stockholm, and a few others in the provinces. Linnaeus' student Eric Gustaf Lidbeck, who supervised the planting of some 36,000 fledgling mulberry trees at Mårtens fälad, outside Lund, even produced a small amount of homegrown silk. Princess Sofia Albertina was presented with one astonishingly beautiful piece, woven from silken threads spun from cocoons tended in the mulberry gardens of Lund University.[135]

Other brocaded silks, patterned with gold threads, were designed by courtiers and woven at the rococo summer palace of Drottningholm, under the supervision of Queen Lovisa Ulrika. As was to be expected from a man who forced his courtiers to wear a "national costume" he himself had designed, Lovisa Ulrika's son Gustav III wore silk coronation robes in 1772 that were thoroughly, and self-consciously, "made in Sweden." As the contemporary label of an extant fabric sample for the robes notes, a little anxiously: "On the whole, the effect is good, and the ground deserves comparison with king Adolf Fredrik's Coronation robe, made by Peyron."[136]

In 1729, Sweden's first cotton printing operation was established, and in 1741, the first cotton weaving shop. By 1746, predictably, Linnaeus struggled to grow cotton in Sweden.[137] Like his projected mulberry plantations, this was a typically cameralist impulse. Mercantilist theory would accept the import of raw materials, so long as a sufficiency of finished goods was exported. As a fall-back scheme, Linnaeus and his students recommended native substitutes for cotton, such as linen and fabric spun from marsh grasses.[138] They here turned to folk knowledge. Had not Linnaeus seen, in 1732, how naked Sami babies were bedded down in moss nests, lined with reindeer fur and "more pleasant than silk clothes"?[139] Traveling through the southern parts of the Russian empire between 1768 and 1774, his student Johan Petter Falck noted that the Bashkirs and Ostyaks wove nettles into cloth and the Cossacks made fabrics from hemp plants. Linnaeus, who advised Alingsås's importation of "Spanish sheep" (merino), also hoped to naturalize American "Cattle with long wool or shaggy hair suitable for spinning and weaving" (buffalo) and "Chinese goats, who have finer, curlier and softer wool than any sheep" (angora).[140]

Another Linnaean project was dye plants. Just how deeply northern Europeans yearned to transform the gray of their wools into the exotic hues of a warmer climate is suggested by the color names of an anonymous Swedish dye recipe book, handwritten sometime in the mid-eighteenth century: "Nutmeg," "Light Coffee Color," "Chestnut Brown," "Lemon Yellow," "Parrot Green," "Gold Yellow," "Olive Color," "Cherry Red," and "Seville Orange."[141] Dye stuffs, imported from abroad, thus conjured up by their colors more weighty exotic imports: coffee, lemons, parrots, gold, and oranges.

Linnaeus hoped to satisfy such yearnings for exotica by homegrown products. In 1759, he condensed his observations from his six provincial journeys into a list of foreign "dye grasses that can be cultivated by us, without being bought from abroad," and dye "grasses and herbs growing at home, which ought to be just as useful as the foreign ones."[142] A year earlier, together with the Swedish consul at Tripoli, he imported a North African beetle that was said to produce as rich a scarlet color as the Mexican cochineal bug. The consul "performed on them several experiments and has already come so far that after a certain manner of killing and drying them, they give an admirable ink in water, hardly worse than the best cochineal."[143]

As a cameralist, Linnaeus hoped most for Swedish tea plantations.[144] In the eighteenth century, tea transformed Eurasian trade. In the seventeenth century, East India companies, assembling complex return consignments, had become a considerable trade presence in the countries around the Indian Ocean. But for tea, they voyaged directly to Guangzhou and purchased their cargo with New World silver. To prevent a tea monopsony, the Chinese government admitted all European nations to this arena of trade. Smaller European countries began to license East India companies (usually run by diaspora merchants) that got into this new, one-stop China tea trade. (The Swedish East India Company, founded in 1731 by English and Dutch traders seeking a flag of convenience, was one such interloper.) Of course, the Eastern trade still remained intricate, since Europeans continued to participate in the Indian Ocean country trade.[145] Yet with the newly simplified exchange of silver for tea, ultimately "the company form of mercantile enterprise was undermined."[146] This well-known historical development, the decline of the national trading companies, meant the involuntary forfeiture of a state-managed foreign trade, and consequently the loss of a key cameralist stratagem. As cameralist theoreticians and Europe's rent-seeking state elites saw it, they were losing control of the Asia trade.

This explains the tone of irony in Linnaeus' contemplating, in 1761, the swap of Latin American silver for Chinese tea and silk: "perhaps some people consider us happier because we have discovered the silver treasures of Potosis [Bolivia], which we pry loose from the innards of the earth with great efforts, ship to Europe with great danger, and then, with no lesser risk, export to the barbaric countries of the distant East Indies and waste them there, bringing home in exchange dry leaves of bushes and thin threads, spun by caterpillars."[147]

Linnaeus believed that this exchange of silver for tea and silk would bankrupt Europe and was outraged that for "just a couple of leaves from a bush," Europe gave up its bullion.[148] (Silver, he felt, had an intrinsic value extending beyond both its use value and its commercial price.) In the 1741 Acts of the Swedish Academy of Science, he asserted "That tea should be able to grow in Europe and Skåne just as well as in China and Japan, there is no doubt any more. But it has never arrived in Europe." That is because tea plants and tea seeds, he explained, died from the equatorial heat during the long sea journey. "If the seed could be brought from China over Russia to Sweden it would grow without fail."[149] Thus there was "still one solution, namely that the seeds and the tree [be] transported with the Caravans that yearly leave China for Russia."[150] As we have seen, Linnaeus planned to send his student Pehr Kalm with these caravans. He also corresponded with the London merchant John Ellis on how "to bring exotic seeds from China, and other distant parts of the world, in a vegetative state. . . . Likewise the best method to preserve the plants alive in such long voyages and so many different climates."[151]

Preparing for the tea seeds' arrival, Linnaeus chose Sweden's southernmost province, Skåne, as the site for his plantations. This region "is so similar to Germany and Denmark that one could say that the sea, as it were, has violently cut [it] off from the southern countries and added it to Sweden."[152] If a plantation of Chinese tea bushes could be established there, he asserted, "I do not doubt that it can tolerate our winters." He added, gloatingly: "Poor Chinese, they will only lose through this more than 100 barrels of gold a year."[153]

In 1746, Linnaeus published an essay on tea in Scandinavia's most widely read publication, the farmer's almanac. He began by warning against tea consumption. "It makes the flesh in the body loose, the nerves flabby, the head stupid, and the body feeble."[154] Upon the observation that tea is none-theless fashionable, and by way of a classic misogynist trope, he then introduced the following syllogism: women are governed by fashion, women govern men, Europeans drink tea. Therefore, "let us bring the Tea-tree here from China! Here there is the same soil, the same Sun, same water, same air. . . . If one used just half the money that in one year alone leaves the Realm for Tea, then one would without fail gain so much Tea through Tea plantations here in the Country that in the future not a pence would leave us for those leaves."[155]

Linnaeus went on to tell his peasant audience how the Dutch, growing coffee in the East Indies, had broken the Arabian coffee monopoly. By breaking the Chinese tea monopoly, he daringly predicted, Europe could become as wealthy as Southeast Asia. Or—for Linnaeus thought in narrowly patriotic and zero-sum terms—Sweden could rob China. "In this way China would follow Arabia's path, losing much of its now flowering bliss, and leaving Sweden not a small part of this its fortune and happiness."[156]

Linnaeus doubted that tea plants could survive the southern sea route. Nonetheless, he instructed the five students he sent to Guangzhou between 1745 and 1751 to bring back tea. All failed.[157] In 1757, however, the Swedish East India Company sent Linnaeus two Chinese tea plants that had arrived in Gothenburg. (They traveled the last leg of their journey, from Gothenburg to Stockholm, with a consignment of herrings from Danzig.) When planted in the Uppsala botanic garden, "at the time of their flowering the treachery of the Chinese first revealed itself; it wasn't the Tea bush but the herb Camelia that one had taken the trouble to transplant from the East Indies."[158]

Three years later, in 1760, John Ellis—in response to Linnaeus' pleas—mailed to the Swedish professor eight tea seeds from London, and Daniel Solander sent two.[159] Hinting at England's imperial strategies for self-sufficiency, Solander wrote his mentor: "I now hope to be the first to send home to Sweden fresh tea seeds. . . . But I couldn't get more than these two because all the others were to be sent to the English colonies in North America."[160]

The London seeds never sprouted, and the next year brought a fresh disappointment. Again the Swedish East India Company reported to Linnaeus that it had brought a live tea plant to Gothenburg. But "the sailors hurried after a long sea journey heedlessly into the city, leaving the tea plant on the table in the captain's room, whereupon it was so eaten by the ship's rats that it died."[161]

Two years later, in 1763, the Company wrote Linnaeus for a third time with news of live tea plants. He scribbled an anxious answer to the East India captain:

> But live Tea Trees? Is it possible? Is it really Tea Trees? Truly if it is Tea, I shall make your name, Mister Captain, more eternal than Alexander the Great. . . . I am old, but if I were sure it was true Tea Trees, I would dare to walk to Gothenburg and carry them myself in my arms to Uppsala. If it is true Tea Trees, I beg you, Mister Captain, for God's sake, for your love of your Fatherland, for the natural sciences, and for all that is holy and famous in the world, treat them with the most tender care. I fear it is a Cassinia, whose leaves are similar to tea and are said, in Dutch Botanic Gardens, to be Tea Trees.

Sensing his dread goddess of fate hovering nearby, Linnaeus added: "My dear, do not tell anyone about this, that Nemesis might not hear."[162] But she must have heard, for the first batch of tea bushes arrived at Uppsala already dead. Linnaeus noted forlornly: "I saw from these earlier dead plants and from their wilted leaves that it was the correct plants, so I can't doubt that anymore, but am now wholly sure and convinced, and that so surely, as 2 times 2 is 4."[163] As a student later remembered, Linnaeus "became inconsolable; his Disciples participated most tenderly in his sorrow."[164]

A reserve lot of ten plants that the captain had held back was then dispatched. "I am so afraid," cried Linnaeus, "that these plants will be destroyed that I dream about it every night."[165] They traveled in a wagon along a ox track, on the knee of the wife of the East India captain (displacing her children). And they arrived in Uppsala still green. Linnaeus got "new life and youth again."[166] Straightaway, and just before the plants perished before his eyes, he wrote his patrons "curious observations about the planting of Tea here in Sweden."[167]

The sad tale of the demise of these plants, which I shall relate in next chapter, is one of the more poignant episodes in Linnaeus' life. In the meantime, Linnaeus argued that the importation of Chinese tea was anyhow unnecessary. We "ourselves own hundreds of kinds of both tastier and healthier plants," such as thyme, oregano, sloe, Arctic raspberries, and bog myrtle.[168] In 1740, he presented some dried leaves of a northern Scandinavian shadow growth, *Linnaea borealis,* "to a high-born lady in Stockholm, under the name of Lapp tea, with which she entertained other ladies who all admired it."[169] The State Commission for Frugality (which was a typical Hat reform effort of the 1740s) recommended lemongrass and speedwell.[170] Pehr Kalm suggested pennyroyal, sassafras, or fresh shavings of cypress wood.[171] Johan Petter Falck added that the Kalmyks and Kazakhs made tea from licorice, stirring milk, salt, and butter into the steaming beverage. More tempting, perhaps, was the Cossacks' tea, brewed from marihuana.[172]

In 1754, the Swedish Academy of Science investigated "samples of a sort of tea that is the most commonly used among the public in China and is called Bat-schia." It was said to grow "heartily" in Finland. The academicians "tasted tea water prepared with this, which they found unpleasant enough."[173] Linnaeus, called in to explain the mystery, noted drily that "the Tea bush is as distinguished from Dwarf Willow as the peacock from the crow."[174] In 1755, Linnaeus again examined a "tea plant." "All and sundry claim to have seen it in Amsterdam and other gardens." This, too, was a "little willow that grows in many places and mostly around Åbo."

The line between tea substitutes and consumer frauds was thin. "Eight years ago," Linnaeus remembered darkly in 1755, "there was a merchant in Stockholm who said that he had gotten a wonderful Tea from China and for a good price. He gave it to me to test. A large Chinese piece of paper with letters of black magic was wrapped around it; among the leaves were buds and flowers of true tea; but when I wetted the leaves, they were Blackthorn."[175]

They may seem odd to us: our hero's scramble to grow rice in Finland or tea in Lapland. Yet Linnaeus' notion of floral acclimatization and his search for indigenous flora underwrote his most basic political goal: to reproduce the economy of empire and colony within his Baltic country, and thus to short-circuit the economic improvements achieved by Holland and England through their international trade. When Linnaeus predicted that Scandinavia's tundra—with its shallow summer melts lying over permafrosts, its glaciers and fields of lichen-covered boulders, its sparse migratory animals and vast seasonal swarms of insects—would flourish with tea plantations, saffron meadows, and cedar forests, all tended by nomads turned farm laborers, he envisioned a future world profoundly different from the one that came about. Now forgotten, Linnaeus' vision provides a more complex vision of our own modernity, and particularly of the ongoing quest of marginalized polities to achieve what I have here termed a local modernity.

CHAPTER 7

"The Lord of All of Sweden's Clams": A Local Life

On 21 July 1748, Baron Carl Hårleman, a member of the Swedish Academy of Science and a cabinet superintendent (*överintendent*), wrote an answer to a now-lost letter from his "most devoted client," Linneus. Penned too hastily to afford a greeting, Hårleman's letter begins abruptly: "Never expect, my master, any reward from a rational government to someone able to manufacture real pearls."

"Father" Hårleman went on to advise his "lovable son" Linnaeus to keep secret his technique for seeding pearls in mussels and not attempt to get a stipend from the cabinet. "As soon as the art is known, the pearls must by necessity become cheap." He added: "and how could it not become known when it will pass through the hands of clerks, who thereby will receive the promised secret."[1] In other words, Hårleman admitted that corrupt civil servants would pilfer Linnaeus' intellectual property: the very process that rewarded technological inventions destroyed their commercial value.

Hårleman's letter must have saddened Linnaeus. He had expected great things from his pearl experiments. Using his standard measure for large sums, he had calculated that Lapland's freshwater pearl fisheries annually earned Sweden "some barrels of gold."[2] In 1761, he also wrote to the cabinet's sitting committee on economic matters, recounting his observations from the Lapland voyage of 1732. "I have seen how pearl mussels are fished at Purkijaur, the most famous pearl fishing place in the Luleå Lapp lands."[3] Linnaeus described how the pearl fishers, balancing on log rafts anchored in the foaming rapids, used wooden tongs to pluck the mussels clinging to the rocks at the bottom of the river. Once the raft was loaded, the fishermen paddled ashore to pry the mussels open, searching for the occasional pearl. "Horrifying masses of shells still left lying witness how many thousands of mussels have been killed in vain; for one often has to break open entire

1000's, yes, often 2000's, without getting any pearl of importance." He concluded that "our so-called pearl mussels" were going extinct.[4]

When Linnaeus moved to Uppsala in 1740, he began experimenting with the mussels in the creek meandering through the city.[5] In this, he disregarded the Younger Rudbeck's conclusion, made after his visit to Purkijaur in 1695, that pearls were "small mussel babies."[6] Linnaeus realized they were the mussel's mechanism for encapsulating foreign objects and thus rendering them harmless. To produce a pearl, he bored a hole through a mussel's shell, poked through it a chalk or gypsum granule, and repatched it with *tuttanego,* an alloy of two parts tin to one part bismuth.[7] He then replanted the inoculated mollusk in the river bed, to layer mother-of-pearl around the grain scratching its soft flesh.[8] The Uppsala rivulet was a tepid and sluggish stream. It watered malarial swamps, and it housed the common river mussel, not the pearl mussel of Lapland's icy rapids. Yet Linnaeus did succeed in seeding pearls in "our miserable lake mussels."[9] In the 1748 edition of *Systema naturae,* he declared his success.

Linnaeus' idea to seed pearls in freshwater mussels was commonly advocated at the time, and aimed at transforming the dwindling wild harvest into a sustainable domestic crop. As he himself confessed, his ability to culture pearls was a small matter. "I have wondered how such a simple thing so long has been hidden from curious people, since luxury has set such a high price on something that in itself is of little value."[10]

The technique Linnaeus used he had learned from the Chinese (who had been seeding river mussels for centuries).[11] By the early 1740s, Swedish East India Company ships had plied the tea route for more than a decade. He might have heard of Chinese pearl plantations from returning travelers. Certainly he had read of them. For when, in 1751, the Swedish Chamber of Commerce queried him about a prospective pearl-fishing inspector, informing him that the applicant had learned to culture pearls in China, Linnaeus wrote back that "I have read in the Travel descriptions about China that were published already more than 15 years ago how the Chinese, who in several areas are superior to the Europeans in imitating Nature, have invented an art to make Mussels bring forth pearls."

Linnaeus added that "he who reads the above-mentioned descriptions . . . cannot help but learn the art, if he gives himself time to do a little experimenting."[12] With his usual optimism, he calculated that three men could inoculate one hundred mussels daily. After twelve years, these would be harvested. "No mussel that is impregnated can fail to bear a pearl, if only she is laid down and allowed to live in her Rapid." Linnaeus anticipated stupendous profits. The only problem, in his mind, was "to hold the knowledge thereof secret, so that foreigners do not get to know about it."[13]

As Linnaeus harvested his first Uppsala pearls in the late 1740s, he hoped to earn money from his technology. The now-lost letter to Hårleman of 1748 indicated this, to judge by the baron's response. As we saw, the opening of the baron's letter crushed Linnaeus' hopes for cash grants from the government. (His aspirations were entirely reasonable. Monetary rewards were common in Sweden during the Era of Freedom, and Linnaeus had seen them distributed to fellow rent-seekers.) But Hårleman continued more promisingly: "However, if he who possesses this art petitions the King for permission to be the sole person allowed to fish and to be the lord over all of Sweden's Clams, this ought not to be refused, since it would be a National profit, for as long as the whole thing remained a secret, this creator of pearls would draw from vain foreigners all the money that could ever be won in that game."[14] What the baron dangled in front of the professor was a promise of an exclusive product license.

With obvious pride in the cameralist economic policy he had helped devise, Hårleman explained why he and his cabinet peers might grant Linnaeus a pearl monopoly. "The Cabinet and the King find it sufficient that the money comes in; it may be under whosoever's care it will happen to be." Mindful of his double loyalties as landowner and civil servant, Hårleman conceded: "the entire property of Sweden shouldn't be solely owned by the Government; what I and my dear Sir own, is not outside of the borders of the country; per the circulation, it will still in the end fall into that Ocean which is the Tax authorities."[15]

Today the baron's cameralist concession—his willingness to keep alive the goose that lays the golden eggs—is a commonplace governmental strategy. In 1748, however, Hårleman rehearsed a novel and complex argument. It needed to be spelled out, as much to curb his own exploitative impulses in his role as civil servant, as to make his intentions clear to Linnaeus.

The magic of money, Hårleman marveled, lies in its capacity to be taxed each time it circulates in the nation's private economy. Private and public interests coincide, in that the richer the commonwealth, the higher the tax revenues. When Hårleman conceded that "the entire property of Sweden shouldn't be solely owned by the Government," he did not articulate a moral theory of limited government. He described the coalescence of the economy of power and the power of the economy.[16]

At the same time, the desire of the cabinet to establish pearl fisheries promised social advancement to naturalists. This was not to the liking of the prospective pearl-fishing inspector we encountered earlier. When, in 1751, the Chamber of Commerce passed Linnaeus' queries on to the applicant for a response, he answered that in their petty-minded insistence on technical details, they insulted a man of his standing, "especially since at my return from China I was graciously permitted to pay humble honor to Their Royal Majs . . . in a Chinese dress, when I was graciously allowed to give a rather long Description about China and my travels."[17]

The Chamber of Commerce did give this China traveler the title of "pearl inspector." When he failed to grow pearls, he was fired. Linnaeus, on the other hand, succeeded as a pearl farmer. Or, rather, he succeeded as a rent-seeker. It took work and planning, however. In a 1761 letter to a cabinet committee on economic matters, he reiterated the statement he had made in his 1751 letter to the Chamber of Commerce that he knew how to culture pearls.[18] But by 1761, he was emphasizing the difficulty of his feat rather than, as in 1751, its ease. He also told none of the anecdotes of pearl-producing tricks that animated his earlier epistle. In 1751, Linnaeus had high-mindedly written that he had kept his pearl manufacture "secret." "I had often thought about submitting it to the Academy of Science, or describing it in a public Dissertation, but . . . I [might] damage many people and especially my fatherland with my curiosity."

He went on to explain why his "curiosity" would hurt his home country. "Pearls in the fatherland as a sturdy Capital nowadays amounts already to a couple of barrels of gold." With the coming of cultured pearls, "the price would fall entirely." Also, "if Pearl plantations were to be common among us, they could not long remain secret before our neighbors in Norway, Russia, and Siberia, who own more stores of Pearl mussels, could thus entirely triumph over us in quantity." He concluded, anxiously: "Yes, even the Nations that live in Southern Europe" would produce pearls.[19]

Sweden was the only European country that still harvested substantial (if dwindling) quantities of wild pearls. The price of pearls would "fall entirely" if pearl farms were introduced, since the technology would be adopted by competitors and in countries without wild pearls. From Linnaeus' zero-sum, cameralist perspective, it was better if the fatherland earned little and other countries nothing, than if all nations earned something. What mattered was not the size of the global economy (that he assumed to be static), but its distribution across nations.

In 1751, Linnaeus cast himself as the selfless patriot, determined to put "entirely to the side" this "curious experiment in nature." In 1761, he almost blackmailed the cabinet instead: "I was intending to tell some foreign power of the art against payment, to ease somewhat my deficit." "But," he wheedled, "if my own dear fatherland would desire to know the art from me, I would possess double happiness, as I am sure that the Estates would rather grant me some appropriate reward than a foreigner, as I wouldn't take it out of the country."

"It seems not unreasonable," he continued, "to earn on something that could be of great use for the public; especially as I can swear on the Bible that I hitherto have never described this art to any mortal."[20] Yet Linnaeus had told his secret to the whole Chamber of Commerce ten years earlier, in 1751. For his epistle of that year briefly explained how to culture pearls, "an art," he had noted on that occasion, "that already is as good as clearly described" in "Travel descriptions from China."[21]

In 1761, Linnaeus also abandoned his argument that that once a "pearl plantation" was established, the art of culturing pearls would become generally known.[22] In 1751, he made this argument the main point of a second letter to the Chamber of Commerce, written four days later. "For all the profit is won in the art, and that will never remain hidden for many years. For each and every man who finds a prepared mussel in the river, or can fill a workman [with aquavit], can quickly learn this simplest of arts."[23]

In contrast, in a pro memoria on "pearl plantation," probably dating from early 1761 and given to a member of the cabinet committee responsible for pearl fisheries, Linnaeus claimed that the secret of pearl culturing was easily safeguarded. "If the porcelain factory in Saxony could have been kept secret, I don't see that it would be any less possible here."

Consider, he wrote, how remote the river at Purkijaur in Lapland really is. "From the Norwegian and the West side, no one can come here without walking over the fjeld mountains, past the church at Quickjock, then be taken an entire 10 miles by boat, walk certain parts, be carried over to Purkijaur." He added: "all of this cannot be done without guides, rowers, and several changes of Lapps, so that a stranger can never pass here without everyone knowing about it."[24] For extra security, he advised that a "faithful mussel guard" be housed in a crofter's cottage by the rapids.[25]

Hearing Linnaeus' 1761 sales pitch, the committee agreed to pay him for his pearl-culturing techniques—if he proved they worked. Linnaeus was anxious to succeed. His purchase of a country estate in 1758 ("for the sake of my little children" and with money borrowed from Uppsala University) threatened to ruin him. In 1759 he had also "dared to stage a coup," and bought "a whole neighboring village of five entire farms or tenants."[26]

The second purchase was financed by loan as well, and also from the university. It worsened the disastrous money troubles he confessed to in a 1758 letter to a friend. "Now, Brother of my heart, I am ready to be hanged. I have always feared debt as snakes, but must now dance at the debtors' prison, which I hadn't expected. This will either make me gray before my time or ghosts will snatch me away with my children."

Not even science could console him. "Now neither correspondence nor news in the realm of letters help; I have left the country, the anchor is raised, I must sail; we'll see if I find a harbor. 200 barrels of grain I get in tax, but the 4 knights in wartime" that Linnaeus' estate was obliged to equip in return for its taxation privileges, "also mean something." Linnaeus found no comfort even in the Chinese rhubarb seeds that a foreign correspondent had sent. "I planted his miserable seeds in a pot, and they are already grown, but all of this doesn't help at all to obliterate my sins and my debt."[27]

It was with high and anxious hopes, then, that Linnaeus boated to Stockholm one July morning of 1761, carrying nine small pearls. Only a few committee members, sworn to silence, saw him display his pearls and demonstrate his pearl-culturing techniques.[28] The chairman asked a goldsmith to break open Linnaeus' pearls. "Ignorant about the purpose of the experiment, he explained it was a pity to destroy any of them."[29] Linnaeus then demonstrated how to seed the mussel with irritants. The chairman declared the pearls real and the technique workable: the meeting was a triumph. Exhilarated, Linnaeus began lecturing on Sweden's need to naturalize foreign plants, and especially rhubarb.[30] Some two weeks later, the committee, fired up by Linnaeus' vision of a fatherland awash in rhubarb, pearls, and sundry other treasures, recommended to the cabinet that the pearl maker be presented a "national reward" of twelve thousand silver thalers.

The cabinet, however, only reviewed the committee's proposal half a year later, in January 1762. It then decided to wait further "until one sees how much money there might be in the future"—which is to say, how acute the ever-present liquidity crisis would be that year.[31] In June 1762, the cabinet recommended to the diet of the national estates that Linnaeus be awarded six thousand silver thalers, as well as the right to appoint his academic successor at Uppsala University. When they convened again that autumn, the estates sold Linnaeus' secret to a Gothenburg merchant and director of the

Swedish Greenland Company. He in turn paid Linnaeus.[32] The estates also granted Linnaeus the right to dispose of his university chair, and ratified the king's recommendation to ennoble him.

Thus Linnaeus was dubbed "von Linné." He submitted to the House of Nobility a sketch of his personally designed coat-of-arms, wherein "my little" *Linnaea borealis* rests above a fertilized hen egg. To his fury, the House's heraldic expert substituted for his cross-sectioned, blood-streaked, and obviously fertilized egg a more conventional, intact, and shell-cased one.[33]

The Gothenburg whaler was given Linnaeus' now-lost manual for pearl production and a perpetual pearl monopoly, subject to a fifty percent tax on net profits. He was going blind, however, and never founded any "pearl plantations." His son later misplaced Linnaeus' instructions. His grandson rediscovered them, and in 1822 successfully petitioned for a renewed pearl-producing license.[34] But yet again, actual pearls were never produced.

This series of failures came as no surprise to Eli F. Heckscher, Sweden's most famous economic historian and co-author of the Heckscher-Ohlin theorem of comparative advantage. In 1949, Heckscher tersely summarized Sweden's effort to produce indigenous pearls during the Age of Enlightenment: "Naturally, the whole apparatus resulted in nothing." From Linnaeus' point of view, however, his nine small pearls, supposedly demonstrating the possibility of domesticating Lapland's wild and foaming rapids—now long since dammed for hydroelectricity—were a rousing success. They resulted in a "national award," a secure academic succession, and noble status. By the autumn of 1762, Linnaeus had indeed become "the lord of all of Sweden's Clams."

Linnaeus was a "lord," yes, but with no material territory of his own. For in another sense, Heckscher was right, of course, to say that Linnaeus' pearl fisheries, and in fact his entire cameralist program, came to nothing. Half of his students perished during their long-distance travels.[35] No wonder that their teacher searched perennially for "sure omens,"[36] and remembered in *Nemesis divina*, his manuscript collection of his spiritual musings, that Pehr Löfling, who had stumbled while saying good-by, "did not come back. Same thing with Forsskål."[37] In bitterly sad memory of Forsskål, who died in Yemen in 1763, Linnaeus named the common nettle *Forsscolia*. He mourned when his students met their untimely deaths, and was quick to blame other people. "Ternström died of chagrin . . . our natural historians always get a miserable cabin and the worst one on the ship where they lie like dogs."[38]

To our eyes, the occasional survival of these European travelers in the tropics may seem more surprising. Carl Peter Thunberg left from Amsterdam to Java in 1771 on a Dutch East India ship which had twenty-one Swedes on board. Only the three officers among them survived the outward journey. In 1777, on returning from Japan to Batavia, Java, Thunberg noticed that of the thirteen guests who had attended his farewell dinner a year earlier, only he and the host were still alive.[39]

Death, including Linnaeus' own slow demise, was then the most serious obstacle to his enterprises. He especially grieved because his students suffered, as he said about Fredrik Hasselquist, a "double death, since not only he has disappeared, but also all his work."[40] Within his own Uppsala circle, he knew of spectacular losses of learning and labor, heartbreaking failures to make public those bodies of knowledge which had been earned with great difficulty.[41] This might explain why he always strove to memorialize his deceased students through posthumous publications, eulogies, and the like; why he himself always hastened to the printer; and why in *Philosophia botanica* he railed against expensive natural history publications which, he charged, even in the rare event that they were completed, remained unavailable because unaffordable.

Yet Linnaeus' students' travel accounts were often fragmentary and late in appearing. As Johann Reinhold Forster's botanist on Captain Cook's second voyage, Anders Sparrman circumnavigated the globe in 1775. Once home he only finished publishing his travel narrative in 1818. Patrons were part of the problem, tending to keep for themselves the booty of travel without feeling obligated to make it public. The Spanish crown simply stored Pehr Löfling's South American collections of 1751–1756. Similarly, the Danish crown locked up Forsskål's Arabian collections of 1761–1763; and the Swedish crown put away Fredrik Hasselquist's Levant herbaria of 1749–1752.[42] Joseph Banks's and Daniel Solander's invaluable Australian and Pacific botanic collections of 1768–1771 were squirreled away at Banks's private residence at Soho Square, London, and were fully published only in the late 1980s.

The Linnaean voyage of discovery was also hampered by poor techniques of storage and preservation. Before Nathaniel Bagshot Ward's invention in 1827 of the so-called Wardian case, a glass box that stabilizes humidity and temperature, live plants rarely survived long sea journeys. Christopher Tärnström purchased live lizards, peacocks, and turtles from Javanese traders and gathered plants from his Cambodian island; none of this reached

Sweden.[43] Pehr Osbeck's Chinese trees died before his ship, *Prins Karl*, reached Cadiz in southern Spain.[44] Carl Peter Thunberg lost his precious live plants from Japan, nursed through a year-long journey at sea, in an autumn storm off the coast of Holland in 1778.

Linnaean survivors were often unproductive upon returning to their homeland, too. Some took up other pursuits. Osbeck, on returning from Guangzhou in 1752, became a country parson, nursing his "hypo-chondriacal ills."[45] Lars Montin became a provincial doctor, and his large Lapland collections of 1749–1751 went unused. Daniel Rolander panicked in tropical, insect-ridden Surinam in 1754, to scramble home, deranged and empty-handed, on the next available ship. Eventually he ended up a beggar. Roland Martin's Arctic Sea gatherings of 1758–1760 were paltry. Barely sur-viving as a private tutor, the erstwhile star student lost a leg and deteriorated into a lonely eccentric, obsessed with idle temperature observations.

Thus it happened that for reasons ranging from high mortality rates, ac-cidental losses of collections, faulty preservation techniques, and penny-pinching patrons, to their own general goals and work habits, the outcome of the Linnaeans' protracted journeys was trifling.

Moreover, though Linnaeus instructed his students to make *oeconomia* the keystone of their foreign travels, they were understandably uncertain about this brief. How was one to go about rebuilding a national economy from scratch, and as a self-sufficient, isolated autarky, in an age of expanding global trade? A vast discrepancy arose between goal and result. For example, during his Arctic Sea voyage, and as he explored northern Norway's coast-line between 1758 and 1760, Roland Martin aimed to "encourage my coun-trymen to [achieve] just as real a profit from fisheries" as that enjoyed by "the foreigner." In the end, however, he found only gray sandstone, "pass-able to sharpen knives on," and some "pretty white pebbles."[46]

The flora, fauna, and technologies that Linnaeus' students *did* manage to import to Uppsala found only small economic use. In a letter of 1765, Osbeck slyly told his aging teacher that his students kept their great finds to themselves.[47] Not so; the real reasons for failure were systemic. As we al-ready noted, the Linnaeans failed in part because they hesitated in assessing which life-forms were genuinely valuable. Rhubarb was not, as Linnaeus believed along with his contemporaries, a "divine medicine" that cured lung disorders, scabies, and fevers.[48] Nor was the "English soil-plant called pota-toes" poisonous, as Linnaeus conjectured, even if it is related to the deadly nightshade.[49] "I don't know why the servants . . . find it so necessary to go on eating" potatoes. "They don't understand how to test their powers either dietetically or botanically, instead judge only by smell and taste."[50] Clearly, the prolixity of science obscured a truth known to the "lower" senses.

In 1749, together with an improver landlord, Linnaeus experimentally "proved" that the period's only major botanical transplant, potatoes, were indeed toxic. He did so by relying on an implicit notion of the natural. "We had a whole heap of swine herded into the courtyard, and in front of them was thrown a sizable heap of potato roots . . . but the naughty swine didn't voluntarily eat a single one but ferreted out bread and other bait, that had been thrown in among the roots."[51]

Ten years earlier, in 1739, Linnaeus participated in a discussion in the Swedish Academy of Science on the potato that shows how unknown it was. Assuming that the potato could only be made into bread, the academicians debated whether to crush the raw potatoes into a mush (as wine-growers do grapes) and bake it into a bread of sorts, or whether to grind boiled potatoes into a baking flour.[52]

The fact that the potato yields far more food per acre than grain, and thrives on poorer and sandier soils, remained lost to the academicians even as cottagers' families grew larger. Hitting upon a novel use at last, the only woman elected to the Swedish Academy of Science before the twentieth century, Countess Eva Ekeblad, took her place in 1748 because she recommended the potato for wig powder and aquavit. (The academicians also hoped that her election would encourage "the entire sex to be attentive in all matters of household economy.")[53]

The Linnaeans also failed to build a domestic textile industry. The winter of 1739–40 killed the mulberries of Ystad, in Skåne; 1759–60 those of Stockholm; 1770–71 and 1771–72 those of Åbo, in Finland; and 1787–88 those of Lund, in Skåne.[54] The Skåne Plantation Bureau, run by Linnaeus' student Erik Gustaf Lidbeck, managed to produce a few pounds of raw silk as late as 1790. But the manufacturer backed out of his promise to buy it because of its dismal quality.[55]

Scandinavia never saw plantations of such fibre plants as cotton, nettles, or hemp. Nor were wild plants such as marsh cotton grasses ever gathered or used in textiles. Linnaeus' hopes for home-grown buffalo wool were also dashed. By 1754, the professor, once optimistic about Baltic cotton, more humbly hoped to show how "foreign seeds can be used with advantage to improve meadows."[56] The textile dye industry made incremental advances.

But here, too, Linnaeus' grander visions came to nothing. His 1754 adventures with cochineal, the Mexican beetle that produces a brilliant red dye, illustrate this failure. "About *Coccionella* I do not wish to speak, never wish to think or remember."[57]

As Linnaeus narrated the painful story, his student Daniel Rolander "on coming home from Surinam sends a potted Cactus with Cochineal. But Linnaeus presides and the Gardener receives the plant, removes all dirt, consequently also the worms . . . and accordingly all hope disappeared to get those he thought could be profitably cultivated in Orangeries. This affected him so that he got Migraine."[58] Linnaeus' substitute cochineal beetles from North Africa survived no better. Proving themselves fragile hothouse creatures, they perished on the instant the humidity exceeded that of Tripoli's deserts.[59]

Linnaeus also failed in his many attempts to grow tea in Sweden. In 1763 he at last received true tea plants. Two specimens lingered for a few years. As he noted in a *vita,* they "flowered in 1765 in Uppsala."[60] The same year, his treatise *Potus theae* asserted that tea grew in Sweden.[61] But while it was being printed, his last tea bushes succumbed to the Nordic winter. As Linnaeus had complained as early as 1746, he saw himself continuing to pay in silver coins "for the Sugar, Tea, and Coffee for those who want to see the botanic Garden."[62]

All that remained to the Linnaeans in their quest for tea was honor and its pursuit. They comforted themselves (falsely) that tea had first entered Europe with "a Swedish Man and in a Swedish ship and to a Swedish harbor."[63] In 1768, one student wrote in this patriotic vein from Paris to console Linnaeus: "Now, Sir, you will get a great revenge on that lie that Tea was first introduced in Europe in Trianon, which once was said in the gazettes."[64]

From Uppsala, Linnaeus dismissed the claim that tea grew in Dutch and French botanic gardens.[65] Yet this was of little comfort to him as he saw his specimens wilt and die, and his achievements ignored by the learned. Attempting to set the record straight, he cautiously summarized his life's work in 1773: "Tea was first seen away from China in the Uppsala Garden; from this others have learned to take care of it in such a way that within a Century Tea will be common in the fields of Southern Europe."[66]

As late as 1791, Linnaeus' student Anders Sparrman advocated European-grown tea. Employing the fashionable rhetoric of participatory and radical politics, he urged "you Europeans! Citizens!" to "abandon the Chinese tea entirely, or plant it yourself" to help abolish slavery. "Europe, through her own forests of tea bushes . . . can avoid the humiliating annual tribute to the barely half-civilized Chinese Nation, of so many hundreds of thousands of measures of silver," which are obtained from ignoble commerce. "Europeans drag away Africa's Children in irons, to bring up in America that silver from the depth while abused weaker Natives are forced to wither away."[67]

Yet the Linnaeans never could grow tea. Nor did they popularize their native substitutes for tea—sloe, bog myrtle, and pennyroyal. Even Linnaeus' exotic "Lapp tea" of 1740 was a poor substitute for real tea. Thirty years after that introduction, the aging naturalist still expected that his namesake, a frail flower some two inches high, would be cultivated as the national beverage.[68] As if to honor this quixotic dream, Sweden's classic Linnaean flora, *Svensk botanik* (1803), opens by describing *Linnaea borealis,* recommending it especially for tea.[69] Yet his son's terse note on brewing it may more appropriately memorialize it: "NB. one shouldn't use too many" *Linnaea* leaves in the water, "for then is rather repulsive."[70]

Linnaeus' ambitious plans for Lapland failed as well. Since lyme grass is a saltwater plant that grows only on sandy beaches, "Lapp wheat" never became a staple grain, as he had predicted to the governor of Västerbotten in 1733. Nor did his projected "millions" of people come to inhabit Lapland or live from its coastal grasses.[71] Saffron, mace, nutmeg, and cinnamon never grew on the tundra; and the permafrost realms never saw larch trees or cedars of Lebanon. Nor did Lapland ever see sables, let alone in Linnaeus' promised "great numbers." To this day, Lapland's treeless tundra remains a vast wilderness, established in perpetuity not as Sweden's garden of cedars, tea, and spices, but as Europe's largest national park.

Linnaeus even failed to transplant Arctic plants into temperate zones. In his 1753 *Demonstrationes plantarum in horto Upsaliensi,* listing one year's plantings in his garden, he remarked that not only did tropic growths fail in Lapland, but "moreover, the Lapp plants don't thrive in our area either as they do in their fatherland."[72] In 1773, in a poignant farewell to his Arctic dreams, Linnaeus "planted on my hill by the Museum a little garden only of Siberian plants that tolerate our winters so well."[73]

Already by 1746, some five years into his acclimatization experiments, Linnaeus suspected that Scandinavia's climate was an insurmountable obstacle to improving the economy. "Yes, if the violent cold here in Sweden

weren't so severe, our Fatherland could feed 10 to 20 times as many inhabitants." He confessed to being sick and tired of "the old drudgery with southern plants."[74] In the 1757 *Calendarium florae* he stressed that "southern" or
"Indian" plants must be housed in greenhouses in winter. By then, he had
become convinced that his colonial cash crops, such as sago, coconut, ginger, and banana, would never acclimatize. It was typical that, when their
rice paddies failed in 1768, Linnaeus and his friend August Ehrensvärd resigned themselves to experimenting with domestic hay grasses.[75]

By 1769, Linnaeus wrote plaintively to the governing body of Uppsala
University that although he had many plants from "the most foreign far-
away places," his "Garden is located in a colder Climate than any other academic" garden. He asked for manure from the university's stables, for cattle
litter, and for a hundred carts of firewood, all to heat his garden. "Life for an
academic Garden is heat above and below ground." If they denied his request, he wrote, "everything is in vain."[76]

Thirteen years earlier, in 1756, when Erik Gustaf Lidbeck was appointed
professor at Lund, 450 miles (700 kilometers) south of Uppsala, Linnaeus
wrote to his former student on a cold March day, from a still wintry Uppsala:
"I would be happy to change places with you . . . if I could have your climate
in Uppsala, I would have almost the whole world, and, if possible, half of the
moon, too. But all Elements fight me. I think we will never get spring. They
say in Lund spring is almost over."[77]

As Linnaeus grew older, he also experienced failures and disappointments in
love. In his mind, these inevitably mingled with the bleak results of his science. For Linnaeus loved his students. They were in fact the only objects of
his love, apart from his small daughters and his animals—his dogs, parrots,
guenon monkeys, and especially his raccoon (named Sjubb).[78] He praised
his students extravagantly and openly compared one to another. He gave
them pet names and called them his "apostles."[79] "Take burning firebrands
and throw them at Professor Kalm," Linnaeus begged a Stockholm friend in
1751, with typical emotionalism, "so that he might come without delay to
Uppsala, for I long for him as a bride for the hour of one o'clock at night."[80]

His students were often good-looking. Johan Peter Falck, for instance,
"was quite tall, upright, muscular, and handsome and the face was oval,
manly, and pleasing."[81] Linnaeus hugged and kissed them.[82] And "it wasn't
unusual to see him examine the outer form of the skull of his most beloved
disciples, to judge each one's memory, intelligence, and inclinations."[83] His
favorite was Pehr Löfling, "grown like Professor Kalm in Åbo, with a long,

straight body, and a manly and pleasant face."[84] Linnaeus dictated *Philosophia botanica* to Löfling "from bed." As he told it, his "dear Löfling" did not want to end this secretarial job.[85] "For as I, in this book, tried to encompass the kernel of all of botany and Mister Löfling never failed to ask me when he didn't understand something, he finally learned the basic tenets so well that I could assert against all false spirits in Flora's Realm, that no one could make him a heretic."[86]

Linnaeus' relation to his students was linked in turn to his relation to his family. He regarded marriages as economic calculations, which men could undertake because women, being fools, marry for love. When he was a bachelor, he knew that he "ought to get engaged to some rich Girl who first could make him happy, and he then her."[87] It was in this spirit that he wrote to Albrecht von Haller about his engagement in 1734 to the girl he would marry in 1739, Sara Elisabeth Moraea, daughter of the town physician at the Falun copper mines in Dalarna. "People probably would not have hesitated to have called this doctor wealthy. Yes, in this very poor region he was truly the richest of all." He added, in a sadder vein, "my father-in-law was, however, quite fond of his money, and didn't want to give up any part of it to his son-in-law."[88]

In his youth, Linnaeus searched for substitute families where he could play the role of a son. He claimed that he was loved as a "son" by his teachers at grammar school, his university professors, his correspondents, and his Dutch patrons.[89] On one occasion, in 1738, the fantasy was of being the son of an eminent professor, and receiving his rightful intellectual patrimony. Linnaeus pretended that he was called to the deathbed of Hermann Boerhaave in Leyden, so that the great Dutch physician could bequeath his duties to the young Swede, transferring them with the seal of a kiss.[90]

The notion of modeling a scientific career on the structures of the family was of course common in the eighteenth century. In Linnaeus' case, however, it may also have reflected ambivalence about his own father. Linnaeus called himself "the prodigal son" or "lost son."[91] As the fifth-generation heir to the parsonage of Stenbrohult, Småland, and as the eldest son, he had been "doomed to be a parson by his parents."[92] To prepare him for this, his father employed a sadistic tutor. Linnaeus attributed his faulty Latin "that he later so often endlessly regretted," to the horrors of that early schooling.[93] In later life, he reiterated his hatred of parents who forced their children to study or punished them with blows, and he made sure that his son was educated kindly.[94]

Linnaeus loved his beautiful mother, who was nineteen when he was born. He often remarked that he "was bred at her own breasts."[95] He especially admired her virtue, and noted that although she had married only to escape a cruel stepmother, she was still a faithful wife. In his own mind, his mother's morals gained merit by his father's failure to please her.

When he visited his home in 1732, Linnaeus noted in a *vita* that he saw his mother. Yet he made no mention of his father, also present. The next year she died, at age forty-four, before he could reach her deathbed. "I in my absence an indescribable anxiety, mourning, and loss."[96] Linnaeus even believed that he had sensed her death from afar. "What is it that makes one anxious when something evil is about to come, when sorrows happen far away, as I in Uppsala, when my mother died in Småland?"[97]

In 1735, two years after his mother's death, Linnaeus revisited the family home. Again his *vita* passes by his father in silence. Instead, he wrote about the memory of his mother. As if to deny a crime he may have committed in his heart, he announced: "I cannot say that I am free of faults, except that I was never a patricide."[98] Perhaps not. Yet in Linnaeus' most extensive *vita*, which describes at length the circumstances and emotions surrounding his mother's death, the reader learns of his father's demise only indirectly, in a brief and ambiguous entry: "Linnaeus had the pleasure of seeing his only Brother take his Father's place."[99]

Linnaeus at once abhorred, and was sexually attracted to, adult women. Arguably, his disgust for women's bodily functions went beyond the ordinary misogyny of his era. Men were the center of his emotional world. Yet he only had two close friends. One was Petrus Artedi, an Uppsala student and his companion in their youth. In 1735, on his way home from a dinner party given by the apothecary Albert Seba, the young man drowned in an Amsterdam canal.[100] Linnaeus was thrown into despair by his loss. Later he met the Swedish court physician Abraham Bäck, who became his "best and most faithful friend," his "brother in flesh," and his "best friend on earth."[101]

Despite his emotional engagement with his students, Linnaeus never became their friend. The reason may have been the way he related their work to his own.[102] In his mind, even the notion of a tributary relation, with its sense of subordination and dependency, was too generous. He believed that he alone should be given credit for the science they served. In 1756, he mourned Löfling's death in part because "*I* was just about to further science through *his* assiduous work." [103] As his funeral orator recalled, the professor "believed himself to have a kind of right to see and describe those plants that his Disciples found, and greatly appreciated that politeness; if he found such a confidence lacking, he could not control his disapproval."[104] Linnaeus thus at once depended on his students and rode roughshod over them. His letters to them mingle reproachful entreaty with haughty condescension. Most of all, they are studded with that frequent imperative of a parent fearful of being forgotten: "Think of me, think of me," he begs Löfling and Thunberg.[105]

Such commandments seem to have had little effect. As soon as his students achieved a position of their own, they abandoned their old teacher. For example, on Linnaeus' recommendation, Pehr Kalm became a professor in natural history and economy at Åbo University in Finland in 1747.[106] When Kalm returned from America in 1751, a voyage for which his old teacher had arranged the financing and supported in many ways, Linnaeus passionately longed to see him. In May of that year, he wrote to Abraham Bäck to ask him to convince Kalm to come to Uppsala.[107] In June, he wrote him four more times, begging Bäck to plead with Kalm.[108] In July, crippled from gout but now doubting that his former student would ever visit him, the professor boated to Stockholm to see the ingrate.[109]

The student whose loss Linnaeus felt most keenly, however, was Daniel Solander. After traveling as Joseph Banks's botanist on Captain Cook's first circumnavigation of the world, Solander settled in London in 1771, as Banks's companion and man-about-town. As Linnaeus complained, he never dispatched specimens or observations to his old teacher from his historic voyage aboard the *Endeavour.* "The ungrateful Solander . . . doesn't send a single plant or insect of all of those he had collected in *Insulus australibus novis.*"[110]

Yet Solander had lodged in Linnaeus' home, had tutored his son, and had flirted with one of his daughters. At one point, Linnaeus even thought of awarding Solander his Uppsala chair over his own son. Linnaeus had patiently endured Solander's many imagined illnesses, advising him, for example, when in 1759 he thought he might have scurvy, "since my gums bleed, when I suck them vigorously."[111] Nonetheless, after 1762 Solander quickly cut his links to Linnaeus. He wrote only once, from Rio de Janeiro in 1768, falsely promising that on returning to Europe he and Banks would travel to Uppsala, "to humbly ask you, Sir, to be the Master Inspector of our recruits".[112] In a gossipy aside, Solander portrayed Banks as an affluent amateur, who spent eight to ten thousand pounds a year on his hobby, but left his studies of natural history unfinished.[113]

But Linnaeus was only advised of Banks's and Solander's reappearance in

Europe by a London acquaintance, the wool merchant John Ellis. The two explorers never journeyed to Sweden, despite Linnaeus' ingratiating references to the new-found fifth continent as "Banksia." Repeatedly raising and dashing the old professor's hopes, Banks wrote Linnaeus in May 1772, again proposing that he and Solander would travel to Uppsala, bringing along parts of their Australian and Pacific herbaria. In a final letter of July 1772, Banks disclosed his and Solander's new plans of touring Iceland instead. He pledged to send botanic specimens instead. None ever arrived at Uppsala.[114] While Banks and Solander sojourned in Iceland, the c. 31,000 sheets of botanical specimens and annotations from the *Endeavour*—including the first Pacific and Australian flora collected by Western naturalists—gathered dust.[115]

Linnaeus' plea of 1771, obliquely addressed to a mutual acquaintance, had gone unheard: "Do but consider, my friend, if these treasures are kept back, what may happen to them. They may be devoured by vermin of all kinds. The house where they are lodged may be burnt. Those destined to describe them may die. . . . I therefore once more beg, nay I earnestly beseech you, to urge the publication of these new discoveries. I confess it to be my most ardent wish to see this done before I die."[116]

In the same year, Linnaeus' former student Andreas Murray wrote him from Göttingen, where he was now a professor: "Misters Solander and Banks negotiate with a Göttingen publisher to publish their observations, issued with many paintings. Mister Solander has asked me to supervise" the publication.[117] Yet this scheme also came to nothing. For as Murray explained, the Göttingen samples of copper engravings "were beautiful, but Mr. Banks wanted them even grander."[118]

The conflict of interest between Banks and Linnaeus continued even after Linnaeus' death. In 1778, Linnaeus the Younger (who had inherited his father's chair at Uppsala) complained: "I know that Banks doesn't want to tell me anything before he has completed his work; but if we could only speak face to face, we would soon agree, since our purposes are different. Mine is only to continue the system, only to determine the plants' genera and species and thereby to preserve the central book already begun in the science [*Systema naturae*]; without having this, one will soon be confused by this plenitude; but Banks, who has money, would illustrate them with descriptions and figures." Linnaeus the Younger placed the blame for the impasse on his compatriots Daniel Solander and Jonas Dryander, now in Banks's employ. "If Banks didn't work with two Swedes, this would probably happen, too."[119]

The son of Linnaeus found especially offensive the underhanded manner in which Banks refused him botanic specimens. In response to the younger scholar's plea, Banks requested "as good a collection . . . as you can spare" of the South American plant specimens sent to Linnaeus the Elder by José Celestino Mutis, promising "to make returns, in things which you cannot easily obtain elsewhere."[120]

Linnaeus the Younger, who was hale enough to travel to England, at least caught a glimpse of Banks's Australian and Pacific treasures. The father, already old when Banks had returned, saw nothing: "If I were not bound fast here by 64 years of age, and a worn-out body, I would this very day set out for London, to see this great hero in botany. Moses was not permitted to enter Palestine, but only to view it from a distance: so I conceive an idea in my mind of the acquisitions and treasures of those who have visited every part of the globe."[121]

Solander's defection was only one of Linnaeus' many personal disappointments. Another occurred when Johan Peter Falck became a professor in botany in St. Petersburg in 1765 and abruptly ceased to dance attendance on Linnaeus. Indeed, scornfully, and with a seeming reference to the muddy stream that ran by (and often flooded) his mentor's Uppsala gardens, he referred to "the Old Guy in the Black Creek." Falck's brother, himself a student in Uppsala, exalted in Falck's appointment. "Now you can ask him to kiss your ass."[122]

The young man went on to gloat: "how much I would enjoy seeing the Old Guy lose his temper . . . for goodness sake don't forget to tell the Old Guy. How shall this not roil his innards?" Five years later, he wrote to St. Petersburg to tell his older brother how the old and lonely professor, "every time He meets me, asks about You. He says he longs endlessly for Letters from You."[123] None came, either from London or St. Petersburg. For it was Linnaeus' tragedy that, as one acquaintance put it in 1764, "hardly anyone loves him, not even here."[124]

Yet Linnaeus loved his students. He acted toward them like a father, which is to say, dysfunctionally. His paternal love was shot through with a demand for recognition of his authority and an open ambivalence that was more raw and direct than people could easily accommodate even in eighteenth-century culture, a time when patriarchal rhetoric dominated social and political discourse and the family was the model of social relations in general. His students, in turn, avoided, rejected, and sometimes hated him. They all sought to escape him. Half-unwittingly, half on purpose, Linnaeus himself provided the model for their flight. For Linnaeus identified with his students and wanted them, in turn, to identify with him. They were his younger self. In Linnaeus' own family, a parsonage had been inherited through five generations, beginning before living memory. He could not picture his university chair any differently. A life without a "son," a designated familial successor, seemed to him an amputated, meaningless existence.

Linnaeus therefore laid down for his phylogenetic science an ontogenetic path, in the form of a model career. His students' travels, imitating his own, were *Wanderjahre,* years literally spent as journeymen. These voyages reflected Linnaeus' artless program of fashioning life as if it were a pre-Romantic *Bildungsroman.* They were also, however, his students' path of escape. En route, they crept into the clergy, found new patrons, became mad, crippled, destitute, or died prematurely. A few found security in university chairs, where they no longer needed to pay homage to their old teacher, and no longer did. This was in some ways the worst fate for Linnaeus, since he wanted his students to remain children, dependent on him in every way. On a conventional and societal level, he could not imagine *not* having a filial successor. On a personal level, he could not imagine having one.

It was perhaps out of pessimism about scholarly continuity and succession—or even about success itself—that in the late 1750s Linnaeus began his testament to "My Only Son," *Nemesis divina* (written from the late 1750s to 1765).[125] It consists of a series of loose sheets that Linnaeus kept in "a pocket similar to those in which one keeps Psalm Books."[126] This secret diary perfectly captures Linnaeus' turn from his early Candidean faith in the material world, to his later and profoundly dark vision of life. His reflections are certainly despairing. In his "pocket" Linnaeus stuffed almost two hundred stories of villainy, sickness, and other tribulations, some gleaned from neighbors and hearsay, others copied from secular history, Scripture, and theological tracts.[127]

In part, the *Nemesis divina* was an emotional safety valve. Here Linnaeus could express himself when inchoate feelings of doubt, fear, and a sense of the uncanny overtook him. Here he could write about sensing the ghosts of his friends, servants, and relatives. "What [is] it that bangs the wall, that haunts, when others die? This is as mysterious as the soul."[128]

Nemesis divina's central thesis, however, and the intuition that governs almost all its entries, is that God punishes sins here on earth. Linnaeus collected his hideous tales as proof of this inverted theodicy. As he saw it, they illustrated his favorite maxim: no one can escape God's vengeance on earth. He told a story, for example, of a young Baltic baron who dug up the grave of an illustrious ancestor in Uppsala Cathedral, "letting the soil run through his fingers to find rings and gold," and "treat[ing] his friends to Rhine wines in the coffin." That retribution would follow from these acts was so apparent to their chronicler that he concluded merely by asking us to "guess what the outcome will be."[129]

To guide us is Linnaeus' oft-stated belief that the nature of the punishment fits the character of the crime. Thus he drew up lists of "whores" who were punished by fire or boiling water, as if that were the end which their inflamed and inflaming passion deserved.[130] Other stories are in a similar vein. A farmer watched gleefully from the shore as his wife crashed through the ice of a frozen lake and drowned. Later his hands rotted. Karl XI confiscated noble estates. Soon thereafter, his castle burnt down. A colonel ran berserk with his sword, slashing at chained and kneeling farmers captured during the 1743 peasant revolt of Dalarna. Eighteen years later, in the Pomeranian campaigns, his face was cut to ribbons.[131]

Another officer, drunkenly steering a coach through Uppsala's cobblestone alleyways, so maimed a young boy that he become a crippled beggar. The reckless coachman returned from the Pomeranian campaign of 1760–61 as a mangled amputee. An Uppsala professor of medicine dissected the head of Linnaeus' maidservant, even though as she lay dying in the city hospital the girl had begged him to bury her intact. Later he fell dreadfully ill. Such tales go on and on, through 203 loose sheets patiently collected through the years.[132]

Nemesis divina thus offers a self-enclosed moral economy in which the immensity of suffering is matched by an immensity of sins. A telling example is the explanation for the political downfall of Linnaeus' patron, Count Carl Gustaf Tessin (who, as we noted, had been the Swedish chancellor, the Swedish ambassador to France, and the tutor of the crown prince).[133] Tessin's diaries witness that he was a cultured and humane man of the Enlightenment. Eager to encourage science in Sweden, he had housed the young Linnaeus in his town palace, introduced him in society, and secured for him the post of chief physician to the Swedish navy. As a measure of his gratitude, Linnaeus had promised himself in 1739 always to thank "God and Co. Tessin" when saying grace at table.[134]

Yet two decades later, in *Nemesis divina*, Linnaeus wrote that Tessin had bricked an English courier—alive and screaming—into a wall of his castle.[135] In a rare moment of distance from his own argumentation, he admitted that this might be hard to believe. He went on to argue, however, that reliable sources claimed a body *was* found when Tessin's castle walls were repaired, and that a courier *was* missing. He himself had noted that from the time this courier had vanished, Tessin had never invited guests. (This Linnaeus viewed as a simple, practical precaution against guests meeting the courier's vengeful and gossipy ghost.) Why did Tessin murder? For nobles, the parson's son explained in a sly aside, such crimes are mere pastimes.

To punish the kind-hearted old count (now recast as a heinous Sadean aristocrat), God removed him from his cabinet post. As Linnaeus saw it, the Deity took a more severe view of crimes directed against Linnaeus himself. During a faculty meeting at Uppsala University, just as he was arguing "an irritating matter" with a "most unfair" theology professor, this professor (Linnaeus gloated in *Nemesis divina*) "falls down under the table, is carried home; never more sees a healthy day." Taking for granted that the entire faculty concurred in his interpretation, he added: "everyone goes home and beats their chests, noting that the Lord sees our intrigues."[136]

Nemesis divina thus traced all pains back to the sins of the victims. To account for the misfortunes of moral innocents, the diarist noted simply that "God seeks revenge on children and grandchildren"—unto the twelfth generation.[137] Linnaeus' theodicy, then, explained suffering by divine vengeance. Perhaps he formulated such pessimistic thoughts in the late 1750s because he had begun to doubt his earlier, sunnier theodicy (which, as we have seen, explained human suffering in a still-Edenic world as the result of a lack of natural knowledge). Perhaps he had begun to wonder about the cold and friendless world he saw—and had created—around himself.

This world was the one Linnaeus inhabited in his decline. In the mid-1770s, he wrote in a nearly unreadable hand that he "limps, can hardly walk, speaks garbled. hardly can write."[138] What we would diagnose as a series of small strokes had rendered the old professor part-paralyzed and speechless. As visitors browsed through his collections and works, he only managed a half-smile, saliva dribbling from his mouth.[139] His once preternatural memory declined to the point where he could not securely recall his own name. In Uppsala, it was whispered that when the old cripple toppled from his chair, his family left him lying on the floor, in his own feces.

The family livelihood now came from his son and the new head of the family, Carl Linnaeus the Younger. From 1759 on, the Younger Linnaeus had been employed as a guide to Uppsala's botanic gardens. In 1763, Linnaeus the Elder had also willed his chair to "My Only Son."[140] Entailing his professorship, he argued, actually meant retaining a public space for natural history: his son's position would allow Uppsala students continued access to Linnaeus' collections of *naturalia*. Yes, Linnaeus admitted, "his son was given an authorized appointment for the chair because of Linnaeus' merits, even though he was only twenty-one years old. The father, however, remained in charge of the chair until the son was able to handle it; through this, his valuable library in natural history, his priceless manuscripts, and his incomparable herbaria could be saved" for Uppsala University.[141] But, having thus secured his son's succession, Linnaeus craftily willed that same collection to his daughters, to ensure their future livelihood as well.

In October 1777, three months before his father's death but years after he had become incapacitated, Linnaeus the Younger was promoted to full professor. Two years earlier, in 1775, the Elder Linnaeus pleaded with the university chancellor to allow the son to take over his duties. "I am old and sickly hardly tolerate the Autumn cold, aged and toothless have difficulties speaking."[142] Sweat pouring down his face, the heir lectured by reading his father's lecture notes aloud. He even spoke of "my" European travels, although he had never left Sweden. Afraid to reveal his ignorance, he refused to admit visitors into the botanic garden and the hothouses.[143]

Linnaeus' students endlessly discussed how "the Pater" had promoted "the young Phoenix L." or "the young Dauphin."[144] An anonymous pamphlet of 1763, asking why Swedes emigrated, even charged that Linnaeus' "Nepotismus" had caused "a Solander, a Falck" to remain abroad. Upon seeing this pamphlet, Falck's brother, whose opinion of Linnaeus we have already noted, wrote to Falck, then inspector of the St. Petersburg Academy of Science's botanic gardens, to congratulate him on being thus described as a national asset. "What do you say about that?"[145]

Linnaeus the Younger was not Uppsala's most despised teacher. (One law professor's "students so spat on him, that the spittle trickled from his coat.")[146] Yet compared to colleagues with scholarly reputations, he was scorned indeed. "The Young Master mostly spends his days amusing himself; doesn't ask so much after Flora as after Nymphs. Has a proud walk, clothes and powders himself à la mode, and pays constant Visits, where beautiful Women are."[147]

Nor was the younger Linnaeus alone in holding a university chair at Uppsala through inheritance. Another such professor by birth, and a contemporary of the Younger Linnaeus, was Christer Berch, son of Anders Berch, Sweden's first professor of cameralism. He too resorted to the stratagem of reading his father's lecture notes aloud. Most days he "sat handsomely powdered behind the curtains" of the windows of his town house, "and watched the people pass in the street." "Beautiful, but enormously fat," and incongruously married to his father's aging housekeeper, Berch the Younger nonetheless seemed to fill his position well.[148]

Linnaeus the Younger, on the other hand, was indecisive and timid. In 1741, his father had been overjoyed at the birth of "such a large fat and well-formed child, and what's more a son, *nota bene* a Carl."[149] But even he came to doubt his only surviving son's talents. In a manuscript of 1776 entitled "Cry from the grave to my once dear wife," which was a memorandum of posthumous wishes, he urged: "My Son ought not to have [the herbarium], who never helped me with it and doesn't have any interest in Botany but wait in case there is any son-in-law who is a Botanist."[150]

No wonder that Linnaeus' daughters were considered tempting prizes. As one Uppsala student pensively wrote another in 1770, after speculating about the content of Linnaeus' will: "3 Misses are unmarried NB."[151] Yet in 1778, and showing some scientific ambition, Linnaeus the Younger bought the herbaria, library, and manuscripts from his sisters, against 6,000 copper *daler* and some other property. The same year, in a letter to a family friend, he noted helplessly that his father's famous herbaria were getting moldy under his care, and that rats and moths gnawed at the specimens. The collections had been disturbed, he believed, when he had moved them from the family's country estate to prevent his sisters from spiriting them away. "Several tricks otherwise could have enticed them to [accept] Banks's cruel offer [to buy them]. No one steals from me a single plant."[152]

Five years after his father's death, in 1783, the Younger Linnaeus himself died, a bachelor of forty-two years of age, and Linnaeus' last surviving son. The Elder Linnaeus' only brother, the Rev. Samuel Linnaeus, wrote in a commemorative letter that of Samuel's own five sons, four "are already dead; but little Carl Samuel, who now is the only one, except myself, who carries the Linnaean name, is five years old, amuses me daily with a quick-wittedness that much surpasses his age."[153] The boy's name, Carl Samuel, was an obvious talisman.[154] Yet like his four brothers, he too died in childhood. Eight daughters remained. But except for forming marriage alliances, women were considered useless for transmitting the family's scientific traditions.[155] With the boy's death, Linnaeus' family died out in the male line and in the noble branch.

When Linnaeus is proudly dubbed "the Lord of All of Sweden's Clams," or when his students are discovered lecturing to drunken children in the provinces, it may seem that my own narrative of an intellectual project has descended into the irredeemably quaint or local. Yet only thus can we chronicle the quite specific profile of the failed and fragmented aftermath of Linnaean cameralism.

As we have seen, the Linnaeans' voyages of discovery were beset by mishaps and disasters, and by a general failure, even on the part of successful (that is, surviving) travelers, to carry on their work after the initial stage of collection. Linnaean import substitution ultimately failed, for reasons of both ideology and science. The economic theory upon which the program was founded (and in particular its producer monopolies) largely benefited rent-seekers. Indeed, the Linnaean ideal of the cameralist state assumed that citizens subordinated their labor, tastes, and fashions to the financial demands of the state apparatus. Linnaeans also assumed that nature remained in Edenic stasis, as a symbiont superorganism created for the benefit of man. But as Löfling's "tragic thesis" of 1749 first suggested, tropical plants could not "grow by us."

In nineteenth-century Scandinavia, Linnaeus' merging of nature and nation was forgotten as an embarrassing and minor intellectual detour made by a man now reconstructed as scientific genius. Yet his physico-theological cameralism deserves to be remembered. It was a guiding preoccupation for Linnaeus. More broadly, its very existence suggests that imperial design may at times be a contingent policy, not a goal in itself.

From c. 1720 to c. 1860, the Swedish state elites hoped for economic independence. Yet import bans failed; colonial crops wilted; and replacement plants proved to be chimera. Perhaps a cameralist economy could only function by means of a tropical tributary? In the 1790s, confronting their triple failure of transplants, substitutes, and import bans, the Swedish state elites did indeed turn to mercantile imperialism. When Linnaeus' last student explorer, Adam Afzelius, left for Sierra Leone in 1792, he was instructed to search out lands for the Swedish flag. In this local case at least, imperialist designs were not in themselves a historical agency. They were one particular expression of the broader ideology of cameralism, or what I have here called the ideology of a local modernity.

CHAPTER 8

"His Farmers Dressed in Mourning": The Fate of Linnaeus' Ideas in Sweden

In the nineteenth and early twentieth centuries, Carl Linnaeus became a Romantic and nationalist icon for Sweden. There, as in his own work, ideas of nature and of nation were perennially in dialogue. In time Linnaeus was thus transformed from a classifier of nature for a future nation into a national exemplar from a mythically natural past.

By the 1770s Linnaeus' legendary career had been secured. But he himself was emaciated and crippled by strokes. As he struggled to describe himself in 1775, he was "so sick that he hardly can talk."[1] That his Uppsala community worried can be seen, for example, in a poem by "J. S.," printed in a local newspaper in 1776, on "Linnaeus' refound health, after he suffered gangrene." It naïvely opens with the query: "But is Linnaeus ill? My sad heart asks."[2]

Neglected and maltreated by his kin, and always in pain, Linnaeus arrived at a "quiet death which ended all tortures, on 10 Jan. 1778, at the age of 70 years and 8 months." As a student would remember, "Linnaeus' funeral in Uppsala Cathedral was the grandest Act I had ever seen, and it deeply impressed me." Linnaeus received what the nineteenth century later came to call a beautiful funeral: "It was a melancholy and quiet evening, and the darkness was only lit up by the torches, flares, and lights of the Procession, slowly marching through town—his farmers, dressed in mourning, followed the carriage with torches."[3]

In 1783, five years after Linnaeus' death, Carl Linnaeus the Younger, the last survivor of the six sons Carl and Samuel Linnaeus fathered and the only one to have lived beyond childhood, died a bachelor and without issue. With the extinction of the Linnaeus family in the male line, and simultaneous with the advent of Romanticism and its emphasis on the singular, autochthonous "great" individual, the interpretation of his achievements shifted away from the familial context in which Linnaeus and his kin had inscribed him. What counted henceforth was the analysis, remembrance, and celebration of one man.

The earliest posthumous biographies of Linnaeus continued the stereotyped and generalized praise offered him during his lifetime at home and throughout Europe.[4] (In Scandinavia, these early hagiographies often took the form of semi-anonymous occasional verse.) Memorial addresses—based broadly on the *vitae* that Linnaeus had lodged in such institutions—were delivered at the Swedish Academy of Science (1778), the Montpellier Academy of Science and the Arts and the Paris Royal Society of Medicine (both in 1779), the French Academy of Science (1781), the Linnean Society of London (1788), and the Société Linnéenne de Paris (1788).[5] Lives and eulogies of Linnaeus were published in science journals, popular magazines, and newspapers all over Europe throughout the 1780s and 1790s.

The Romantic interpretation of Linnaeus began with the *minne* written in 1778 by his younger brother (but published much later). In his richly evocative memorial, which became a key source for later Scandinavian nationalists and Romantics, the Reverend Samuel Linnaeus depicted his elder brother's life as the unfolding of a destiny, and his childhood as a series of auspicious signs. Yet in Sweden, tributes to Linnaeus tapered off soon after his death. After all, Linnaeus had promised the state elites economic independence through import substitution, and this cameralist program had come to nothing. Alongside the ascendancy of German Romantic morphology within what could now be constituted as the life sciences, this political disappointment may explain why, during the first half-century after his death, Linnaeus was hardly memorialized in his homeland.

How the political climate changed in Sweden in the later eighteenth century, in Linnaeus' old age and after his death, is exemplified by an annual prize question offered by the Swedish Academy of Science and its aftermath. In 1763, the (baseless) query was: "What can be the causes of such a multitude of Swedes emigrating each year?" The Academy added a second question that typifies the eighteenth-century faith in the possibility of a perfectly managed economy: "And through what laws and ordinances can this be prevented?"[6]

Many of the more realistic respondents, however, traced the cause to Sweden's "law and ordinances" themselves, that is, to its cameralist economic legislation. It placed, they claimed, an impossible stranglehold on liberty and commerce. So common was this subversive and (to the Academy) surprising argument, that the Academy—that veritable beehive of rent-seekers and champions of cameralism—refused to publish the responses, contrary to established custom. This in turn occasioned a huge scandal. A war of pamphlets was launched. For the first time, Swedes publicly queried the fundamental morality, as well as the economic efficiency, of cameralism. Two years later, in the 1765–66 parliamentary session, and to some extent because of this debate, the Hat government fell after a twenty-seven-year reign.

The new Cap government immediately abolished subsidies to manufactures. As a result, the number of workers fell by some 37 percent over the decade. Alingsås textile manufactures collapsed, going from 1,100 to 175 laborers.[7] The Hat party's economic policies, with which Linnaeus was so closely linked, were now widely blamed and criticized as unfair and ineffective. Nonetheless, if only because of his aging students' academic positions, Linnaean cameralism lingered in the universities around the Baltic Sea. As we saw, Linnaeus had helped establish chairs in "practical economics" at Åbo University in 1747, at Uppsala University in 1759, and at Lund University in 1760. All three first occupants had been students of Linnaeus'. Pehr Kalm at Åbo University and Erik Gustaf Lidbeck at Lund University even specialized in botanic acclimatization experiments.

As late as 1784, in a last-ditch effort to retain cabinet subsidies for Lidbeck's Skåne Plantation Bureau, the chancellor of Lund University revived Linnaeus' botanic transmutationism. He wrote the king that greenhouse-reared mulberry saplings could be trained to tolerate the harshest winters.[8] Unimpressed, the king cut further funding. For in the new climate of criticism, educated Swedes increasingly questioned the economic utility of both cameralist legal measures and botanic transplant schemes. After all, Linnaeus and his students had repeatedly, indeed spectacularly, failed to naturalize exotic plants.

By the 1780s, Linnaeus and his students had become the butt of parody in newspapers, novels, and magazines. One such spoof, published in the newspaper Stockholms-Posten in 1781, presented an imaginary Linnaean botanist, "Henric Durr," who on his travels supposedly discovered many more plant genera than even Fredrik Hasselquist had on his journey of 1749–1752 to the Levant. "And since Hasselqvist only discovered 100, Mr. Durr's value compared to Hasselqvist's is like 1906:100 or as around 19:1. Happy the Botanists who possess such a well-defined measuring rod to measure their genius."[9]

Interestingly, the anonymous creator of "Durr" had already forgotten the economic impulse behind Linnaean botany. As he ironically put it: "As long as Botany is studied in this old praiseworthy way, to enumerate, describe, and draw plants, without bothering about a heap of details, such as their use in Medicine, their use in the Manufactures, etc., it cannot but win important progress, precious for humanity."[10]

Around the same time, former Linnaeans themselves turned into ironizing Romantics. They now mocked their erstwhile teacher's once fashionable excursions around Uppsala. Aging converts to the virtues of self-parody, they described their former selves as a "Grass-hunter troop under the so very famous Sir and Knight von Linné." They recalled their botanical ramblings thus: "We damaged the meadows so much by our tramping that they turned into a few bales of hay, about which no one dared to complain so as not to be named an enemy of the natural sciences, which in those days was considered high treason at the university, even though no opportunities were missing to find a thousand plants with less trouble in the botanic garden."[11]

The easiest target for post-Linnaean gadflies was Linnaeus' erstwhile student Carl Peter Thunberg, the old Japan traveler who now held Linnaeus' chair at Uppsala. In Fredrik Cederborgh's hugely popular novel *Uno von Trasenberg* (1809–10), one character reads aloud from a "newspaper article" made up of "observations" pasted together from Thunberg's ponderous *Resa* (published 1788–1793). Cederborgh's hatchet job achieved cult status. Manfully attempting to incorporate a sense of light irony in their national repertoire of intellectual skills, educated Swedes gleefully quoted such Thunbergisms as: "Water is that element which makes sea journeys both outside and inside of the Netherlands so nimble and comfortable."[12] Quoted in Sweden even today, is Thunberg's immortal comment upon first entering France: "To me it could not but seem both strange and ridiculous to hear Burghers and Farmers all speak that, in other places so noble, language."[13]

Pehr Osbeck and Olof Torén, those earnest and heavy-footed China travelers, were similarly mocked. Osbeck bitterly commented that during his journey, "when a Captain happens to see a stereus humanum on board" and jokingly demands that the resident naturalist examine it, "then it isn't strange that times are tedious and the trip seems longer."[14] Torén and Osbeck were also pilloried in public by a fellow East India chaplain, Jacob Wallenberg, whose comic travel diary of 1769 became a publishing sensation in 1781. The author begs a fictive Linnaeus not to make him into a disciple. "I must kneel for his majesty of the kingdom of plants, duke over crocodiles and mermaids, etc., and lord of quadrupeds, birds, and insects, our great knight Linnaeus, asking most humbly to be freed from these stony excursions."[15]

Linnaean natural theology, too, was mocked. When an aging professor of economics at Lund University wrote a typically Linnaean dissertation entitled "On God's Wonders of Nature," the *Stockholms-Posten* review simply noted: "among which this work however hardly can be counted."[16] Further afield, Linnaeus' classificatory pedantry was turned to other, sometimes more political ends. In Austria in 1783, Ignaz von Born published an anticlerical *Monachologia*, which was translated into German, French, English and Italian, and set out "the Natural History of the various Orders of Monks after the manner of the Linnaean System."[17]

As the century drew to a close, there were probing criticisms of Linnaean economics as well. In 1804, traveling through Scandinavia, the German Romantic Ernst Moritz Arndt acutely targeted the acclimatization schemes that underwrote Linnaeus' cameralist enterprise. "One was supposed to believe that Sweden suddenly had become Asia Minor and Sicily." He especially attacked Sweden's "unnatural" silk factories. The Linnaeans, he charged, "did not realistically face the geographic situation and economic realities of their country, and foolishly they worked against nature itself."[18] Mocking the sunny natural theology that supported Linnaean economics, Arndt noted that God had given Scandinavia herrings but no salt. Her manufacturing, he continued, would always remain insignificant compared to her North Sea fisheries.[19]

Arndt perceptively analyzed Linnaeus' cameralism in terms of a rent-seeking society. "On the whole, one can probably assume that the projects and hopes were partly fantastic and chimerical; that individuals perhaps needed them to enrich themselves. . . . Much money was wasted, trade confused, profits [made] by means of fraud, and the desire to cheat promoted through unnatural subsidies; and the character of the people destroyed because of the many coercive laws."[20]

Arndt's criticism extended to the intellectual and material decay of Sweden around 1800, too. Even the purely scientific collections of the Linnaeans were sorely neglected. This became clear to him from meeting Linnaeus' last student, Adam Afzelius, who had explored West Africa between 1795 and 1796. As Arndt noted, Afzelius was given no space to exhibit his magnificent Sierra Leone ethnographia. Arndt simultaneously marveled at the collection—"how many household goods, weapons, metal works, and leather works are not better made than in London or Paris"[21]—and lamented its pitiful state of preservation.

In 1798, the English traveler Edward D. Clarke observed a similar state of affairs upon meeting Linnaeus' student Anders Sparrman, who had accompanied Captain Cook on his second voyage, and had explored southern Africa between 1775 and 1776, and Senegal and Cape Verde in 1787. On returning to Sweden, Sparrman became curator of the natural history collections at the Swedish Academy of Science, and later, professor at the Stockholm College of Medicine. Yet Clarke was scandalized to see that after two decades, Sparrman's African and Pacific collections remained stuffed in their shipping crates.[22]

Even the anthropological study of these objects, apart from the Linnaeans' hopes to use them as vehicles for economic instruction, had come to naught. Visiting Uppsala in 1804, Ernst Moritz Arndt met Linnaeus' student and successor, Carl Peter Thunberg, who had explored southern Africa, Ceylon, Java, and Japan between 1770 and 1779. He was now an old professor, "a great curiosity," who "still" lived in Linnaeus' house amid the botanic gardens, "now falling into ruins."[23] The interiors of his home, Arndt gossiped, were chaotically cluttered, with wooden crates piled on top of each other. Despite the aged man's pleas, Uppsala University provided no exhibition space for his Japanese and Sami ethnographia.[24] Together, the graybeard and the youngster rummaged through boxes in search of "Lapp glories," which Arndt deemed, alongside Thunberg's Japanese collections, as "in parts quite amusing."[25]

Well into the nineteenth century, Thunberg dutifully authored dissertations on "Foreign Trees, Bushes and Flowers, that can Tolerate the Swedish Climate" (1820) and "Native Trees and Bushes, that are Worth Cultivating" (1821). But these take their place next to his 291 other, and non-economic, dissertations. Thunberg was a distinguished botanist. His real interest lay in replacing Linnaeus' sexual system of plant classification with a natural order, and his botany remained contiguous with his erstwhile teacher only in that both sought to identify the fauna and flora of the Bible. (Both, too, employed their sons as demonstrators in the botanic garden.)

Yet despite Thunberg's labors, around 1800 Scandinavian natural history was in shambles. Clarke noted this when he attended an Uppsala lecture by Thunberg in 1798. Six slovenly, impoverished-looking boys, age fourteen and under, turned up for the event. While Thunberg rambled, the boys broke into a fistfight. As Clarke explained to his English readers, this was no surprise, seeing how much aquavit these children drank.[26] (Swedish historians have interpreted their behavior more probingly, by noting that it may have been a seminar open solely to noblemen.)

In 1809, the young Carl Adolph Agardh, later a noted biologist, similarly reflected in his diary on the decay of Linnaean natural history when he first visited Uppsala's botanic garden: "Only 4 flowerbeds with plants. The greenhouse wasn't rich either. . . . In the menagerie only a little baboon, sick from tapeworms." Agardh also enumerated Uppsala's puny collection of stuffed animals: "A Gnu. Gazelles, one bear, one Tiger, one Reindeer, etc."[27] By 1812, another English tourist noted that much of Linnaeus' old botanic garden was now a potato patch. The greenhouse windows were smashed, the stone floors buckled and cracked, and everything overgrown with weeds.[28]

In the spring of 1826, Carl Adolph Agardh, now a professor of natural history at Lund University, presided over the university's doctoral festivities. As he mounted the speaker's rostrum—he used the bishop's pulpit in Lund's great Romanesque cathedral—the forty-one-year-old academic could look back on a typical, even exemplary, Linnaean career. Like Linnaeus' students a generation before, Agardh came from a simple family and a provincial milieu. His father was a small-time merchant in the Skåne fishing village of Båstad. As a boy, Agardh took private lessons in natural history from Linnaeus' student Pehr Osbeck. Osbeck, who traveled to Guangzhou between 1750 and 1752, ended his career as a country parson at Hasslöf, three hours walk from Båstad. There, the naturalist—who later became the model for the much-beloved Swedish stereotype of the Linnaean parson—doctored and lectured his farmers, tutored boys, grew chestnuts and mulberries, arranged his modest collections of local insects and Chinese *naturalia*, and wrote the Scandinavian equivalent of Gilbert White's *Natural History of Selbourne* (1789), the *Sketch of a Description of Laholm's Parsonage* (first published in 1922, and an immediate success).[29]

Upon finishing his training with Osbeck, Agardh supported his university studies and financed his European travels by serving, as so many of his predecessors, as a tutor in a noble household. In 1809, he also embarked on a voyage of discovery through central Sweden that perfectly recapitulated the provincial journeys Linnaeus and his students undertook from 1732 on.[30] Agardh's higher research, too, looked orthodoxly Linnaean. In 1820, 1826, and 1828, he published installments of an all-encompassing survey of the world's algae, *Species algarum*.[31] His vernacular economic treatises have a familiar ring: "On the Fodder-beet's Use for Sugar" (1812); "On the Species of Sea-weed Found by the Beaches of Gothenburg and Bohuslän Provinces, and about the Way and Value of Using Them in Agriculture" (1816); and "On the Improvement of Tobacco Plantations" (1819).[32]

Agardh's institutional career was also typically Linnaean. His Lund professorship in natural history and economics, which he received in 1812, had been founded on Linnaeus' recommendation in 1756. The chair was initially financed by a public industrial investment fund (*manufakturfonden*). Its first holder was Erik Gustaf Lidbeck, the pupil who took part in Linnaeus' Västergötland journey of 1746, and whom we have encountered as the creator of the Skåne Plantation Bureau and as a producer of raw silk.

If anything, Agardh's career looks belated. It closely approximated that of Linnaeus' friend and contemporary Johannes Browallius, in its turn a model for social advancement by way of natural knowledge in eighteenth-century Sweden. In 1737, Browallius was named professor in natural history at Åbo University. Two years later, he was ordained, and took on a parish, a job for which—as colleagues noted "with wonder and astonishment"—he learned Finnish, merely in order to be able to speak with his flock of farmers.[33] In 1746, he switched to a chair in theology. And three years after that, he was made a bishop. Agardh, too, joined the church early. In 1812, he was made a professor. Four years later, he took orders. And in 1835, he was made a bishop. If anything, Agardh was more old-fashioned than his predecessor. Browallius devoted his mature energies to proving that Gothic (old Swedish) and Finnish "are the mothers of all other European languages." Agardh spent his old age persecuting Baptists.[34]

Agardh's social background, formal education, early publications, and institutional career thus all seem quintessentially Linnaean. But their spiritual impulses differed profoundly. The Linnaeans were utilitarians, handymen, as it were, of the Enlightenment. Agardh was a conservative Romantic. Together with his friend, the famous Romantic poet and professor Esias Tegnér (also later a bishop), Agardh saw himself as a universal genius, engaged in higher learning for its own sake. He was at once a botanist, mathematician, politician, and theologian. His youthful essays in economic botany, which follow the Linnaean recipe of minute observations of *naturalia* contained within a grand scheme of industry, were career-minded gestures, never to be repeated later on in his life.

Instead, Agardh interested himself in plant anatomy and plant physiology. For philosophical inspiration, he turned not to Linnaeus' classificatory schemes but to Friedrich Wilhelm Joseph von Schelling's idealist and Romantic philosophy of nature. And the task he set himself, and upon which his science foundered, was in the end a universal natural system. His *Species algarum* was less a taxonomic list than a Romantic fragment, that is, the intimation of a self-consciously infinite project that not only was never finished but that *could* never be—and indeed was hardly intended to be—finished. Dozens of volumes were projected. Only one and a half were ever published.

Even though these volumes look back to Linnaeus for their focus on classification, Agardh's outlook as a scientist differed profoundly from that of eighteenth-century "natural historians." He had already begun to absorb the revolutionizing perspective of what would come to be termed "biology": biology that emphasized the function of individual organisms (how their parts fit together to maintain a living existence), rather than the study of the organism's kinship relations to other organisms and to inorganic nature (as visualized by the curiosity cabinet, which brings together minerals, plants, animals, and ethnographia under the purview of the superdiscipline "natural history"). Agardh, as a biologist, was convinced that what he studied was the mystery and specificity of life itself.

Given such broad and irreversible transformations in the life sciences after around 1800, in which Agardh participated in his own small way, it is not surprising that the topic of his lecture at Lund University's doctoral commencement in 1826 should have been the profoundly belated subject of *antiquitates Linnaeanae*. Agardh did not offer his audience new research carried out in the footsteps of Linnaeus, something which his audience could have expected, and for which his career and published work would have prepared him. Instead, he presented his mighty predecessor's work as something truly past—not as an ongoing science, but as the object of antiquarian interest.

He did so with devastating effectiveness, simply by reading aloud from the bishop's pulpit some letters by Linnaeus. Before he crowned the doctoral candidates with laurel wreaths, he concluded his remarks with what he explained was Linnaeus' last letter, addressed to his "dearest," "very dearest Sir and Brother, Abraham Bäck."[35] Written on 5 December 1776, its author misdated it by twenty years, as "5 Dec. 1756."[36] Linnaeus had written, folded, addressed, and sealed this letter, but then left it on his desk. Perhaps he forgot it. Perhaps he no longer recognized what it was, a letter ready for the post. After his death, his son had found it and sent it on. More than a year after it had been written, it had thus greeted Linnaeus' old friend from the grave.

Wracked by apoplectic strokes, Linnaeus had scratched a few lines intended to comfort Bäck for the death of his only son, a seventeen-year-old youth and Linnaeus' "little Brother," and to bid farewell to "my best friend in this world." For Linnaeus, too, was dying and knew it.[37]

Half a century later, Agardh used the same lines to bid farewell to Linnaeus. Standing in the bishop's pulpit that May day in 1826, he began reading in mid-letter. It made what he framed as Linnaeus' last words sound even more uncanny: "God has decided to dissolve more than half of those ties that keep me in this world. *Vale*—dissolve—only way out—that way out be—be dissolved—Live well, I am my Brother's; Brother is mine; I am my—my Brother's constantly unto death faithful Brother Linnaeus."[38]

With these lines, Agardh bade farewell not only to a particular naturalist or to a particular form of knowledge, but to an entire moral and political dream. For by 1826, Linnaeus' project itself, as I have outlined it in this book, would certainly have seemed like an *antiquitas:* the cameralists' hope for a self-sufficient Swedish nation with an imperial economy complete within its bounds.

Indeed, the last formal expression of this dream was a farce. In 1832, Sweden's first Linnaean Society (founded on a Baltic island in 1808) inaugurated a Stockholm acclimatization garden while dining next to "a Nature temple around an altar to Linnaeus."[39] At the time, Sweden was still a backward agricultural nation. The material circumstances favorable to Linnaeo-cameralism remained in place. So did its imaginary fancies. We recognize the hopes that this Linnaean Society now condensed out of a half-forgotten past: its "Tables" recommended "Satyr-Apes" as factory workers and servants to "fetch water, rotate steak spits, pound spices." It advocated "House-dogs" and "Turkish Skunks" for wool; cats, moles, and weasels for furs; and kangaroos, gerbils, and anteaters for flesh. Once more, a future order was projected in which mongoose and skunks chase rats; beavers and seals catch fish; and camels and llamas pull carts.[40]

Again, gradual climate acclimatization was envisioned, although now over generations and on the basis of the inheritance of acquired characteristics. Taking chinchillas as an example, the Society suggested they first be housed in a hothouse, in the next generation in mild heat, "and their off-spring finally in an unheated room. A manner of rearing that ought to be followed in domesticating all animals from hot Zones."[41]

The Society also, typically, hoped to domesticate indigenous flora and fauna. They wished to tame species that are protected as game today, but were shot by poachers then, such as European moose (*älg*) ("since this Grand Animal will soon become extinct") and red deer (*kronhjort*) (which "soon will no longer exist in our country").[42]

This Linnaean Society unwittingly engineered its own downfall by plotting to charge admission to a public park in Stockholm, Humlegården, where they gardened and reared their animals—beavers, parrots, goldfish, guinea pigs, and hares. Incensed locals smashed the busts of Linnaeus and Flora, chopped down the dwarf almond trees and the jasmine bushes, and vandalized the greenhouses where grapes and pineapples grew. By 1838, the Society had sold its carousel, abandoned its "Hermit House," leased its hay-meadows and orchards to an apothecary, and given up on its "little houses to sell sweets and fruits from."[43] When the society formally dissolved itself, in January 1842, only four members were present. Neither the chairman nor the secretary turned up. After a long wait, the deputy secretary took it upon himself to bid "a sad FAREWELL!"[44]

During the Romantic era, Sweden became for the rest of Europe what Lapland had been for Sweden in the Enlightenment: a holiday land for the ethnographic tourist and a natural utopia. In a newly industrializing world, Sweden's dependence on Europe and its marginal status were made ever more obvious, while post-Linnaean Lapland became an exotic stage set for Romantic travel and for antics and jokes. When the Englishman Edward D. Clarke met his first Sami, an old woman in Luleå, he promptly kissed her. One Sunday in July 1799, at Enontekis parish church, three days north of Torneå by reindeer-sled, he flew a seventeen-foot white satin balloon—the first, he prided himself, to float above the Arctic Circle.[45]

In the Scandinavian journeys of Clarke, Mary Wollstonecraft (1796), Giuseppe Acerbi and Bernardo Bellotti (1798), Thomas Malthus (1799), and Ernst Moritz Arndt (1804), Linnaeans even found *themselves* to be the objects of foreigners' natural histories. But first and foremost, these ethnographic tourists were interested in what they cast as surviving Viking customs. They were thrilled, for example, that in northern Sweden farmers still used rune staffs. Clarke took notes on folk dancing. Acerbi, inspired by

Scandinavian folksongs and by Sami formulaic oral poetry (*jojk*), arranged for a quartet to perform his own "runic" melodies in Uleåborg, a northern Finnish town.[46]

Arndt recorded folk superstitions and admired Scandinavia's light-haired children. He noted that her oldest aristocratic families "are all blond," and argued that farmers who homesteaded near pagan sacrificial groves, and therefore supposedly were Viking descendants, had hair in "the old highly honored golden yellow color."[47] In a similar racialist vein, Clarke found the supposedly mixed-blood Swedes below the fifty-ninth north latitude (equivalent to the southern tip of Greenland) to be dirty and dishonest. But he enthused about the supposedly purebred Lapland homesteaders, and compared their horn-blowing milkmaids—he meant this as a compliment to their sensuality—with "South Sea savages."[48]

Included as tourist attractions of the north were the Linnaean naturalists of Europe's most Arctic outposts. Both Acerbi and Clarke were amused by how the drunken, slovenly appearance of the Lapland parsons contrasted with their fluent Latin. Even in the remotest parts of Lapland, they noted, every doctor, apothecary, and cleric possessed a natural history cabinet. Like other Romantic tourists in search of sentimental souvenirs, they bought herbaria and Linnaeana from these aging and impoverished students of the master.[49]

This had become a fashion since 1783, when a young and wealthy Englishman, James Edward Smith, had purchased Linnaeus' natural history collections, library, and manuscripts, and shipped them to London. (Smith received Linnaeus' correspondence for free, since Linnaeus' thrifty widow used it as padding in the shipping crates.)[50] While Linnaeus' treasures thus found their way to London, the book-boxes of North Scandinavia's itinerant booksellers, as the tourists of around 1800 noted, contained only Bibles, prayerbooks, and almanacs. As Clarke remarked, by 1798 Linnaeus' *Flora Lapponica* was as likely to be found in Lapland as the Koran.[51]

Lapland's Romantic travelers and ethnographic tourists were followed by big game hunters, such as Paul du Chaillu, discoverer of the African gorilla and African pygmy people; Prince Roland Bonaparte; and the English captain Alexander Hutchinson, author of the briskly titled *Try Lappland* (1870).[52] By the 1880s, the Baedecker guides asked English polite society not to take offense at the democratic familiarity of Norwegian peasants when shooting snow grouse on the Lapland tundra. It was wiser, they noted, to treasure as a quaint remnant of Viking custom the fact that the gunbearers sat down to eat together with the shooting party, shook their hands, and refused tips.

By the late nineteenth century, too, Scandinavian farmers were presented alongside Sami nomads as northern savages. In the 1878 Paris world exhibition, Swedish peasant women were displayed next to the Sami. In the 1893 Chicago world exhibition, Swedes were even paraded *as* Sami. And despite vigorous protests from the National Federation for Preserving Swedishness Abroad, in the 1911 Nordland exhibit on the Kurfürstendamm in Berlin, Swedes were exhibited alongside Inuits, Sami, and Samoyeds.[53] Scandinavians themselves, in their quaint backwardness, had become part of Arndt's rustic "Lapp glories," "in parts quite amusing." Perhaps we can pinpoint the closing date for such ethnographic delights at 1923, when the city of Gothenburg planned for its tricentennial celebration a "Lapp camp." When protests by both Sami and Swedes forced the municipal council to cancel the project, it instead constructed a "Lilliput Town"—peopled by Danish dwarfs.[54]

Given the failure of the Linnaean project and the general backwardness of Sweden around 1800, it was no wonder that soon after his death, Linnaeus was largely forgotten in his homeland. His first public monument, which remained his sole official commemoration until 1811, was a modest funeral plaque placed in Uppsala Cathedral in 1798, eight years after Jardin des Plantes in Paris acquired a bust of him, and twenty years after his death.[55]

Admittedly, Gustav III (ruled 1772–1792), the Swedish monarch who most closely modeled himself after an enlightened despot, planned to erect a botanic auditorium as a shrine to Linnaeus as early as 1787, when he donated the Uppsala Castle gardens to the university. The king and his French architect, Jean-Louis Deprez, imagined a grand Doric structure for Linnaeus' Temple. Its central hall was to be decorated with a frieze showing Linnaeus led through nature by a Goddess of the Enlightenment, and with a monumental statue of the naturalist as a muscular Greek hero.[56] Plans were shelved in 1792, however, when the king was murdered by a disaffected aristocrat and "lover of liberty." A much scaled-down auditorium was inaugurated only in 1807, one century after Linnaeus' birth.[57]

The Linnaeus statue that Gustav III intended as the building's centerpiece was commissioned in 1822 and finished in 1829. By then, this project was a historicizing gesture toward a distant past. It was initiated by Uppsala's student fraternities and one of Sweden's greatest poets, Erik Gustav Geijer, a leading member of the pan-Scandinavian and Romantic Gothic Association (*götiska förbundet*).

A few days before the unveiling ceremony in 1829, the governor of Uppland county complained that Linnaean devotees "fawned and fussed about some sort of party when Linnaeus' statue is to be uncovered." But, he wrote, "since most of the Botanic Auditorium where the statue is set up is filled with recently arranged stuffed animals," guests would have to be shoehorned in among the beasts on display. "It will become a ridiculous celebration."

On the night of the festivities, when the resident burghers of the little university town were let in to view the statue, the botanic auditorium become so thronged that the archbishop himself, a candle in each hand, had to shepherd the crowds out through a back door. Even so, the county governor, responsible for policing the affair, admitted some success. "The mass of students and the little old ladies von Linné, the spinster among whom nearly swooned as the statue was uncovered, did stir one's feelings."[58]

The Uppsala statue of 1829 was the first major representation of Linnaeus commissioned by a Swedish institution since his death in 1778. Arguably, it marks the beginnings of Sweden's Romantic nationalist (*nationalromantiska*) cult of Linnaeus. It was uncovered in the din and clatter of 120 canon shots, hurrahs, and the chant of a song entitled "God Preserve Our King." During the dinner that evening, numerous toasts were proposed. As was typical of the period, each was rounded off by a song written especially for the occasion. The guests—local notables, blood relatives of Linnaeus', Uppsala University professors, and representatives of various national institutions—drank to the king, to Linnaeus' students and relatives, to the sculptor, to Uppsala University, and so on through the night. The key toast, however, was not offered to Linnaeus himself or to his natural history. It was to something once removed: the *memory* of Linnaeus. The 1829 Uppsala celebration thus accords nicely with the tenor of the 1826 speech at Lund University by Carl Adolph Agardh, which similarly bracketed Linnaeus and his science as "Linnaean antiquities."[59]

After the mid-nineteenth century, that memory of Linnaeus enjoyed a renaissance in Scandinavia. Linnaeus' birthplace was made a museum in 1866; his country estate in 1879; his father's parsonage in 1935; and his Uppsala town house in 1937. "Linnaeus liqueur" and "Linnaeus cream cakes" were manufactured and sold in cafés. The cakes, glazed in baby-blue sugar and topped with a buttery profile relief of Linnaeus, are still served to-day in one Uppsala coffee house.[60] Temperance lodges, youth clubs, local cultural associations (*hembygdsföreningar*), children's magazines, and babies were all named after the great Swede.[61] Hagiographic pamphlets, medals, exhibitions, songs, statues, festivals (and once even a ballet) were produced. From around 1900, May 23 was often celebrated as Linnaeus' Day. Flags were flown, schools closed, parades and dances arranged, and "Linnaeus flowers" peddled. In addition to local panegyrics, typically written by par-sons and schoolteachers, the poet Carl Snoilsky's "Prince of Flowers" was read out in town squares and school auditoriums.

The cult of Linnaeus reached its apotheosis in the bicentennial jubilee of 1907, which was celebrated across Sweden with huge if incongruous "Linnaeus and home county [*hembygd*] festivities."[62] Uppsala University threw a grand party. The archbishop, crown prince, university president, and student representative all made (as one newspaper reporter nervously reported) "striking speeches in ringing Latin." The most prolonged applause, however, was reserved for when the representative of Finland spoke—in Swedish.[63]

On May 23 of that year, the railroad tracks of Småland were choked with extra trains traveling to Linnaeus' birthplace. In towns all over the country, "hundreds of schoolchildren with small Swedish flags in their hands" jostled student choirs singing "patriotic songs."[64] To prepare the ground, that morn-ing's newspapers had admonished their readers: "The public is requested to participate in the Linnaeus celebration by means of a general flying of flags! All flags to the top of the flagpole today!"[65] Newspapers also urged their readers to support the schoolgirls who would peddle "little, cute" paper flowers in honor of Linnaeus Day and for charitable causes such as tubercu-losis sanatoriums. "Don't forget to buy Linnaeus flowers!"[66]

In one southern town, Helsingborg, almost five thousand "children in na-tional dress, the girls with flower garlands in their hair and flower festoons on their dresses and the boys with blue-yellow marshal's ribbons," marched through the town center, behind a sea of flags. Such parades were staged in most Swedish towns. Even hamlets celebrated. In one typical scene, in Råå village, Raus parish, a village schoolteacher named August Chronquist "in-terpreted in a brief speech the flower king's deeds and noble personality," had the children sing "patriotic songs" and shout hurrah for Linnaeus, and closed the school for the afternoon.[67] Other schoolchildren were marshaled for celebratory weeding of potato fields, or planting of pine forests.

It should be recalled here that in the late nineteenth century, Swedes emigrated in huge numbers to America. In the 1880s, at the height of the outflow, about one percent of the population left each year. At that time close to eighty percent of the population still worked in agriculture. Indeed, Sweden was approaching a precarious "Irish" state of affairs, with potato-eating peasants farming tiny plots of land. No wonder that as "America letters" began arriving in the home parishes, describing a New World where land was free and where it was legal even to form Bible study groups, others began packing their "America chests." Returning for visits, some "America Swedes" even sported that dazzling sign of success: gold teeth.

But Linnaeus, being a patriot, had never emigrated to Minnesota, as a bishop pointed out at one Linnaeus festival in 1919, expressing the anxieties of a state elite that for the last half-century had watched as the disenfranchised lower orders voted with their feet. Throughout his rambling speech, the bishop emphasized the children's "duty" to remain in Sweden. He closed his sermon by reading aloud the defensive last lines of the Swedish national anthem, itself a nineteenth-century "invented tradition" aggressively positioning itself and its higher spiritual values against the mere material well-being of the American Midwest. "May all of us always want to exclaim: 'Yes I want to live, I want to die in the Nordic countries!'"[68]

Around the turn of the century, at the height of the Linnaeus cult, the man was even presented as a saint. Supposed "folk myths" about the scientist were presented as biographical facts even in scholarly journals. One typical such "legend" found its way into Swedish newspapers in 1907 (its sentimentality is a sure sign of its fabrication). One Sunday, the story went, "little Carl" was missing from his home. Suddenly, his parents remembered how he had prattled about rare plants growing on some steep cliffs nearby. Rushing there, they found the toddler "having fallen asleep just at the very edge of the rock face, but with a rare plant between his hands, folded as if in prayer. Now the father did not have the heart to scold him, but lifted the little boy with the prophetic cry: 'You will for sure become a flower king!'"[69]

Tales of Linnaean cultists' personal experiences, too, took on a mystical character. In 1920 one Linnaean tourist described in a newspaper article his visit to the cloister ruins that once housed Sweden's two principal medieval saints, Eric the Holy and Saint Bridget. At a heightened pitch he described how he then entered Linnaeus' Uppsala garden, and at last "felt that I stood on *holy ground,* and it was a *ver sacrum*—holy spring."[70]

An Uppsala professor struck this same tone when he described in a 1924 newspaper article how, traversing Småland, he finally ascended to a small wooden hut, Linnaeus' place of birth, as if to the birthplace of Christ: "Unconsciously we hasten our steps, our cheeks grow red, our eyes shine, our hearts beat faster. Only a few more steps and we stand on a place holy to all Swedish minds, we contemplate that wonderful daylight against which the newly born Carl Linnaeus for the first time opened his eyes at that first dawn of that May morning."[71]

A poem of 1907, framed and hung on the wall of this "holy" curate's cottage, describes the environs in analogous terms.

> Between the spruce the birch stands,
> and speaks Swedishness autumn and spring
> and on each birch trunk,
> the name of Carl Linnaeus is carved."

The poem then closes by echoing a well-known Swedish Christmas hymn: "To us was born a flower king."[72]

Around the turn of the century, then, Sweden's political conservatives fashioned the commemoration of Linnaeus into a nationalist spectacle. At a time of rapid industrialization, mass emigration, and social unrest, they launched the Enlightenment naturalist as a compromise figure of royalty, a "flower king."[73] This industry of memory culminated in the bicentennial of 1907, where each keepsake and souvenir (penny prints, jubilee programs, and special newspaper runs), and each gesture and performance (school songs, workers' dances, and public speeches), was configured as a benediction to class unity and national greatness.[74]

In the Romantic nationalist conception, Linnaeus foreshadowed a second and less bellicose Era of Greatness (*storhetstid*), when Sweden would recapture, through science, "the honor of victory" that her two other national heroes, Gustav II Adolf and Karl XII, had supposedly garnered in the Thirty Years War (1618–1648) and the Great Northern Wars (1700–1718). Only this time, the victory would be "without blood."[75]

One frequently cited model, which inspired the founding of the Swedish Linnaean Society in 1918, was the way Germany honored Goethe. Other models were England's Shakespeare societies, and Sweden's own Carolingian associations. Like Goethe in Germany, Linnaeus was the emotional and symbolic focus of a nationalist modernization myth, emphasizing social harmony over class justice. As one newspaper put it in 1919: "The memory must be able to elevate and unify all classes and parties—bring them to forget for one moment all that divides and disrupts."[76]

Such hopes for national unity by way of the memory of Linnaeus had been the central theme in the jubilees of 1878 and 1907. When in 1885 a Linnaeus statue in central Stockholm was unveiled in a public ceremony, a leading newspaper celebrated how "everywhere one saw famous faces, heroes, and veterans in the service of science and politics, teachers and students, workers, poets, people from all social classes. They now spread into all parts of the city after having been, for a short moment, unified in a matchless triumphal march to celebrate the purest, finest, and perhaps greatest memory that Scandinavia possesses."[77] In the same vein, one speaker at the 1907 jubilee celebrations at Gothenburg University pointed out that Linnaeus Day "is the most universal of Swedish memories. Not even the commemoration of Gustav II Adolf, which the Swedish people know as their greatest national memory, has the same power to speak to different races and social groups."[78]

Swedish nationalists thus believed that the commemoration of Linnaeus could "draw together these splintered ways" (that is, political parties) into "a national day for the entire people" (folk). It would also tear the "manual worker" away from his "soulless pleasures" of aquavit and socialism. Or so one journalist argued in 1918, as he outlined a model "Linnaeus Festival" dominated by patriotic speeches and folk music.[79] Other, more acute observers, such as the Socialist Bengt Lidforss, writing about the bicentennial celebration of Linnaeus in 1907, noted that "it is, after all, a fact that most of the poor in our country remain entirely unmoved by the revelry."[80]

Over time, the Linnaeus cult grew increasingly associated with racialist ideologies. Linnaeus described himself as "not big, not small, thin, brown eyes."[81] Contemporaries noted mainly that he was small and dark. By the late nineteenth/early twentieth century, however, conservative opinion-makers imagined that this "genuine son of the Swedish people," of the "Småland yeomanry that for centuries have tilled the inherited soil," was of "genuine Swedish inheritance" and "the most healthy and pure blood." Therefore he must have been blond.[82] A celebrated oil painting of 1846 showed Linnaeus as a blond and blue-eyed boy dressed in a spurious yellow and blue folk costume.[83] In 1874, a popular magazine described him as "a mild youth with blond curls."[84] Sweden's ABC, Läsebok för folkskolan, which taught virtually every Swedish child of the period to read, referred to the "light-haired, lively, highly cultured scientist."[85] Around 1900, the fashion was to drop the effeminate "von" while retaining the more Nordic-sounding "Linné" over "Linnaeus," and for spelling Linnaeus' first name "Karl." Laudatory poems ended in military barks such as "Hail, Karl Linné! Hail, Karl Linné!"[86]

In 1926, when the Swedish Linnaean Society journeyed to Linnaeus' birthplace in Småland on its annual outing, the notables of the small rural parish invited their visitors to dinner. One member of the society, a professor, expressed his thanks to the municipal chairman, a farmer, as coffee was served on the hotel terraces overlooking the lake: "I heard the chairman's manly words [of a dinner invitation], I looked into the men's steel-blue eyes, and felt the same feeling as Linnaeus, who knew that at home at Stenbrohult there lived the old Goths."[87] The professor went on to describe how, in the eighteenth century, "the new realm came. It was the Goths, the blue-eyed Germanics, who created the realm, the spiritual realm, where Linnaeus himself was one of the princes." The speaker then cheered "for the future of our fatherland," and sang the national anthem "in unison" with his new friends.[88]

The next Tuesday, the board of the Stenbrohult municipality was formally to accept its responsibility for payment for this drunken feast celebrating "blue-eyed Germanics." Although the bill came to about a thousand Swedish crowns (an industrial worker's annual salary), its processing was expected to be a routine affair. After all, many municipal board members had themselves attended the party. One political party, however, refused to pass the motion. "The Social Democrats stressed that it is unacceptable to charge this bill to the municipality, since all the guests considered their visit to be a pleasure trip."[89]

The builders of the new Sweden had no intention of financing the pleasures of the old regime. When the local Social Democrats refused to foot the bill for the Linnaean Society's annual outing, they acted not only on their deeply held principles of clean government and social justice. They also expressed their special distaste for "the flower king." For by 1926, Linnaeus was so closely linked to conservatives that socialists had little sympathy with him.[90] During the interwar period, the newspapers and magazines of the Social Democratic Party were profoundly committed to educating their working-class readers. They often carried articles on Swedish culture, science, and history. Yet these hardly ever mentioned Linnaeus. It was exceptional when articles on Linnaeus appeared in the metal union magazine *Metallarbetaren* (in 1932) and the party daily *Social-Demokraten* (in 1934). Predictably, both concerned his description of working conditions in the Swedish metal industry.[91]

When, in 1932, the Social Democrats were elected to govern (they remained in power until 1976), they had no intention of making May 23, or Linnaeus Day, an official Swedish tradition. After all, Linnaeus Day had originally been set up to counter May 1, the union-sponsored International Workers' Day. In the new Sweden, attempts to use Linnaeus as a unifying national figure petered out, as earlier initiatives wound down, and as national chauvinism generally, and especially after 1945, went out of fashion. Admittedly, in 1935 Linnaeus' childhood home was inaugurated as a museum with all the hackneyed props and devices: youths in folk dresses, children clutching flags, and choirs singing "Our Land," "Happiness to the Country," "Sweden Is Everything in the Whole World for Me," and "My People." The day was even broadcast by radio. Despite the rain, thousands lined the road to Linnaeus' home, "bowing and curtseying to the Crown Prince and his wife, who arrived by car."[92]

This was, however, a local celebration. On a national level, a dislike for Linnaeus seems to have become fairly general, and especially because of the way he had been invoked in public schools. As one newspaper put it in 1948: "Linnaeus' role in the school schedule has become so tragic that one could cry. . . . stop rote learning of stamens and pistils and include instead a bit of modern genetics." The editorial concluded with a poignant plea: "Abolish all the old stuff that has lost all interest long ago."[93] Or, as Sweden's most beloved comic writer, Frans G. Bengtsson, put it, "A curse on Linnaeus!"[94]

The results of such attitudes became obvious a decade later, in 1957, the year of the largely uncelebrated 250th anniversary of Linnaeus' birth. In a radio competition designed to identify the country's best educated grammar school students, the finalists lost on a question of stamens and pistils. As a newspaper commented: "The strangest thing was perhaps the uncertainty about Linnaeus' sexual system, his place of birth and his year of birth, since otherwise he is the great prophet for scientists."[95] The same year, a newspaper questionnaire informally tested the man in the street's knowledge of why Linnaeus was famous. The answers alarmed old-fashioned educators: "The flowers, he gave them names." "He discovered the Linnéa, Småland's heraldic flower." "For the flowers, and as a teetotaller."

Suggesting that propaganda among schoolchildren is largely useless from the point of view of the indoctrinators,[96] a full twenty-five percent of the respondents only knew "in general that he was a flower king," without being able to specify what this in turn might mean.[97] In response, education specialists produced several radio programs, one television program, and even—as part of a "'Made in Sweden' *goodwilloffensiv*"—a short film about Linnaeus.[98] As opposed to the turn-of-the-century Linnaeana, however, these productions centered on science rather than on patriotism. "Let botany become modern again," one newspaper urged. "Don't walk around like a sheep in nature!"[99]

Despite a few such belated efforts, after the Second World War Linnaeus was largely forgotten in his home country. Indeed, when his classic journey from Öland and Gotland was published in 1962, only seventy copies sold.[100] True, he was mentioned at the Seventh International Congress of Botany, held in Stockholm in 1950. His grave was visited by two Japanese, even, "who came here after five days of air travel via Bangkok, Calcutta, Cairo and Rome." The Russian delegation, too, traveled to Uppsala. As bemused local reporters noted, they observed that absolute silence we now know as characteristic of subjects of totalitarian states. Anyhow, they had little time for history. The Seventh Congress was dominated by the debate over Lysenko.[101]

Seven years later, in 1957, one Swedish newspaper aimed a below-the-belt blow at a small and insecure nation. "'You're a Swede—then you surely know what this flower is called,' foreigners used to say. They won't in twenty years!"[102] Despite such warnings, however, the Swedish image of Linnaeus continued to fade or, rather, to shrink. This is neatly measured by statues of the man. If we line up Sweden's public sculptures of Linnaeus in chronological order, we find that the Uppsala statue (1829) and the Stockholm statue (1885) depict Linnaeus as an old man, the Lund statue (1938) shows him as a university student, and the Älmhult statue (1946), as well as the Stenbrohult statue (1948), present him as a child.[103]

Moreover, both these last statues were commissioned by Småland municipalities, illustrating how the Linnaeus cult moved from national to regional. This is true for post-war pageantry, too. It was local, *and* it localized Linnaeus. In 1945, for example, the Farmers' Union's annual convention (*riksting*) met in Jönköping, Småland. As entertainment, and in front of 12,000 people, a prominent local farmer and parliamentarian, Gustaf Svensson i Vä, played Linnaeus. "A bouquet of twelve small girls dressed as *Linnaea* flowers formed a ring around the flower king."[104]

In 1957 another Småland town, Växsjö, staged a "flower march" to celebrate Linnaeus, its most famous grammar school graduate. "Flower floats" were drawn by tractors driven by female members of the Swedish Automobile Association. On one such float, Linnaeus (played by a schoolboy) shared space with mythic and fairy-tale figures such as Saint Sigfried, the Viking amazon Blenda, and the magic midget Nils Holgersson. Further crowding the float were two "Lapps" (played by Sami), and "a great number of small schoolchildren more or less disguised as flowers."[105]

This "flower march" did little to kindle interest in Linnaeus. The midget, riding a goose, seems to have stolen the show. Nor was a second Linnaean renaissance inaugurated by the breathless tabloid rubrics of the 1960s, such as "Linnaeus TV Hero—If TV had Existed," "Sexologist Linnaeus Amazingly Modern Man," "Linnaeus—Our Greatest PR-man?" and "Not Only Flowers for the Sex-Radical Linnaeus."[106]

The Romantic nationalist process of deification that prompted cultural conservatives of the late nineteenth century to position Linnaeus as "flower king" and compromise national symbol, better able than Gustav II Adolf to battle the specter of Marx, can be understood as an "invented tradition" mythicizing histories and practices invented by nineteenth-century state elites to inculcate their citizenry with patriotic sentiments.[107] Yet despite extensive promotion by the press and by public institutions, the Linnaeus cult failed to root itself in the collective consciousness of Scandinavians. This was a case of an "invented tradition" that resoundingly failed.

Over the course of two centuries, quite different reasons were proposed for Linnaeus' historical importance, too. As we saw, the man himself understood his naturalist knowledge and his fame through the lenses of his biblical faith, his Gothicism, and his notion that a university chair was akin to a family parsonage and was therefore a patrimony for a civil servant caste. Linnaeus' continental followers, on the other hand, appreciated his sexual system and binomial nomenclature. Indeed, his floral codices made botany a popular hobby among notables in the later eighteenth century. Moreover, Linnaeus' rejection of rhetoric and of the manners and ornaments of courtly culture struck a chord in the later 1700s, an era appreciative of virtues such as authenticity and simplicity.

With the advent of Romanticism and Romantic tourism, and in the context of Sweden's continued backwardness relative to the Continent, Swedes, and in an ultimate irony the Linnaeans themselves, came to be regarded by more advanced Europeans in the same manner as the Linnaeans in the past had regarded non-Europeans and especially the Sami—as ethnographic "curiosities."

Nineteenth-century Swedish conservatives fashioned Linnaeus into a "flower king," embodying the military and racial virtues of Sweden's seventeenth-century empire, but pointing to another realm—modernity and science—where such virtues could be reasserted. This Romantic and nationalist icon of science was demolished in the 1930s, with the advent of Social Democratic governance. Linnaeus dwindled into a local hero dimly recalled as a "famous teetotaller" and listed in Smålandian tourist brochures alongside Viking amazons and magic midgets.

"Without Science Our Herrings Would Still Be Caught by Foreigners": A Local Modernity

Our trajectory of Linnaeus' reception in his homeland began with Linnaeus' own conviction of his redemptive role as a "Moses" of science, proceeded through his status as a patriot genius of nineteenth-century Sweden, and wound up in Småland, in the regionalists' pride in their only world-class celebrity. By a process of historical reduction, the "second Adam" became Småland's greatest son (*bygdens störste son*).

This trajectory resembles a funnel: when the magisterial Enlightenment classifier of the earth's minerals, animals, and plants passes through it, he shrinks into a bronze statue of a boy, gazing at a flower by the grass-roofed cottage where he was born. Yet this image, growing ever more localized and localizing, reflects a profound truth about him. Linnaeus was a quintessentially local man.

This book has been an attempt to write a history of that localness. As I have conceived that task, it means situating Linnaeus as he situated himself. And that in turn means accepting as central the question of how he linked the universal with the local, or, to use the terms of this book, nature with nation. As I now come to a close, I want to address briefly the broader context of Linnaeus' economic projects.

Throughout the body of this book, I have used the word "cameralism" in a narrow sense. I have meant by it the theoretical elaboration of fiscal and economic governance by Scandinavian and German courtiers and civil servants from c. 1650 to c. 1780. But the term can be productively employed also as a shorthand notation for the political goal of a rationalistically governed autarky.

Earlier, I contrasted this ideal to the classical economists' aim of an ungovernable yet self-regulating global modernity, and to the Romantic antimodernists' hope for an infinitude of custom-governed, local communities. This broader definition of cameralism is not rigorous, but it is strategically useful. It serves as a coded communiqué, hinting at broad similarities of thought, or an ideational complex that has endured over the last three hundred years. Indeed, I believe that if we ignore for a moment Luddite and Romantic traditions of thought, modernity's most central fault line runs between global modernizers (or Smithian liberals) and local modernizers (or cameralists).

Eli F. Heckscher, in his standard work on early modern economic doctrines, argued that in the mid-eighteenth century, cameralism went "underground." (As he saw it, it later emerged in Nazism).[1] I will leave aside the vexed question whether cameralism, thus broadly conceived, should be interpreted in terms of descent and genealogy or in terms of convergent evolutions.[2] But for Scandinavia, it seems more accurate to say that instead of going "underground," cameralism relocated from the social to the natural sciences.

After all, Linnaeus was a principal, if failed, guarantor of a modernization policy by import substitution, and his strategy in turn hinged on a botanical premise here termed "acclimatizationism" or "adaptationism." His *science* underwrote a fresh political solution to the Asia trade (which, for complex theoretical reasons, the cameralists regarded as becoming unmanageable). The idea was that science would create a miniaturized mercantile empire within the borders of the European state.

Toward the end of the eighteenth century, cameralist economic doctrines were again challenged, and now by the doctrines of classical economics, England's industrial revolution, and the cameralists' own failure to improve their local economies by means of either their standard legal measures or by Linnaeus' more imaginative imperial botany. In response, cameralism transmogrified into, or reappeared as (depending on whether you employ a homological or analogical narrative), a defensive, almost Romantic creed. It again relocated, to moral philosophy and the social sciences. By 1841, the German cameralist Friedrich List accused Adam Smith of what German conservatives now considered deadly sins: "boundless cosmopolitanism," "dead materialism," and "disorganizing particularism and individualism."[3]

Broadly, then, seventeenth-century cameralism was a policy of legal reforms of the economy. It was radical in the sense of being administratively transformative, but not in the sense of being democratizing or egalitarian. In the eighteenth century, cameralism remained (in this sense) a progressivist improvers' creed. And it continued to turn to the natural sciences for its methods and evidence. Then, around 1800, it was submerged in the politics of reaction.

In our own century, it reappeared in new garbs within the economic doctrines of anti-imperialist nationalisms: Stalinist socialisms-within-one country, neo-Marxist dependency theory, and even, arguably, neo-orthodox development theory. I hasten to add that these modern economic philosophies also profoundly differ from each other, and from earlier forms of cameralism. Neo-Marxist dependency theory attempts a historical explanation and ethical judgment of global economic interdependency. It is a hermeneutics. As such, it remains within the purview of the German tradition of thought, which from the Romantic era on became entangled in historicist teleologies of class and race.

Neo-orthodox development theory claims to be a science in a positivist sense. An ahistorical creed, it attempts a technological blueprint for national economic independence. It emerged in the Anglo-American academic world, and drew from the experience of the Great Depression,[4] technical criticisms of neo-classical economics,[5] and the widespread belief in the West from c. 1930 to c. 1980 that centrally planned economies did, would, or could achieve higher economic growth, more advanced industrial technologies, and more equitable distribution of social goods.[6]

At the minimum, neo-orthodox development theorists assumed that, because of their supposedly higher rates of capital accumulation, state-governed economies could step efficiently from traditional to transitional economies. More recently, and with the collapse of the Occident's last colonial power, the Russian empire—a collapse that originated precisely in the realm in which it had proclaimed its superiority, the economy[7]—this belief in a Soviet model was replaced by a related belief that the economic achievements of the Far East are due to the protectionist and statist aspects of their economies. More recently still, the financial collapse of those countries was taken as evidence of the need of more protectionism.[8]

But despite their differences, neo-orthodox development theory and neo-Marxist dependency theory still resemble each other, and resemble older forms of cameralism. (I again bracket the question of descent or convergence). In this broader sense, and whether it is set in a traditional, transitional, or modern economy, cameralism is a "catch-up" modernization doctrine. It measures a state's economic success in terms of national self-sufficiency, capital accumulation, and the development of advanced indigenous technologies.

From the 1650s to the 1990s, cameralists and neo-cameralists have dismissed service and finance industries both in moral terms and as an economic force. They bracket them together as "speculation." This in turn they oppose to a category that they embrace with exceptional moral fervor, namely "production." Over their three centuries of existence, cameralists have persistently inclined toward a missing-component type of macro-economics, too. One marked family resemblance among cameralists is how they make a fetish of the latest high technology and of the latest state-supported enterprise devoted to it. This holds true from Johann Joachim Becher's spice-trading German East India Company (1660s) and Carl Gustaf Lidbeck's silk-producing Skåne Plantation Bureau (1750s), to Friedrich List's German railroad ventures (1840s) and Edith Cresson's "hyper-industrial" French computer company, Groupe Bull (1990s).[9]

In all of these phases, too, cameralists favor an activist state, protecting such ventures. Typically they use production subsidies on exports, along with tariffs, commodity controls, currency exchange and capital transfer restrictions, cheap loans and seed-money, and infrastructure investments. Typically, too, they create domestic producer cartels or monopolies (by tacit agreement, state licenses, or state ownership). Cameralists—who are typically civil servants and monopoly producers—thus seek to replace profit-seeking with rent-seeking. More fairly, because their ideology is not only an emanation of self-interest, they base their economic philosophy on the idealistic premise that bureaucratic discretion over the distribution of monopoly rents will not occasion lobbying, bribes, or other rent-seeking activities.

From the seventeenth century to the present, cameralists have promised a local modernity.[10] They have promised, that is, a state whose rationalized, centralized power structures will constitute the condition of a self-sustained economy. They attempt to deliver this local modernity by means of extending political power over the local economy. Consequently, they support protectionism, and/or state-governed means of production (which, in turn, comes to mean *also* protectionism). Their most central argument is this: the monopoly structures created by the confluence of political and economic power, which neoclassical economists believe hinder economic growth, are in fact economic growth's favored precondition.[11]

For much of the post-World War II era, neo-orthodox development theory and neo-Marxist dependency theory inspired poor economies' attempts to grow. After c. 1990, most development specialists abandoned these neo-cameralist creeds. Many turned instead to public choice theory, especially of rent-seeking societies. (This subset of economics has been particularly useful in explaining the spectacular failures of Western development aid.)[12] But since academia provides a harbor of last resort for the ideas of yesterday, cameralist forms of reasoning still live on in the professional writing of history, and especially in the historiography of the growth of global economy.[13]

In this book I have not attempted to place Linnaeus within a supposed trajectory of modernization. Bristling with arrows of causality, such an enterprise would have underwritten the metaphysics of much modern historiography. Instead I have situated Linnaeus within the problematic he himself deemed paramount. In doing so, I have found unhelpful the teleological historiography on eighteenth-century economic modernization and international relations. Indeed, to use it would constitute a lack of critical distance. For ultimately these schemes trace their own pedigrees to the cameralist thought-world to which Linnaeus belonged.

Instead of turning to earlier historiography, then, I have attempted to dislocate our interpretative habits (or what we might think of as our vernacular understanding of the economy), by recovering a sense of how Linnaeus located himself and his theories of utility and of improvement, which is to say, how he understood his science and its principal tasks. This more modest attempt may in turn inspire a more radical estrangement of our presumably natural languages in matters of economics. (To choose an example at random, it may help dispel our vernacular belief that the French state's subsidies of the computer firm Groupe Bull in the 1990s followed an evidentiary protocol, rather than simply reflect those same colloquial conventions).

Linnaeus' acclimatization project brilliantly highlights the contingencies of economic improvement. At the same time, it becomes predictable, or at least coherent, once we recognize that it was predicated on the idea of a zero-sum international economy, and on the concomitant notion that the tertiary sector is parasitical. Given these premises, or axioms, the following conclusion holds true: if a polity aims for *both* economic and political independence (which comes to roughly the same thing under this purview) *and* a multifunctional, complex economy, then either it must conquer economically and technologically diverse territories, or it must make its homeland economically and technologically diverse.

In eighteenth-century agrarian economies this latter choice in turn meant, essentially, making that homeland more ecologically diverse. As we would put it today, Linnaeus believed that the key incentive to trade is the existence of different endowments of natural resources. He thought that people trade over "ecological divides."[14] But this does not mean he accepted international trade as inevitable. Instead he radically reconceptualized what he saw as the natural root of the problem. Science would overcome ecology.

* * *

The task of this book has been to unearth the archival and primary sources of how Linnaeus conceptualized these problems of natural diversity, national independence, and economic progress, and thus how he linked his concepts of nature and nation. A central argument throughout has been that Linnaeus believed he could re-create a colonial economy within his Nordic homeland because of the relationship he assumed between a human science, or economics, and a divine order, or theology. He regarded his natural history as at once the investigation of Edenic nature, and as a science auxiliary to cameralist economics.

Linnaeus' cameralism and natural theology were both conventional, mature creeds. Indeed they were first developed nearly a century earlier. Linnaeus did not so much elaborate them as reposition them. Or rather, he explored their relation to one another, selectively blending elements from each into a new syncretic creed. This seemingly incremental alteration of the philosophies of nature and society that he had inherited had radical implications. For in bringing together conventional assumptions about nature and nation in an unexpected manner, Linnaeus hit upon novel arguments to legitimize a particular set of governmental economic policies, just at the moment when those policies were becoming ever more difficult to sustain—which is to say, at the emergence of the transnational world of the eighteenth century.

We have also seen how, by meshing natural history and economic theory, and by casting himself as a guardian of this new science, Linnaeus carved out an exalted and lucrative role for himself as a governmental adviser. In his own eyes, he was first and foremost an improver. He understood his science to be an applied technology that was to serve state economic needs, and he valued his economic work as highly as his classificatory schemes.

More generally, Linnaeus' natural knowledges represent an early instance of the confluence of science and governance (and its theorization) that is both a hallmark of modernity and the focus of its antimodern and postmodern critiques. Linnaeus asked a question that, in various permutations, state elites ask to this day: can the natural sciences alter the terms of the trade-off between political independence and economic growth?

Must one chose between national autonomy (as incorporated in a closed-off economy governed by nationals), and material riches and economic growth? Or can native subjects, using only local means of production, build a complex and complete local economy, incorporating contemporary technologies, and functioning as a microcosm of the global economy?

This question is typically posed by local elites that have vested interests—political power, social prestige, and economic rents—in retaining a state-controlled economy roped off and isolated from larger economic regions. In turn, these state elites often turn toward the natural sciences as a way to displace the theoretical onus of their political economy from the management of people to the constitution of the natural world.

The Linnaean voyage of discovery is a paradigmatic example of this. For it was conceived as a device to transfer foreign life-forms and technologies, and thus to build national autarkies complete with local variants of the Asian goods that Europeans—as a matter of practical politics—were no longer willing to forgo.

Indeed, the Linnaeans traveled with the belief that, through their efforts, Europeans would no longer need to venture out of their own countries, let alone their small Western peninsula of the Eurasian landmass. In reconstructing this short-lived project, this book memorializes a local attempt at a local modernity, a now-forgotten future of the past.

Chronology of Linnaeus and Linnaeana

1529	Swedish Lutheran Reformation.
1555	Olaus Magnus' *Historia*.
1560	Death of Gustav Vasa, the king who first managed a national unification of Sweden and masterminded Sweden's Reformation. Himself elected king by nobles, he was able to introduce a hereditary kingship in Sweden.
1632	Lycksele Seminary established, for converting the animist Sami to Lutheranism. Gustav II Adolf, main architect of the Swedish empire in the German lands and the south and east Baltic, is killed in battle, at age 38. His only child, Kristina, age 5, is anointed future regent of Sweden.
1648	Peace of Westphalia ends Thirty Years War.
1654	Kristina renounces her throne and converts to Catholicism. Her cousin Karl X Gustav made king.
1660	Karl X Gustav dies.
1673	Johannes Schefferus' *Lapponia*.
1673	Lapland Settlement Act formally opens Swedish Lapland to colonists.
1674	Linnaeus' father, Nils Ingemarsson (Linnaeus), is born to a Småland farmer family.
1679	Olof Rudbeck the Elder's *Atlantica* begins to be published.
1681	J. F. Regnard travels to Lapland.
1688	Linnaeus' mother, Christina Brodersonia, is born to the family of the parson of Stenbrohult, Småland.
1695	Swedish royal astronomical expedition to Lapland. Olof Rudbeck the Younger accompanies as botanist.
1697	Karl XI dies, at age 42. The 15-year-old Karl XII becomes king of Sweden.
1700	Great Northern Wars begin.
1702	The great Uppsala fire: end of publication of *Atlantica*.
1707	Carl Linnaeus born, first child to Christina Brodersonia and Nils Ingemarsson Linnaeus, curate of Stenbrohult, Småland.
1710	Linnaeus' father is appointed parson of Stenbrohult, on death of previous holder, Linnaeus' maternal grandfather. Uppsala Science Society founded.

1710s	Widespread famines in Scandinavia.
1710–11	Bubonic plague in Scandinavia.
1711	Henric Benzelius travels to Lapland, sponsored by the Uppsala Science Society.
1716	Sara Elisabeth Moraea, future wife of Linnaeus, is born to the town physician of Falun, Dalarna.
1718	Karl XII is murdered, at age 36. Effective collapse of the Swedish empire. Linnaeus' brother Samuel is born.
1719	Sweden's Era of Freedom, parliamentary near-absolutism, begins. Ulrika Eleonora, sister to Karl XII, cedes throne to her husband, Fredrik I. Parliament reduces power of king in favor of *riksrådet,* including sixteen members from all four estates, and *sekreta utskottet,* a smaller body responsible for war, finance, and diplomacy, which excludes farmers, and has a majority of nobles. The new prime minister is Count Arvid Horn.
1719–20	Russians launch punitive scorched-earth invasions of Eastern Swedish seaboard.
1720	Great Northern Wars end.
1721	Peace of Nystad formally ends the Great Northern Wars. Sweden loses most of her south and east Baltic and German lands.
1724	Alingsås textile manufactures founded.
1726	Rörstrand porcelain manufactures founded.
1729	Sweden's first cotton-printing manufacture founded.
1731	Swedish East India Company founded (it closes in 1813). Linnaeus first drafts his sexual system of plant classification, in a ms., *Hortus Uplandicus.*
1732	Linnaeus travels in Lapland, sponsored by the Uppsala Science Society.
1733	Linnaeus' mother, Christina Brodersonia, dies, at age 45.
1734	Linnaeus travels in Dalarna, sponsored by the regional governor.
1734–35	Around New Year Linnaeus becomes engaged to Sara Elisabeth Moraea; they met during his Dalarna voyage.
1735	On 23 June, Linnaeus receives his doctorate in medicine at Harderwijk university, Holland. His best friend, Petrus Artedi, drowns. *Systema naturae* first sets out the parameters of Linnaeus' global classifications.
1735–38	Linnaeus lives in Holland and travels to Paris and London.
1736	*Fundamenta botanica* first elaborates the sexual system's rules in print.
1736–37	Maupertius heads Lapland expedition to test whether the globe is a sphere flattened at the poles.
1737	*Hortus Cliffortianus:* Linnaeus' first tropical and subtropical flora, listing the contents of his patron George Clifford's gardens and green houses. *Flora Lapponica:* the West's first sub-Arctic and Arctic flora, and the first significant test-case of Linnaeus' sexual system of plant classification.
1738/39	The government of Count Arvid Horn is forced out. Power goes to mercantilist anti-Russian Hat party. Power shifts from high to low aristocracy.
1738	Linnaeus returns to Sweden. He marries Sara Elisabeth Moraea and opens a private medical practice in Stockholm specializing in venereal disease.

1739 The Swedish Academy of Science is founded: Linnaeus is co-founder. He is appointed chief physician to the Swedish navy.

1741 After a protracted tenure battle, Linnaeus is appointed professor of medicine at Uppsala University. His first child, Carl, is born. Linnaeus travels on the Baltic islands of Öland and Gotland, sponsored by *Manufakturfonden.* Sweden's first cotton-weaving manufacture is founded.

1741–43 War against Russia. Sweden begins, and loses.

1743 Dalarna peasant revolt. Linnaeus' first daughter, Elisabet Christina, is born.

1744 Linnaeus' second daughter, Sara Lena, is born: she dies in fifteen days.

1746 Linnaeus travels in Västergötland, sponsored by the national estates.

1747 Åbo University chair in practical economics founded on Linnaeus' advice.

1748 Linnaeus' father, Nils Ingemarsson Linnaeus, dies at age 74. Pehr Kalm leaves for North America. The only woman member in the Swedish Academy of Science before the twentieth century, Countess Eva Ekeblad, is elected.

1749 Linnaeus travels in Skåne, sponsored by *manufakturfonden* and parliament. Pehr Löfling's *Gemma arborum. Pan Svecicus,* a pamphlet on cattle fodder, first uses a binomial nomenclature. Fredrik Hasselquist leaves for the Levant. Linnaeus' third daughter, Lovisa, is born.

1751 *Philosophia botanica,* a rewrite of the 1736 *Fundamenta botanica,* publishes rules for a true binomial nomenclature for the first time. Adolf Fredrik is crowned king. Pehr Kalm returns from North America.

1750 Pehr Osbeck leaves for Guangzhou (Canton) as East India Company ship's chaplain.

1751 Linnaeus' fourth daughter, Sarah Christina, is born. Pehr Löfling leaves for Spain.

1752 Fredrik Hasselquist dies in Smyrna at age 50. Pehr Osbeck returns from China.

1753 *Species plantarum* is the first major Latin taxonomical work to apply binomial nomenclature to flora consistently. Linnaeus is the first Swedish scientist to receive the Order of the Polar Star (*Nordstjärneorden*).

1754 Linnaeus' second son, Johannes, is born.

1756 Failed coup d'état by Adolf Fredrik, supported by his wife, Lovisa Ulrika, and farmers. Widespread Scandinavian famines. *Flora alpina.* Pehr Löfling dies from tropical fevers in present-day Venezuela at age 25.

1757–62 Intermittent wars between Prussia and Sweden.

1757 Linnaeus' fifth daughter, Sophia, is born. Johannes (his second son) dies, at age 3.

1758 The 10th edition of *Systema naturae* for the first time applies binomial nomenclature to fauna in a major Latin taxonomical work. Linnaeus purchases country estate of Hammarby, outside Uppsala.

1759 Uppsala University chair in practical economics founded on Linnaeus' advice. Linnaeus purchases village of Säfja next to Hammarby. Carl

Linnaeus the Younger is appointed demonstrator in Uppsala University's botanic garden at age 18.

1760   Lund University chair in practical economics founded on Linnaeus' advice.

1760–61   Pomeranian campaigns.

1761   Pehr Forsskål leaves for the Ottoman empire and the Arabian peninsula.

1762   Linnaeus is ennobled and takes the name "von Linné," is awarded 6,000 silver thaler, and is given the right to chose his own successor to his Uppsala chair.

1763   Linnaeus wills his Uppsala chair to his son, Carl Linnaeus the Younger. Pehr Forsskål dies in Yemen, at age 31.

1765   Cap party gets into power for first time. Censorship abolished. Hat subsidies for manufactures abolished, and Hat inflationary politics stopped, followed by deflation and bankruptcies of previously state-supported manufactures.

1768   Daniel Solander joins Captain Cook's first circumnavigation of the globe (1768–1771) as Joseph Banks's botanist. Johan Petter Falck joins the Orenburg expedition to explore the southern parts of the Russian empire.

1769–71   Hats get back in power. Promise a revised constitution and stronger monarchy but do not follow up.

1770   Carl Peter Thunberg leaves for southern Africa, Ceylon, Java, and Japan.

1771   King Adolf Fredrik dies, at age 61.

1772   19 August. Successful coup d'état by Gustav III, son of Lovisa Ulrika and Adolf Fredrik. The Era of Freedom ends; enlightened royal despotism follows. Anders Sparrman leaves for southern Africa, where he joins Captain Cook's second circumnavigation of the globe (1772–75).

1773   Widespread famines in Scandinavia.

1774   Johan Petter Falck dies in Kazan, at age 42.

1777   Linnaeus the Younger is appointed full professor at Uppsala.

1778   Linnaeus dies, at age 71, after a series of debilitating strokes.

1779   Carl Peter Thunberg returns from Japan and takes up a position as botanic demonstrator at Uppsala university.

1782   Linnaeus' first daughter, Elisabet Christina, dies at age 39.

1783   Linnaeus' first son, Carl Linnaeus the Younger, dies at age 42. The "von Linné" branch of the Linnaeus family becomes extinct. Carl Peter Thunberg takes over Linnaeus' university chair.

1784   James Edward Smith buys and ships to London Linnaeus' natural history collections, library, and scientific manuscripts. Linnaeus' correspondence arrives too, since his widow used it for crate-filling material.

1788   The Linnean Society of London is founded, among others by James Edward Smith.

1792   Gustav III is murdered by a disgruntled minor noble. Dietrich Heinrich Stöver, a German historian, writes the first book-length biography of Linnaeus.

1797	Linnaeus' only brother, Samuel Linnaeus, pastor of Stenbrohult, dies at age 79. The Linnaeus family becomes extinct in the male branch.
1798	First public monument to Linnaeus in Sweden is erected: a funeral plaque in Uppsala cathedral.
1799	Widespread famines in Scandinavia.
1806	Linnaeus' wife, Sara Elisabeth Moraea, dies at age 90.
1807	An auditorium in honor of Linnaeus is completed at Uppsala University.
1808	Sweden's last war to date, with Russia and Denmark.
1809	In a democratizing revolution initiated by a coup d'état by aristocratic officers, Sweden receives a new constitution strictly limiting the role of the monarchy. Jean Baptiste Bernadotte, son to a French lawyer and one of Napoleon's generals, is crowned Karl XIV Johan in 1818. Finland, previously part of Sweden, is ceded to Russia.
1829	A statue of Linnaeus is inaugurated at Uppsala University.
1830	Linnaeus' fifth daughter, Sophia, dies at age 73.
1835	Linnaeus' fourth daughter, Sarah Christina, dies at age 84.
1839	Linnaeus' third daughter and last living child, Lovisa, dies at age 90.
1856	Scandinavian famine: confined to Swedish Lapland.
1866	Linnaeus birth home, Råshult curate cottage, Stenbrohult, Småland, is made a museum.
1867	Paris International Botanic Conference adopts an international code of plant nomenclature taking as its starting point *Species plantarum* (1753).
1873	Lapland Settlement Act allowed to lapse.
1875	Linnaeus is first condemned by international scholar, the German historian of science and plant physiologist Julius Sachs.
1878	Centennial jubilee of Linnaeus' death.
1879	Linnaeus' country estate, Hammarby, is made a museum.
1903	The standard biography of Linnaeus, Th. M. Fries' *Linné. Lefnadsteckning*, is published.
1905	International code of taxonomic nomenclature for fauna takes the 10th edition of *Systema naturae* (1768) as its starting point. Norway, previously part of Sweden, becomes independent.
1907	Bicentennial jubilee of Linnaeus' birth.
1918	The Swedish Linnaean Society is founded on the model of Germany's Goethe societies. It begins publication of Linnaeus' correspondence and other important manuscripts.
1932	Social Democrats elected to government in Sweden.
1935	Linnaeus' childhood home, Stenbrohult parsonage, is made a museum.
1937	Linnaeus' Uppsala town house is made a museum. The botanic garden is restored.
1943	The half-finished publication of Linnaeus' correspondence ceases.
1951	Linnaeus is first condemned by Swedish scholar Sten Lindroth.

Biographical References

Adanson, Michel (1727–1806). French botanist and explorer of Senegal. Correspondent of Linnaeus.

Adlér, Carl Fredrik (1720–61). Linnaeus' student. Traveled to Java and Guangzhou (1748) and several more voyages. Swedish East India Company ship's surgeon (*överfältskär*). Financed by Carl Gustaf Tessin. Died outside Java.

Adolf Fredrik (1710–71). King of Sweden. Preceded by Fredrik I. Succeeded by his son Gustav III.

Afzelius, Adam (1750–1837). Linnaeus' student. Traveled to Sierra Leone 1792–93 and 1794–96, financed by the English Sierra Leone Company. Assistant (*e.o.*) professor at Uppsala.

Agardh, Carl Adolph (1785–1859). Osbeck's student. Professor of natural history at Lund University, expert in algae, eventually bishop of Värmland.

Agardh, Jakob Georg (1813–1901). Professor of botany and expert on algae. Son of Carl Adolf Agardh.

Alströmer, Claes (1739–94). Owner of Alingsås textile manufactures, noble (*friherre*), and government adviser (*kansliråd*). Son of Jonas Alströmer.

Alströmer, Jonas (1685–1761). Owner of Alingsås textile manufactures, co-founder of the Swedish Academy of Sweden in 1739, and government adviser (*kommerseråd*).

Artedi, Petrus (1705–35). Lutheran parson's son, Uppsala University student, zoologist, and Linnaeus' best friend in his youth. Drowned in Amsterdam. See also Albert Seba.

Bäck, Abraham (1713–95). President of the Swedish College of Medicine and physician to the Swedish court. Correspondent of Linnaeus and his closest friend after Artedi.

Backmansson, Anders (1697–1772). Swedish mercantilist in the English tradition, and author of *Arcana oeconomiae* (1730).

Bagge, Peter Samuelsson (1710–79). Gothenburg merchant, director of the Swedish Greenland Company, and member of the estates (*riksdagsman*) for the burghers. Bought Linnaeus' pearl-culturing technique from the estates in 1762 and received a state-licensed monopoly of pearl culturing.

Baltimore, Frederick Calvert, Lord (1731–71). Owner of Maryland, America, and amateur natural historian. Correspondent of Linnaeus.

Banks, Sir Joseph (1743–1820). President of Royal Society of London. Employs Linnaeus' student Daniel Solander as his botanist on Captain Cook's first circumnavigation of the world on the *Endeavour* (1768–1771), and himself travels on it.

Barsch, Johannes (1709–38). German-born Dutch botanist and Surinam doctor. When he was in Holland, Linnaeus was offered Barsch's colonial position. He turned it down and recommended Barsch, who died soon after arrival in Surinam.

Bauhin, Caspar (1560–1624). Swiss botanist, professor of anatomy and botany in Basel, and author of *Pinax theatri botanici* (1623). Frequently cited by Linnaeus.

Becher, Johann Joachim (1635–82). German cameralist and author of *Politische Discurs* (1668). In later life Becher rejected trade and monetary instruments as evil, and turned to alchemy and anti-Semitism.

Beckmann, Johann (1739–1811). Linnaeus' student. Professor of economics in Göttingen.

Benzelius, Henrik [Henric] (1689–1758). After making a Lapland journey (1711), he traveled to join Karl XII in Bender, where he was captured by the Russian army in 1712. He was eventually freed and traveled home through Asia Minor, returning to Sweden in 1718. He became professor in Oriental languages at Uppsala, then a theology professor, a bishop, and in 1747 an archbishop.

Berch, Anders (1711–74). Sweden's first professor of cameralism, appointed 1741 at Uppsala. Author of *Sätt at igenom politisk arithmetica* (1746) and *Inledning til allmänna hushålningen* (1747).

Berch, Chister (1735–92). Son of Anders Berch. Professor of cameralism at Uppsala by inheritance.

Bergius, Peter Jonas (1730–90). Linnaeus' disciple, professor of natural history in Stockholm, physician, and economic improver.

Berlin, Anders (1746–73). Linnaeus' student. Traveled to Guinea Bissau and Senegal 1773, financed privately.

Bielke, Sten Carl (1709–53). Swedish noble (*friherre*), courtier (*kammarherre*), judge, and improver landlord. Pehr Kalm's patron and Linnaeus' friend.

Bjerkander, Claes (1735–95). Linnaeus' student. Country parson.

Bjerkén, Pehr af (1731–74). Linnaeus' student. Public doctor of Stockholm.

Björnståhl, Jacob Jonas (1731–68). Linnaeus' student. Appointed professor in Oriental languages at Uppsala in 1776 and at Lund in 1779. From 1767, Björnståhl traveled abroad. He died in Saloniki. Correspondent of Linnaeus.

Blackwell, Alexander (1700–47). English doctor employed at Swedish court as royal physician. Executed in Stockholm for espionage. Husband of Elisabeth Blackwell, herbalist author.

Boerhaave, Hermann (1668–1738). Professor of botany, medicine, and chemistry in Leyden. Linnaeus' patron during the 1735–38 Holland stay, and together with John Ray one of his scientific idols.

Bonnet, Charles (1720–93). Natural historian, entomologist, and philosopher at Geneva.

Borgström, Eric Ericsson (1710–70). Swedish owner of an iron works, and the donor of the Uppsala chair of "practical economics" sponsored by Linnaeus. Borgström also gave money towards the experimental farm on which this professor was to be housed, and towards an assistant teaching position linked to this chair (*docentur*).

Brander [ennobled Skjöldebrand], Erik (1720–1814). Swedish consul in Tripoli, 1753–65. Correspondent of Linnaeus, and amateur naturalist.

Brodersonius, Samuel (c. 1658–1707). Linnaeus' maternal grandfather. Parson at Stenbrohult.

Browallius, Johan (1707–55). Professor and later bishop in Åbo. Linnaeus' friend.

Buffon, Georges-Louis Leclerc de (1707–88). French count and possibly the Enlightenment's most famous natural historian. Linnaeus detested him as a Frenchman, noble, rhetorician, and nominalist.

Burman, Johan (1706–79). Professor of botany in Amsterdam. Linnaeus' friend.

Burman, Nicolaas Laurens (1734–93). Son of Johan Burman. Linnaeus' student. Professor of botany in Amsterdam.

Camerarius, Rudolph Jacob (1665–1721). Professor of medicine at Tübingen and German botanist. Linnaeus studied his theories of plant sexuality.

Candolle, Alphonse de (1806–93). Together with his father Augustin, the formulator of nineteenth-century rules of taxonomic names.

Candolle, Augustin Pyramus de (1778–1841). Together with his son Alphonse, the formulator of nineteenth-century rules of taxonomic names.

Cederborgh, Fredrik (1784–1835). Swedish Biedermeier parodist; also mocked the Linnaeans.

Celsius, Olof the Elder (1670–1756). Polyhistorian, professor of theology at Uppsala. Linnaeus' teacher. Employed Linnaeus to help him botanize around Uppsala, 1730 and 1731. Invented the Celsius thermometer, but used 100 to denote water's freezing temperature, and 0 for its boiling: Linnaeus reversed it to the present standard.

Celsius, Olof the Younger (1716–96). Son of Celsius the Elder. Librarian and professor of history at Uppsala, and later bishop in Lund.

Child, Sir Josiah (1630–99). English mercantilist.

Chydenius, Anders (1729–1803). Parson, representative in the 1765–66 parliamentary session, and key to the Hat government's fall. Author of many works, among them *Källan til rikets wan-magt* (1765), *Den nationale winsten* (1765), and *Tankar om husbönders och tienstehions naturliga rätt* (1778). Although he did not read English and lived in isolation in northernmost Sweden, by the early 1760s Chydenius had developed a full-fledged laissez-faire economic philosophy. He remains the most original thinker of the Swedish Enlightenment.

Clerck, Carl Alexander (1709–65). Entomologist, Stockholm civil servant (*kommissarie*).

Clifford, George (1685–60). Dutch banker of English descent. Linnaeus' most important patron during his Holland stay.

Cook, James, Captain (1728–79). English navy captain, voyager of discovery. Explorer of Australia, discoverer of much of the Pacific (e.g. New Zealand, Hawaii). Daniel Solander, Linnaeus' student, accompanies him on his first circumnavigation of the globe, 1768–71.

Dahlberg, Carl Gustaf (?-1775). Swedish officer (*överstelöjtnant*), left Sweden after he was captured by rebels and forced to shoot at fellow officers in the 1743 Dalarna peasant revolt. Member of Dutch merchant marine and Surinam planter. Correspondent of Linnaeus and host of Linnaeus' explorer-student Daniel Rolander.

Dillenius, Johan Jacob (1684–1747). Professor of botany at Oxford. His post was founded by the English consul at Smyrna William Sherard (1651–1722), and his stated task was to complete Bauhan's *Pinax*. Early critic of Linnaeus.

Djurberg, Daniel (1659–1736). Swedish pietist and professor of theology at Uppsala University. Djurberg became interested in converting the Jews while traveling on the Continent. On returning to Sweden (where Jews were forbidden to settle), he instead sought to convert the Sami.

Dryander, Jonas Carlsson (1748–1810). Linnaeus' student. Joseph Banks's librarian, and later librarian to the Linnean Society, London.

Duhre, Ander Gabriel (c. 1680–1739). Swedish mathematician. Duhre in 1738 proposed to the estates an "Economic Society" which would own all means of production and be run by civil servants trained as naturalists.

Ehrenswärd, Augustin (1710–72). Count, landowner, engineer, and fortifications field marshal (*fältmarskalk*). Correspondent of Linnaeus.

Ekeberg, Carl Gustaf (1716–84). Captain at the Swedish East India Company, member of the Swedish Academy of Science. Correspondent of Linnaeus. Sponsor of Linnaeus' explorer-students leaving for China.

Ekeblad, Eva (1724–86). Countess, born De la Gardie. Elected to the Swedish Academy of Sciences in 1748 for her recommendation that potatoes be used for distilling aquavit and making flour and powder.

Ellis, John (1711–76). London merchant and amateur natural historian. Correspondent of Linnaeus.

Elvius, Pehr (1710–49). Secretary of the Swedish Academy of Science.

Fabricius, Johan Christian (1745–1808). Took a private seminar on the natural system given by Linnaeus in 1764 to a small group of students, including Adam Kuhn from Philadelphia as well as students from Stettin, Denmark, Holstein, and Sweden. Professor of economics and natural history at Kiel.

Falck, Anders (1740–96). Astronomer, brother of Linnaeus' student Johan Petter Falck.

Falck, Johan Petter (1732–71). Linnaeus' student. Traveled to Russia, Caucasus, Kazan, and West Siberia, 1768–74, as part of the Orenburg expedition. Curator of the botanic garden and professor at St. Petersburg. Opium addict; slit his throat in Kazan.

Forsskål, Petter (1732–63). Linnaeus' student. Traveled to Egypt and the region of present-day Israel, Jordan, Saudi Arabia, and Yemen, 1761–63, financed by the Danish crown. Appointed professor of botany but died before he could take up the post. See also Michaeli and Carsten Niebuhr. Linnaeus named the common nettle *Forsskohlia* in memory of his "bitter death."

Forster, Johann Georg Adam (1754–94). Professor and eventually librarian in Mainz. Son of Johann Reinhold Forster. Participated in Cook's second circumnavigation of the globe (1772–75).

Forster, Johann Reinhold (1729–1798). German natural historian. Occasional correspondent of Linnaeus and translator of his students' works into German. Employed Linnaeus' student Anders Sparrman as botanist on Captain Cook's second voyage, on which he was the chief naturalist.

Fredrik I (1676–1751). King of Sweden. Preceded by his wife and Karl XII's sister, Ulrika Eleonora (r. 1719–20), succeeded by Adolf Fredrik.

Fries, Elias (1794–1878). Professor of biology in Uppsala.

Fries, Thore Magnus (1832–1913). Son of Elias Fries. Professor in biology in Uppsala. Author of the standard Linnaeus biography.

Gadd, Pehr Adrian (1727–97). Åbo professor in chemistry, and Linnaean botanic transmutationist.

Geer, Charles de (1720–78). Swedish noble (*friherre*), courtier (*hovmarskalk*), owner of Löfsta estate, and distinguished entomologist. Correspondent of Linnaeus.

Geijer, Erik Gustaf (1783–1847). Professor at Uppsala. Leader of Gothicist and Pan-Scandinavian movement and famous poet. Promoter of 1829 Linnaeus statue at Uppsala, together with the student fraternities.

Georgi, Johann Gottlieb (1729–1802). German-born natural historian, member of the Russian Academy of Science in St. Petersburg. Correspondent of Linnaeus.

Giseke, Paul Dietrich (1745–96). Linnaeus' student. Professor of natural history in Hamburg.

Gledisch, Johan Gottlob (1714–86). Professor of medicine and natural history in Berlin. Correspondent of Linnaeus.

Gmelin, Johann Georg (1709–55). German-born natural historian employed by the Russian government to explore Siberia (1733–1743). Professor at St. Petersburg and later Tübingen. Correspondent of Linnaeus.

Grew, Nehemiah (1641–1712). English natural historian and botanic experimenter. Linnaeus studied his theories of plant sexuality.

Grill, Claes (1705–65). Director of the Swedish East India Company. One of the leading men in the Hat party. Correspondent of Linnaeus.

Gronovius, Johann Frederik (1690–1762). Senator in Leyden, natural historian. Participated in Linnaeus' 1737 Dutch "club" discussing *Systema naturae*. Together with Lawson (see below), Gronovius paid for the publication of the first edition of Linnaeus' *Systema naturae* (1735). See also Swieten.

Gronovius, Laurenz Theodor (1730–77). Son of Johann Frederik Gronovius. Senator in Leyden, natural historian.

Gustav II Adolf, or Gustavus Adolphus (1594–1632). As warrior and king of Sweden, he vastly expanded the Swedish empire's domains. Killed in battle during the Thirty Years War.

Gustav III (1746–92). King of Sweden. Son of Lovisa Ulrika and Adolf Fredrik. His successful coup d'état in 1772 brought the Era of Freedom to an end. Killed by a disgruntled minor noble.

Gustav IV Adolf (1778–1837). Ascended to throne in 1796, at age 18. Preceded by his father Gustav III, succeeded by Karl XIV Johan (founder of the Bernadotte monarchy). Disposed of in 1809 coup d'état—lived in exile, insane, as "Colonel Gustavsson."

Gyllengrip, Gabriel (1687–1753). Swedish noble (*friherre*), county governor of Västerbotten, Lapland. Correspondent of Linnaeus.

Haller, Albrecht von (1708–77). Swiss botanist, anatomist, and Alpinist. Professor of medicine at Göttingen. Correspondent of Linnaeus.

Hallman, Johan Gustaf (1726–97). Linnaeus' student. Medical doctor and spy, who in 1749 was sent abroad by Chamber of Commerce (*kommerskollegium*) to study silk production technologies. Hallman returned in 1754 and planted some 25,000 mulberry saplings outside Stockholm. Nothing came of the project, and with the fall of the Hats in 1765–66, the state cut all subsidies to this enterprise.

Hårleman, Carl (1700–53). Swedish noble (*friherre*), high civil servant (*överintendent*). Correspondent of Linnaeus.

Hasselquist, Fredrik (1722–1752). Linnaeus' student. Traveled to Egypt and the region of present-day Israel, Lebanon, and Turkey, 1749–52, financed by Uppsala University, the Swedish Levant Company, and private donations. Teaching fellow in medicine at Uppsala (*med. adjunkt*). Died in Smyrna, at age 30.

Höpken, Anders Johan von (1712–1789). Swedish noble (*greve*), head of government department (*kanslipresident*), Chancellor of Uppsala University, co-founder of the Swedish Academy of Science in 1739. Correspondent of Linnaeus.

Hornick, Philipp Wilhelm von (1638–1713). Austrian author of the cameralist tract *Österreich über alles* (1684).

Jussieu, Antoine Laurent de (1686–1758). Professor of botany in Paris. Brother to Bernard de Jussieu.

Jussieu, Bernard de (1699–1777). Botanist in Paris. Brother to Antoine Laurent de Jussieu.

Justi, Johann Heinrich Gottlob von (1717–71). A German cameralist of a later date, more humanitarian than his fellows, author of *Staatswirthschaft* (1758).

Kalm, Pehr (1716–79). Linnaeus' student. Traveled in northwest Russia, 1744–45, and northeast America, 1748–51, financed by Sten Carl Bielke, the Swedish Bureau of Manufactures, and others. Professor of economics and natural history in Åbo.

Karl X Gustav (1622–60). King of Sweden. Preceded by queen Kristina and succeeded by Karl XI.

Karl XI (1655–97). Swedish king. Preceded by queen Kristina and succeeded by Karl XII.

Karl XII (1682–1718). Swedish king. Preceded by Karl XI.

König, Johan Gerard (1728–1785). Linnaeus' student. Traveled in the region of present-day Thailand, Sri Lanka, and Tranquebar (India) 1767–85, partly financed by the nabob in Arcot. Was Danish Tranquebar trade post physician, remaining there until his death.

Kristina (1626–89). Queen of Sweden. Preceded by a caretaker government during her minority and earlier by her father Gustav II Adolf and succeeded (upon her conversion to Catholicism in 1654) by her cousin Karl X Gustav.

Laestadius, Lars Levi (1800–61). Botanist, parson, and a charismatic preacher of part Sami descent. Founder of the Hebraic revivalist movement in nineteenth-century Lapland.

Laffemas, Barthélémy de (1545–1612). French mercantilist.

Lagerström, Magnus (1691–1759). Director of the Swedish East India Company. Correspondent of Linnaeus. Thanks to Lagerström and Linnaeus, the company preferentially hired ship's chaplains and surgeons who were educated in natural history.

Låstbom, Johan Andersson (1732–1802). Professor in economics and natural history at Uppsala (1759), chosen by the chair's donor Eric Ericsson Borgström. Later Uppsala cathedral parson (*domprost*).

Lawson, Isaac (?-1747). Scottish physician and botanist. Participated in Linnaeus' 1737 Dutch "club" discussing *Systema naturae*. Together with Gronovius, Lawson paid for the publication of the first edition of Linnaeus' *Systema naturae* (1735). See also Swieten.

Lidbeck, Anders (1772–1829). Son of Erik Gustaf Lidbeck. Docent of natural history at Lund university and later professor of aesthetics there.

Lidbeck, Erik Gustaf (1724–1803). Linnaeus' student. From 1756 professor of natural history and economics at Lund University. Expert in botanic transmutationism, and founder of the Skåne Plantation Bureau (*skånska plantageverken*). Correspondent of Linnaeus.

Linnaeus, Carl the Younger (1741–83). Linnaeus' son. Professor of botany at Uppsala University by inheritance.

Linnaeus, Nils Ingemarsson (1674–1748). Linnaeus' father. Parson in Stenbrohult, Småland.

Linnaeus, Samuel (1718–1797). Linnaeus' only brother. Parson of Stenbrohult, Småland, in the fifth generation, and well-regarded as a beekeeper.

Löfling, Pehr (1729–56). Linnaeus' student. Traveled in Spain and Spanish South America, 1751–56, financed by the Spanish crown. Died of fever in present-day Venezuela.

Lovisa Ulrika (1720–82). Queen of Sweden. Sister of Frederick the Great of Prussia. Married to Adolf Fredrik. Linnaeus arranged her and her husband's curiosity cabinets.

Martin, Anton Rolandsson (1729–85). Linnaeus' student. Traveled in Norway and the Arctic Sea, 1758–60, financed by the Swedish Greenland Company. Degree in medicine (*med. kand.*).

Martin, Roland (1726–1788). Linnaeus' student. Professor of anatomy in Stockholm.

Maupertius, Pierre-Louis Moreau de (1698–1759). French scientist who in 1736 led a French expedition to Lapland to measure the length of a degree along the meridian, and thus to test Newton's hypothesis that the globe is a sphere flattened at the poles. Maupertius brought back to Paris two "Lapp" sisters (daughters of a Swedish grocer).

Meldercreutz, Jonas (1715–85). Captain, professor of mathematics at Uppsala.

Merian, Maria Sibylla (1644–1717). German-born entomologist, painter, and engraver, voyaged on a scientific journey to Surinam, 1698–1701. Linnaeus frequently cited her works on European and Surinamese insects.

Michaelis, Johan David (1717–91). Professor in Oriental languages at Göttingen. Correspondent of Linnaeus. Pehr Forsskål's teacher.

Miller, Philipp (1691–1771). Gardener and author in Chelsea, London. Correspondent of Linnaeus.

Monson, Anne (1714–1776). English amateur botanist. Occasional correspondent of Linnaeus.

Montchrétien, Antoine de (c. 1576–1621). French mercantilist.

Montin, Lars (1723–85). Linnaeus' student. Traveled in Lapland, 1745–51, financed by private donations. Regional public doctor (*provincialläkare*) in Halland.

Moraea, Sara Elisabeth (1716–1806). Linnaeus' wife.

Moraeus, Johan (1672–1742). Linnaeus' father-in-law. Public doctor (*stadsläkare*) of copper-mining town of Falun, Dalarna.

Mun, Thomas (1571–1641). English mercantilist and a director of the English East India Company.

Murray, Johan Andreas (1740–91). Linnaeus' student. Professor in medicine and botany at Göttingen. Correspondent of Linnaeus.

Mutis, José Celestino Bruno (1732–1808). Spanish clergyman, from 1760 physician to the viceroy of New Granada in South America. His *Flora de Bogotá o de Nueva Granada,* with over 6,000 illustrations drawn by indigenous American artists, was not published until the twentieth century. Correspondent of Linnaeus.

Niebuhr, Carsten (1733–1815). Danish voyager of discovery (with Forsskål), and the Arabian expedition's sole survivor. Member of Swedish Academy of Science.

Osbeck, Pehr (1723–1805). Linnaeus' student. Traveled to Guangzhou, 1750–52, as Swedish East India Company ship's chaplain. Carl Gustaf Tessin's castle chaplain, and later country parson at Hasslöv, Halland.

Pallas, Peter Simon (1741–1811). German natural historian, professor of natural history in St. Petersburg, member of the Swedish Academy of Science. Correspondent of Linnaeus.

Paulli, Simon (1603–80). Professor of medicine at Copenhagen and author of a 1648 *Flora Danica.* Paulli also recommended bog-myrtle as a substitute for tea.

Polhem, Christopher (1661–1751). Famous Swedish inventor and mechanic, civil servant (*kommerseråd*).

Ray, John (1628–1705). English parson, natural historian and natural theologian. Fellow of Trinity College, Cambridge, and naturalist of Black Notley, Essex. Linnaeus admired him and borrowed his natural theology and much of his botany from Ray.

Retzius, Anders Johan (1742–1821). Professor of natural history in Lund.

Reuterholm, Nils Esbjörnsson (1676–1756). Regional governor for Dalarna. Sponsored Linnaeus' Dalarna journey (1734).

Roberg, Lars (1664–1721). Professor of medicine in Uppsala.

Rolander, Daniel (1725–1793). Linnaeus' student, specializing in entomology. Voyaged to Surinam (Dutch Guyana), 1755–56, as the client of Carl Gustaf Dahlberg and Charles de Geer. On returning to Europe he sold his collections to two Copenhagen professors. Became insane and lived on public charity until his death. See also Carl Gustaf Dahlberg and Martin Vahl.

Rosén, Nils von Rosenstein (1706–73). Professor of anatomy and practical medicine at Uppsala. Co-operated with Linnaeus in their campaign for breastfeeding babies.

Rothman, Göran (1739–78). Linnaeus' student. Traveled in Algeria and Morocco, 1773–76, financed by the Swedish Academy of Science and the Tripoli ambassador to Stockholm.

Rothman, Johan Stensson (1684–1763). Linnaeus' Växsjö grammar schoolteacher. Introduced Linnaeus to Tournefort.

Royen, Adriaan van (1705–79). Professor of botany in Leyden. Correspondent of Linnaeus.

Rudbeck, Olof the Elder (1630–1702). Professor in botany and anatomy at Uppsala, main Gothicist ideologue in the later Carolingian empire, and author of *Atlantica* (1679–1702). His and his son's illustrated edition of Bauhin's *Pinax*, with several thousands of readied but never printed woodcuts, was lost in the 1702 Uppsala fire.

Rudbeck, Olof the Younger (1660–1740). Son of Rudbeck the Elder. Professor in medicine and botany at Uppsala, succeeding his father, and Gothicist ideologue. Voyaged to Lapland, 1695. Linnaeus' teacher and patron at Uppsala. Linnaeus lived in his house and tutored his sons.

Sachs, Julius (1832–97). German plant physiologist and historian of botany. Famously first to condemn Linnaeus for having retarded the science of botany.

Salvius, Lars (1706–1773). Stockholm printer and Linnaeus' Swedish publisher.

Sauvage, François Boissier de la Croix (1706–67). Professor of medicine in Montpellier.

Scheffer, Carl Fredrik (1715–86). Swedish count and government member (*riksråd*). Introduced the physiocrats in Sweden.

Schefferus, Johannes (1621–79). German-born professor of law and rhetoric at Uppsala from 1648, and author of *Lapponia* (1673), an account of the Sami based on Lutheran missionary reports, and the West's first anthropological monograph of a single people.

Schreber, Johan Christian Daniel von (1739–1810). Linnaeus' student. Physician, naturalist, and professor in Erlangen.

Seba, Albert (1665–1736). Amsterdam apothecary and famous collector of *naturalia*. After a dinner at his house one evening in 1735, Linnaeus' best friend Petrus Artedi drowned in a canal.

Sibthorp, Humphrey (1713–97). Professor of botany at Oxford.

Siegesbeck, Johan Georg (1686–1755). Medical doctor, botanical demonstrator in St. Petersburg. Protests Linnaeus' theory of plant sexuality, partly on moral grounds.

Sloane, Sir Hans (1660–1753). English physician and collector of *naturalia*, instrumental in founding the British Museum in that he arranged to sell his collection to the state after his death, for 20,000 pounds.

Smith, Sir James Edward (1750–1828). English gentleman and natural historian, purchased Linnaeus' *Nachlass* in 1784 and founded the Linnean Society in London.

Sofia Albertina (1753–1829). Swedish princess, sister to Gustav III.

Solander, Daniel (1733–1782). Linnaeus' student. Traveled in Lapland (1753); later, as botanist, accompanied Joseph Banks on Captain Cook's first circumnavigation of the globe (1768–71), and afterwards worked as Banks's librarian. Honorary doctor in medicine and law at Oxford.

Sparrman, Anders (1748–1820). Linnaeus' student. Traveled around the Cape of Good Hope and joined Captain Cook's second circumnavigation of the globe, 1772–75, as the Forsters' botanist. Professor of natural history and pharmacology in Stockholm. Avid abolitionist. Traveled to West Africa in 1787–88 to investigate the possibility of Swedish African colonies. See also Georg and Johann Reinhold Forster.

Stobaeus, Kilian (1690–1742). Professor of history at Lund University and a physician. Linnaeus' teacher 1727–28.

Stöver, Dietrich Heinrich (1767–1822). German historian who wrote the first book-length biography of Linnaeus in 1792.

Swieten, Gerhard van (1700–72). Dutch doctor; reorganized the education of Austrian physicians; personal physician to Maria Theresia of Austria. Participated in Linnaeus' 1737 Dutch "club" discussing *Systema naturae*. See also Johan Frederik Gronovius and Isaac Lawson.

Tärnström, Christopher (1703–1746). Linnaeus' student and parson. Traveled to Java and Cambodia, 1745–46, as Swedish East India Company ship's chaplain financed by the Swedish Academy of Science. Died on voyage out, off Cambodia.

Tegnér, Esias (1782–1846). Professor at Lund University, later bishop. One of Sweden's most famous poets, a neo-Gothicist, and a Pan-Scandinavian. Helped initiate the national Romantic cult of Linnaeus.

Telander, Johan (1694–1746). Linnaeus' first tutor, later a parson. Linnaeus remembered him for his sadistic beatings.

Tessin, Carl Gustaf (1695–1770). Swedish count and landowner. Tessin, the son of the French architect who built the Stockholm royal palace, was a key figure in the Hat party, a Speaker in the House of Nobility, a Chancellor (*riksråd*), Sweden's ambassador to Paris, 1739–1742, a tutor of the crown prince Gustav III, and a member of the Swedish Academy of Science. Sweden's most enlightened leader, he was also Linnaeus' main Swedish patron. He employed Pehr Osbeck as chaplain at his country estate Åkerö after Osbeck's return from China.

Thunberg, Carl Peter (1743–1828). Linnaeus' student. Traveled to southern Africa, Ceylon, Java, and Japan, 1770–79 as Dutch East India Company surgeon; is regarded as a key figure in Japanese history of medicine. Professor of medicine and botany at Uppsala on the death of Linnaeus the Younger in 1783.

Tilas, Daniel (1712–1772). Swedish noble (*friherre*), regional governor (*landshövding*), government mining adviser (*bergsråd*), and *riksheraldiker* for the House of Nobles: in this role he quarreled with Linnaeus over his proposed heraldic shield.

Torén, Olof (1718–1753). Linnaeus' student. Traveled to Guangzhou, 1750–52, as Swedish East India Company ship's chaplain.

Tournefort, Joseph Pitton de (1656–1708). French botanic taxonomer and voyager of discovery. Linnaeus borrowed most of his plant genera from him.

Triewald, Mårten (1691–1747). Captain in the fortification division of the Swedish army, popularizer of Newtonian mathematics and physics. Encouraged the founding of the Swedish Academy of Science in 1739.

Uddman, Isaac Hansson (1731–1781). Linnaeus' student. Doctor in Uppsala.

Ulrika Eleonora (1688–1741). Queen of Sweden 1719–20. Preceded by Karl XII, succeeded by her husband Adolf Fredrik.

Vahl, Martin (1749–1804). Linnaeus' student. Professor of botany in Copenhagen. Vahl used some of Daniel Rolander's Surinam collections in his own work; Rolander had sold them previously to a Copenhagen colleague who then passed away.

Vaillant, Sébastien (1669–1722). French physician and botanist. Linnaeus' theory of plant sexuality derived mainly from Vaillant.

Wargentin, Pehr (1717–83). Statistician, Uppsala teaching fellow (*adjunkt*), fellow-founder and secretary of the Swedish Academy of Science.

Abbreviations

Archival abbreviations

UUB	Uppsala University Library
HBL	Hunt Botanical Library, Pittsburgh
LSL	Linnean Society, London
KB	Royal Library, Stockholm

Abbreviations for Linnaeus' works

AU	*Adonis Uplandicus* (1888 edition)
BS	*Bref och skrifvelser*
CD	*Collegium diaeteticum* (1907 edition)
CPS	*Catalogus plantarum rariorum Scaniae*
DN	*Diaeta naturalis* (1958 edition)
EA	*Egenhändiga anteckningar* (1823)
FD	*Föreläsningar öfver djurriket* (1913)
FL	*Flora Lapponica* (1905 edition)
HU	*Herbationes Upsalienses* (1952 edition)
IL	*Iter Lapponicum* (1913 edition)
LD	*Linnés disputationer*
LN	*Lachesis naturalis* (1907 edition)
ND	*Nemesis divina* (1968 edition)
PB	*Philosophia botanica* (1751 Stockholm edition)
SB	*Spolia botanica* (1888 edition)
Skrifter	*Skrifter* (I, 1905; II, 1906; IV, 1908; V, 1913)
TL	*A Tour in Lapland* (1971 facsimile of 1811 print)
Ungdomsskrifter	*Carl von Linné's ungdomsskrifter* (1888)
Valda	*Valda avhandlingar* (various dates)
Vita	*Vita* (1957 edition)

Abbreviations for other frequently used sources

K. Vet. Akad. Handl.	*Kungliga Vetenskapsakademins Handlingar*
TMF	T. M. Fries, *Linné* (1903)
SLÅ	*Svenska Linnésällskapets Årsskrift*
SBL	*Svenskt biografiskt lexikon*
COR	Smith, *Correspondence of Linnaeus*

Notes

All translations, unless otherwise noted, are mine. Punctuation, spelling, and capitalization of works appear as in the originals.

Introduction

1. BS I:2, 59. Linnaeus to the Academy of Science. Uppsala, 10 January 1746.
2. Heckscher, *Mercantilism,* vol. 1, 22–25; Whittaker, *History of Economic Ideas,* 280–319.
3. Heckscher, *Mercantilism,* 240–254; Mun, *England's Treasure by Forraign Trade.*
4. For an alternate dating see Small, *Cameralists,* 21.
5. Briggs, "The Académie Royale des Sciences and the Pursuit of Utility," 40; Ambjörnsson, "'Guds Republique,'" 4–6.
6. Heckscher, *Mercantilism,* 15–21.
7. This was especially true for backward nations, like the German principalities and Scandinavia. See Hildebrand, "Economic Background," 19.
8. Heilbroner, *Worldly Philosophers,* 29.
9. Small, *Cameralists,* 107–108. See Heckscher, *Mercantilism,* vol. 1, 40, and Rydberg, *Svenska studieresor,* 136, for how, in the 1740s, Sweden's first professor of cameralism clamored for tropical colonies where slaves and criminals would purchase Swedish woollen cloth and silk brocades.
10. Daniel Solander to Linnaeus. London, 5 February 1761. LSL.
11. Von Schulzenheim, *Åminnelse-Tal,* 23.
12. EA, 61. See also TMF II:97.
13. Sachs, *Geschichte der Botanik;* Daudin, *Les méthodes de classification;* Foucault, *The Order of Things;* quotation from Delaporte, *Nature's Second Kingdom,* 191; Atran, *Cognitive Foundations,* x.
14. Sachs, *Geschichte der Botanik.* Foucault, *Order of Things.* Quotation from Hagberg, *Carl Linnaeus.*
15. See *Dagens Nyheter,* 14 December 1951, Sten Lindroth, "Linné och hans profet," review of Knut Hagberg, *Den linneanska traditionen* (Stockholm: Natur och Kultur, 1951). See reply by Folke Isaksson, *Morgontidningen,* 30 December

1951; and for similar charges see Sten Lindroth, *Dagens Nyheter,* 29 October 1964.

16. Lindroth,"Legend och verklighet."

17. See, for example, Ritterbush, *Overtures to Biology,* 9; Lanham, *Origins of Modern Biology,* 123; Bodenheimer, *The History of Biology,* 107.

18. See Sheets-Johnstone, "Why Lamarck Did Not Discover the Principle of Natural Selection."

19. Worster, *Nature's Economy,* 553; Londa Schiebinger, "The Private Life of Plants: Sexual Politics in Carl Linnaeus and Erasmus Darwin." In Benjamin, ed., *Science & Sensibility,* 121–143.

20. Sernander, "I Linnés fotspår," 48. In Germany, Linnaeus was seen to have started "modern" (Nazi) race science. See Möller, "Die Rassenforschung in den nordischen Ländern." On the Jewish issue, see Felix Bryk, "Var Linné svensk?" review of a 1942 article by the Swedish Nazi scholar Ernst Alker, in *Deutsche Vierteljahrschrift für Literaturwissenschaft und Geistesgeschichte,* which "proved" Linnaeus was Jewish, since he was short, brown-eyed, and an Old Testament fundamentalist. See also "Linné dålig som german. Svensk forskare gör honom i tysk tidskrift till jude!" *Trots Allt,* no. 20, 1942. Sweden's most famous literary critic, the *Mitläufer* Fredrik Böök, defended Oscar Levertin in *Svenska Dagbladet,* 23 October 1939, but also repeatedly hinted that Levertin was Jewish.

21. For example, Petander, *De nationalekonomiska;* Schauman, *Studier i Frihetstidens;* Wiselgren, "'Yppighets nytta.'"

22. Especially Heckscher, "Linnés resor"; Frängsmyr, "Den gudomliga ekonomien," 217–244; Hildebrand, "Economic Background"; Heckscher, *Sveriges ekonomiska historia;* and Lindroth, *K. Vet. Akad. Hist.* See also Gunnar Eriksson, "Rudbeck och Linné—enmansinstitutioner," in Gunnar Broberg and Karin Johannisson, eds., *Kunskapens trädgårdar—om institutioner och institutionaliseringar i vetenskapen och livet* (Stockholm, 1988), 17–31, and Sörlin, "Apostlarnas gärning," esp. 79.

23. See Small, *Cameralists.*

24. Smith, *Inquiry;* Heckscher, *Mercantilism.* This school was started by Schmoller, *The Mercantile System,* 1896. Small, *Cameralists,* also regards cameralism as the ideological forerunner of Wilhelmine Germany, but emphasizes cameralism's social welfare legislation.

25. See, however, Tautscher, *Staatwirtschaftslehre;* Dittrich, *Kameralisten.*

26. These older histories assume that national economic independence is a self-evident common good. I investigate it as a specific and historically contingent ideology. I have also been inspired by public choice theory's notion of rent-seeking as formulated by A. O. Kreuger, "The Political Economy of the Rent Seeking Society," *American Economic Review* 64:3 (June 1974), and Jagdish Bhagwati, "Directly Unproductive Profit-Seeking Activities," *Journal of Political Economy* 90 (1982).

1 "A Geography of Nature"

1. Jonas Hallenberg, who was the Småland's nation's history lecturer at Uppsala University and who collected *Linnaeana* for a planned biography, asked for the memoir. He published a genealogical tree; see Hallenberg, "Stam tafla öfver Linnaeiska Slägten."

2. On its printing history, see Dahlgren, "Philosophia Botanica," 2. One extant copy of the work, now at the Linnaean Society in Uppsala, is inscribed "D. Cn Solander 1750." As for *Fundamenta botanica*, the print run was earlier than the imprint date. See also Soulsby, *Catalogue*, 39–42.

3. Linnaeus, *Den lärda världens omdöme*. On the student riot, see TMF I:37–39.

4. EA, 215.

5. Hofsten, "Linnés dubbla bokföring."

6. A. J. Wilmott, "Systematic Botany," 36.

7. The tier "family" was inserted between order and genus only at 1800, and the tier "phylum," situated between kingdom and class, even later. My thanks to Peter F. Stevens for explaining how, in terms of the Aristotelian essence-concept Linneaus relied on, varieties did not signify as a separate rank. See also Linnaeus, *Philosophia botanica*, aphorisms 93–97, on the *Numero, Figura, Proportione, Situ*.

8. Linnaeus, *Fundamenta botanica*, aphorism 159; Stearn, "Species Plantarum," 36.

9. Atran, *Cognitive Foundations*, 170–171, 180; Mayr, *Growth of Biological Thought*, 179; Larson, *Reason and Experience*, 48. See also BS I:8, 149. Marshall of the Court Anders Johan von Höpken to Linnaeus. Stockholm, 31 October 1751.

10. Gledhill, *Names of Plants*, 8–47.

11. Ibid., 18.

12. Quoted from Tessin's diary of 6 July 1751, in EA, 245.

13. Description by Johann Beckmann from 1765, quoted and translated in Sernander, "Hårleman," 81.

14. BS I:1, 90. List of merits, autograph, undated, probably written in March 1767. See also Svenska vetenskapsakademien *Protokoll* I:5 (2 June 1739), first meeting.

15. EA, 206, 209, 210, 212, 221, quote from 212.

16. Tullberg, *Linnéporträtt*, and Soulsby, *Catalogue*, 199–216.

17. *Famam extendere factis;* sometimes *famam extollere factis.*

18. *Vita* II, 58.

19. The letter is reprinted in Virdestam, "Samuel Linnaeus," 120.

20. Lindroth, "Legend och verklighet," 115. In a 1763 Academy prize question, Linnaeus competed under the transparent anagram C. N. Nelin, or Carl Nilsson Linné.

21. Samuel Linnaeus, 26 February 1778, quoted in TMF I:2, n3 and in Virdestam, "Samuel Linnaeus."

22. Letter written by M. E. Forsander, 11 August 1823, quoted in TMF I:2, n3.
23. Almer, *Linné om skaparen*, 21.
24. Sahlgren, "Linnés talspråk"; TMF I:4.
25. Quoted in TMF I:6–7, n4. The death scene is also described in Samuel Linnaeus's *minne*, reprinted in Virdestam, "Samuel Linnaeus," 120.
26. Linnaeus, "Rop ur grafwen."
27. BS I:2, 127. Linnaeus to Elvius. Uppsala, 29 April 1749. See also BS I:4, 75. Linnaeus to Bäck. Uppsala, 3 June 1748.
28. *Vita* II, 60.
29. Samuel Linnaeus' *Minne*, here quoted from Hagberg, *Carl Linnaeus*, 33.
30. The letter is reprinted in Virdestam, "Samuel Linnaeus," 117.
31. Ibid.
32. TMF I:4.
33. The letter is reprinted in Virdestam, "Samuel Linnaeus," 117.
34. Ibid., 122.
35. PB, aphorism 159.
36. Linnaeus, *Deliciae naturae* (1772), 1939, 108.
37. Linnaeus, "Tal," (1759), 1939, 95.
38. Lindroth, "Legend and verklighet," 66. Linnaeus assumed that 2 Sam. 7:9 and 1 Kings 17:8 described him.
39. *Vita* III, 147.
40. "Progressus botanices," Lin pater, Bot., LSL. See also EA, 204.
41. *Vita* III, 146–47.
42. Albrecht von Haller, anonymous review of Linnaeus's *Flora Svecica*, *Göttingen gelehrten Anzeigen* (1746): 670, partially reprinted in Stöver, *Leben des Ritters Carl von Linné*, 214.
43. Linnaeus, *Deliciae naturae*, 108; *Vita* III, 146–147.
44. Roberts, *Swedish Imperial Experience*, 70–75, esp. 71.
45. Erasmus Darwin said that Linnaeus "may be said *to have formed a language*, rather than to have found one," in the preface to Carl Linnaeus, *A System of Vegetables, according to Their Classes Orders Genera Species with Their Characters and Differences. . . . By a Botanical Society, at Lichfield*, 2 vols, sel. trans. from 13th ed. of Linnaeus, *Systema vegetabilium*, and Linnaeus the Younger, *Supplementum plantarum*, vol. 1, preface.
46. COR II:258. Linnaeus to Haller. Hartecamp, 8 June 1737.
47. Linnaeus, *Deliciae naturae*, 108.
48. Goethe, quoted in Friedenthal, 306.
49. Goethe, *Schriften zur Naturwissenschaft*, II:9 A, 1977, 270–271, reprinted bookbinder's bills and bibliographical references.
50. COR II:552. Jean-Jacques Rousseau to Linnaeus. Paris, 21 September 1771.
51. Quoted from one of Bois-Reymond's academic orations, in Müller, "Abraham Bäck," 108.
52. See also BS I:5, 322. Linnaeus to Nils Dahlberg. Uppsala, 25 October 1771. Dahlberg accompanied Gustav III on his European grand tour, as his physician.

They met Rousseau in Paris in 1771, and talked about Linnaeus. On 21 September 1771, Rousseau wrote a flattering letter to Linnaeus (now kept in LSL).

53. PB, aphorism 296.

54. Ibid. See also aphorisms 297 and 298.

55. Linnaeus, *Den lärda världens omdöme* (1740), 1952, also partially reprinted in EA, 96–97.

56. *Monthly Review* (1750) III:205, cited in EA, 244.

57. Linnaeus, *Philosophia botanica*, 1751, aphorism 93–97, and Atran, *Cognitive Foundations of Natural History*, 180.

58. Stevens, "Metaphors and Typology," 172.

59. COR II:345. Albrecht von Haller to Linnaeus. Göttingen, 25 August 1740.

60. Knight, *Ordering the World*, 44–46.

61. COR II:91–97. Dillenius to Linnaeus. Oxford, 16 May 1737.

62. Ibid. See also Dillenius to Linnaeus. Oxford, 18 August 1737.

63. Ibid., 152. Dillenius to Richard Richardson. Oxford, 25 August 1737.

64. Georges-Louis L. de Buffon, "Initial Discourse," *Histoire naturelle*, vol. 1 (1749), trans. John Lyon, in Lyon and Sloan, *From Natural History*, 106.

65. Sloan, "The Buffon-Linnaeus Controversy," 238, 356–375; Larson, "Linné's French Critics."

66. BS I:6, 292–293. Linnaeus to Joh. Otto Hagström. Uppsala, 9 February 1774.

67. Cited from Hagberg, *Carl Linnaeus*, 265. Compare ibid., 111.

68. EA, 67.

69. Gillespie, *Science and Polity*, 161, 151. Stearn, *Botanical Gardens*, lxxxvii-xcii.

70. COR II:229. Linnaeus to Haller. Hartecamp, 3 April 1737.

71. Schuster, *Linné und Fabricius*, from the author's "Nachwort über das natürliche System," IX. In the sixth edition of *Genera plantarum* (1764) he reduced them to fifty-eight.

72. PB, aphorism 77. See also Stearn, *Botanical Gardens*, xc; Dahlgren, "Philosophia botanica," 14; Schuster, *Linné und Fabricius*, IX, from "Nachwort."

73. *Vita* III, 135; PB, aphorism 77 and 155, quoted in Stafleau, *Linnaeus and the Linnaeans*, 45, 61. Schuster, *Linné und Fabricius*, XIX, quoting from Giseke, *Caroli a Linné*; Atran, *Cognitive Foundations*, 174; Linnaeus the Younger, "Linné den Yngres brev," 160, quoted from Linnaeus the Younger to Abraham Bäck, 1778.

74. Fabricius, "Einige nähere Umstände," unpaginated. See also Giseke, *Caroli a Linné*, for Fabricius' notes on Linnaeus' 1764 lectures; and Schuster, *Linné und Fabricius*, XII, from "Nachwort," on how Linnaeus lectured on this topic in 1767 and in 1771.

75. Schuster, *Linné und Fabricius*, XVII-XIX, from "Nachwort."

76. Ernst Mayr, *Towards a New Philosophy of Biology*, 181.

77. *Vita* IV, 170.

78. COR I:369. Alexander Garden to John Ellis. Charlestown, 13 January 1756.

79. Atran, *Cognitive Foundations*, 174. See also Schuster, *Linné und Fabricius*, XVII-XIX, "Nachwort."

80. COR II:232. Linnaeus to Haller. Hartecamp, 3 April 1737.
81. Atran, *Cognitive Foundations,* 175. I have not found primary evidence to either support or disprove this view. See also Mayr, *Philosophy of Biology,* 181.
82. COR II:258. Linnaeus to Haller. Hartecamp, 8 June 1737.
83. Ibid., 229. Linnaeus to Haller. Hartecamp, 3 April 1737.
84. Ibid.
85. Ibid., 232. Linnaeus to Haller. Hartecamp, 3 April 1737.
86. Ibid., 258. Linnaeus to Haller. Hartecamp, 8 June 1737.
87. Linnaeus, *Herbationes Upsalienses* (1747), 1952, 47. See also Gledhill, *Names of Plants,* 8.
88. Knight, *Ordering the World,* 55.

2 "A Clapper into a Bell"

1. Ms. now kept in UUB, D.82a. It is a vernacular text, authored in February 1729 by Petrus Artedi. Also cited in TMF I:67. It discussed the plants around his parents' home.
2. Hagberg, *Carl Linnaeus,* 36.
3. Florin, *Trädgården,* 5–6.
4. Gledhill, *Names of Plants,* 6–8.
5. Hagberg, *Carl Linnaeus,* 51. See also Ashworth, "The Persistent Beast," p. 54–57, on how Linnaeus, in his 1760 *Anthropomorpha,* borrows from Aldrovandi and Gesner in describing *Homo caudatus,* tailed man.
6. *Vita* II, 61.
7. TMF I:166. Cited from summary in the protocols of Småland's *nation.*
8. Ibid., 28; quoted from Linnaeus.
9. Ibid., 48–49.
10. Juel, "North American Flora." Seven years later, in *Hortus Cliffortianus,* which tallies the gardens of a rich Dutch banker, Linnaeus listed c. 170. In 1753, he grew c. 270. For *Species plantarum* of the same year, he examined c. 500 dried specimens.
11. Eriksson, "Olof Rudbeck d.ä.", 84; TMF I:48; Juel, "North American Flora," 65. This herbarium had been assembled by Joachim Burser (1583–1639).
12. *Vita* II, 61.
13. Linnaeus, quoted in TMF I:66. See also *Vita* II, 61–62, and TMF I:55. In 1727 Linnaeus used Kilian Stobaeus' library. Moving to Uppsala in 1729, Linnaeus used Olof Celsius the Elder's private library and botanic notebooks. In summer 1730 Linnaeus moved to Olof Rudbeck the Younger's house and used that family's holdings.
14. Veendorp, Becking, and Baas, *1587–1937.*
15. On the destruction of forests around Uppsala in Linnaeus' times, see BS I:1, 158–160. Linnaeus to Uppsala University's rector. Undated, but filed at 11 February 1756.
16. Linnaeus, *Skånska resa* (1749), 1975, 197, 126, for how he climbed the highest

point of south-west Skåne, Rameleklint, and sees the "Tottenska Skavsjö, Silfversköldska Häckebiera, Piperska Björstorp" forests. The now forested ice-age ridge was "green meadow." Lower down, the grain field were dotted with oaks. "The lords do not allow the farmers to cut them down."

17. TMF I:56, quoted from Linnaeus' diary, April 1729.

18. Ibid., 31.

19. Preface to *Spolia botanica*, quoted from Hagberg, *Carl Linnaeus*, 53.

20. Linnaeus to Kilian Stobaeus, 8 November 1728, quoted in TMF I:42. There were no zoological specimens available to students at Uppsala.

21. Gertz, "Olof Celsius," 53. By contrast, in 1730 the botanic garden of Leyden housed over seven thousand live plants.

22. Sernander, "Linnaeus och Rudbeckarnes," 133, 147 (quoted from a contemporary annotation by Lars Roberg). Linnaeus' first *Hortus Uplandicus* lists 454 plants, the second 417, distributed according to Tournefort (the only classificatory system Linnaeus used, apart from his own).

23. Gertz, "Olof Celsius," 53; TMF I:56.

24. *Vita* II, 63. See also Uggla, "Species Plantarum," 12.

25. Eriksson, "Olof Rudbeck d.ä.," 84; quote from TMF I:38; *Vita* III, 64, and commentary, 204.

26. Gertz, "Olof Celsius." See also Olof Celsius, "Plantarum circa Upsaliam sponte nascentium Catalogus." *Uppsala Vetenskapssocietets handlingar*, vol. 2 (1732, printed 1735), and Olof Celsius, "Tilökning på några örter, fundna i Upland, sedan *Catalogus plantarum Uplandicarum* utgafs år 1732," *K. Vet. Akad. Handl.* vol. I, 1740. Linnaeus published excerpts of Celsius' manuscript "Flora Uplandica" as part of the first edition of *Flora Svecica* (1745).

27. Quoted in Gertz, "Olof Celsius," 46.

28. TMF I:64, 69.

29. Hagberg, *Carl Linnaeus*, 69–71.

30. *Vita* II, 64.

31. Mayr, *Growth of Biological Thought*, 158–179.

32. Later printed as Vaillant, "Sermo de structura florum." Linnaeus read a review in 1729, in *Acta Eruditorum* (Leipzig: 1719). For his reaction, see TMF I:59–60; Stöver, *Leben des Ritters Carl von Linné*. For how Rothman mentioned Vaillant to Linnaeus, see Stearn, *Botanical Gardens*, lix; and Hagberg, *Carl Linnaeus*, 66.

33. TMF I:59.

34. EA, 35. See also COR II:212. Bernard de Jussieu to Linnaeus. Paris, 20 July 1740.

35. On how Linnaeus ranked botanists see *Vita* III, 153–154, "Florae officiarii." Linnaeus equipped and sent four mounted soldiers from his two country estates to fight in Sweden's 1757–1762 Pomeranian War. He named them Säfbom, Säfgren, Hammar, and Hammarblom. See Torén, "Carl von Linné som rusthållare," 48–56, esp. 40. On Linnaeus's cows Lilja, Blomstra, Sommarros, Gullgumman, Hammarros, and Gonos see Forsstrand, "Linnés ekonomi," 70–89, esp. 79, and Sigurd Wallin, ed., "Ukkunder rörande familjen Linnés ägor,"

SLÅ 33–34 (1950–1951), 67–94, esp. 84. At his death, Linnaeus owned fourteen cows, as well as pigs, horses, turkeys, hens, geese, "old sheep," and "young Dito." While Linnaeus's library was valued at 166 *riksdaler,* his "200 Barrels of Rye and Barley" was valued at 333 *riksdaler.*

36. *Vita* II, 76.
37. Carl Linnaeus, autograph ms., "Calendae [sic] botanicae perpetua," now kept in the Linnean Society, London, Lin. pater, Varia. Remains of a seal indicate the sheet of paper was fastened to a wall.
38. *Vita* IV, 160.
39. *Vita* V, 181.
40. *Vita* III, 136, 151; IV, 166.
41. TMF I:60–63; Linnaeus, *Praeludia* (1729), 1908.
42. Ibid.
43. Linnaeus, *Hortus Uplandicus . . . methodum propriam et novam a sexu* (13 May 1731), 1888. See also PB, 93–97. TMF II:67, for "authopsien."
44. TMF I:66–67, 74–75.
45. See Wachenfelt, "Lapplandsporträtt." The portrait discussed here is kept in the Linnaean museum in Uppsala.
46. Carl Linnaeus, *Systema naturae* (1735); *Musa Cliffortiana florens Hartecampi 1736* (Lugduni Batavorum: 1736); *Genera plantarum* (Lugduni Batavorum: Apud Conradum Wishoff, 1737).
47. Here Linnaeus had the painter put in an imaginary book.
48. Not listed in Soulsby, *Catalogue.*
49. Quoted from Linnaeus, "Linnés företal till Species plantarum," (1753), 1952, 3.
50. Carl Linnaeus, *Termini botanici,* resp. Johannes Elmgren (Uppsala: 1762), and the "editio nova avctior" (Leipzig: 1767). First edition also reprinted in Linnaeus, *Amoenitates academicae,* vol. 6, 217–246. See also Soulsby, *Catalogue,* 139–140.
51. BS I:8, 149. Letter from Marshall of the Court Anders Johan von Höpken to Linnaeus. Stockholm, 31 October 1751.
52. COR II:174. Johann Frederik Gronovius to Richard Richardson. Leyden, 1738.
53. PB, 77, also cited in Stafleau, *Linnaeus,* 45; PB, 155, also quoted in Stafleau, *Linnaeus,* 61.
54. Linnaeus, *Caroli Linnaei libellus amicorum* (1734–), 1958, 22, quoting Johan Browallius's preface, Uppsala, 15 February 1735.
55. Quoted in TMF I:67.
56. LD, 200. Quote from Linnaeus' self-review in *Lärda Tidningar* of his dissertation *Incrementa botanices,* resp. Jacobus Biuur (Uppsala: 1753). See also Heller, *Studies,* chap. VII, "Linnaeus on Sumptuous Books."
57. Linnaeus, *Deliciae naturae* (1772), 1939, 108.
58. Linnaeus, "Skaparens afsikt med naturens verk" (1763), 1947, 77.
59. Linnaeus, "Tal, vid deras Kongl. Majesteters höga närvaro" (1759), 1942, 98.
60. PB, 291–299.
61. TMF I:66, and 58. Quote from Linnaeus.

62. Linnaeus, *Herbationes Upsalienses* (1747), 1952, 5.

63. *Vita* III, 120.

64. PB, 293; Linnaeus, *Herbationes Upsalienses* (1747), 1952, 5. See also Arvid Hj. Uggla's preface to Linnaeus, "Herbationes Upsalienses" (1747), 1950–51, 97–106.

65. Linnaeus, *Herbationes Upsalienses* (1747), 1952, 7, quoted from J. G. Acrel, "Tal, vid praesidii nedläggande i Kungliga Vetenskaps Akademien," 1796.

66. *Vita* III, 120.

67. Daniel Melanderhielm, memorial about Uppsala University written 1783, reprinted in Annerstedt, *Uppsala universitets historia,* bihang V, 137–138, and partially quoted (in modernized transliteration) in preface to Linnaeus, *Herbationes Upsalienses* (1747), 1952, 9.

68. *Vita* III, 120. See also *Vita* II, 79, also quoted in Uggla's preface to Linnaeus, "Herbationes Upsalienses" (1747), 1950–51, 102, from EA.

69. Linnaeus, *Botaniska exkursioner* (1753), 1921, 17, 22, 29.

70. Carl Hårleman to Linnaeus, 28 July 1748, quoted in Uggla's preface to Linnaeus, "Herbationes Upsalienses" (1747), 1950–51, 102. See also Sernander, "Hårleman."

71. *Vita,* II, 79, also quoted in Uggla's preface to Linnaeus, "Herbationes Upsalienses" (1747), 1950–51, 102, from EA.

72. Stearn, "Background"; Atran, *Cognitive Foundations;* Heller, *Studies,* 41–75.

73. PB, 365.

74. Linnaeus, *Miscellaneous Tracts,* 1775, 125, from his 1763 speech at Uppsala University's annual conferment of doctoral degrees; Linnaeus, "Linnés företal till Species Plantarum 1753," (1753), 1952, 3; Mayr, *History of Biology,* 72; May, "How Many Species," 42–48.

75. PB, 162. See also 316.

76. Linnaeus the Younger, "Linné den Yngres brev," 160. Quoted from an undated letter by Linnaeus the Younger to Abraham Bäck in 1778.

77. Linnaeus, *Cynographia,* 1756.

78. COR II:232. Linnaeus to Haller. Hartecamp, 3 April 1737.

79. Quoted from Linnaeus, *Miscellaneous Tracts,* 1775, 125, from his 1763 speech at Uppsala University's annual conferment of doctoral degrees. For other calculations see Linnaeus, "Linnés företal till Species Plantarum," (1753), 1952, 3.

80. TMF I:48; Selander, "Linné i Lule Lappmark," 18.

81. Linnaeus, "Skaparens afsikt med naturens verk" (1763), 1947, 74.

82. Ibid., quote from 77. In the eighteenth century a proper name was called a "trivial name." "Specific name" meant a diagnostic phrase of 8–10 words. See Linnaeus, *Fundamenta botanica,* 1736, aphorism 256.

83. TMF I:56.

84. Linnaeus, *Deliciae naturae* (1773), 1939, 108.

85. Most early modern naturalists abbreviated these diagnostic tags, and skimped on congenera name-changes. On synoptic phrasenames, see Stearn, *Botanical Gardens,* lxx-lxxii; Stearn, "Background"; Larson, *Reason and Experience,* 7–11.

86. Heller, *Studies,* 57; Gledhill, *Names of Plants,* 18.

87. Svanberg, "Turkic ethnobotany," 60.
88. BS I:1, 4. Linnaeus to the Swedish king. Undated, but written during the famine of 1756. Arrived 9 March 1757.
89. Ibid., 5.
90. Ibid.
91. TMF I:25, "Inledningsord vid Linnaei första botaniska föreläsning i Stockholm maj 1739."
92. Linnaeus, *Hortus Upsaliensis,* 1748.
93. Linnaeus' Spanish, Italian, German, and Dutch correspondents wrote to him in Latin. The French used French, and the English, English.
94. *Vita* III, 113. See, however, Stearn, *Botanical Gardens,* civ and cv for student comments on Linnaeus' German. I find Linnaeus' own testimonial more convincing.
95. BS I:1, 4.
96. Heller, *Studies,* 43.
97. Ms. *Species plantarum* (ca 1746–748), now kept in LSL, dated by Hulth, "Linnés första utkast."
98. The copy now kept at the Linnaean Society at Uppsala is inscribed "D. Cn Solander 1750." Linnaeus often paid students who did secretarial work with copies of his works. He probably gave Daniel Solander an early print-run copy.
99. Heller, *Studies,* 43.
100. Quoted in ibid., 55.
101. Uggla, "Species Plantarum," 15.
102. PB, 293; also quoted in Linnaeus, *Herbationes Upsalienses* (1747), 1950–51.
103. *Vita* III, 153–154, list of "Florae officiarii."
104. Torén, "Om Olof Torén," 20–21.
105. LSL. Pehr Osbeck to Linnaeus. Gothenburg, 30 September 1752.
106. Ibid. Pehr Osbeck to Linnaeus. 26 February 1751, "Skeppet Prins Carl wid Padrie."
107. Ibid. Pehr Osbeck to Linnaeus. Cadix, 26 February 1751.
108. Ibid. Rolander to Linnaeus. 18 January 1755. Amsterdam.
109. Ibid. Pehr Osbeck to Linnaeus. Gothenburg, 30 September 1752.
110. Ibid. Pehr Osbeck to Linnaeus. 26 February 1751. "Skeppet Prins Carl wid Padrie."
111. Ibid. Pehr Osbeck to Linnaeus. "Skeppet Prins Carl den 21. Martii 1751. Swänska stylen"; Gothenburg, 21 November 1751; [Gothenburg] 18 August 1752; [Gothenburg] 26 August 1752; see Pehr Osbeck to Linnaeus. N.d., n.p., but written in Gothenburg, autumn 1752.
112. Pehr Osbeck to Linnaeus. 18 August 1752; 18 October 1752; 11 November 1752; and ibid. Undated, but written in autumn 1752.
113. COR, 267. Linnaeus to John Ellis. Uppsala, 22 October 1771.
114. LSL, Pehr Osbeck to Linnaaeus. Undated, but written in the autumn of 1752.
115. Ibid.
116. Heller, *Studies,* 76. Linnaeus also guarded his right, as he saw it, to set his own

binomials. See Daniel Solander's anxious comments, as he sends Linnaeus an insect collection, in letter from Daniel Solander to Linnaeus. 24 May 1760. Helsingör. LSL.

117. Quoted in Uggla, "Species Plantarum," 15.

118. Dahlgren, "Philosophia Botanica," 6. The student was the twenty-two-year-old Pehr Löfling, resident in Linnaeus' Uppsala home as the tutor for Carl Linnaeus the Younger.

119. Ibid., 14.

120. Linnaeus, "Linnés företal" (1753), 1952, 3.

121. Quoted from a footnote on p. 481 in Linnaeus' 10th edition of *Systema naturae* in Heller, *Studies,* 72.

122. Stearn, *Botanical Gardens,* lxxii. It lists 1,434 plant species.

123. Juel, *Hortus Linnaeanus,* 3; Stearn, *Botanical Gardens,* lxix-lxx.

124. COR I:31. Peter Collinson to Linnaeus. London, 20 April 1754.

125. Ibid., 35. Earl of Bute to Peter Collinson. No date.

126. LSL, Pehr Osbeck to Linnaeus. Stockholm, 25 April 1757.

127. *Vita* IV, 161. Quoted and translated in Heller, *Studies,* 56.

128. Carl Linnaeus, praes., *Reformatio botanices,* resp. Reftelius (Uppsala: 1762), reprinted in Linnaeus, *Amoenitates academicae,* vol. 6, 315, quoted and translated in Heller, *Studies,* 74.

129. EA, 207.

130. Linnaeus, *Nomenclator botanicus* (Leipzig: June 1772).

3 "The Lapp Is Our Teacher"

1. P. C. C. Garnham, "Linnaeus' Thesis on Malaria in Sweden," *SLÅ* 1978: 80–97, including trans. of Carl Linnaeus, *De febrium intermittentiuum causa,* Ph.D. Harderwijk University, 23 June, 1735; Stearn, *Botanical Gardens,* lix; Lindroth, "Legend och verklighet," 101. Linnaeus claimed that "intermittent fevers" were caused by clay particles dissolved in drinking water. His defense expounded on Hippocrates, bk. VI, 14 and 15. See also TMF I:207. Most members of Sweden's Collegium Medicum had degrees in medicine from Harderwijk.

2. Bäck, *Åminnelse-Tal,* 73-75.

3. TMF I:145.

4. *Vita* IV, 165; Linnaeus, *Egenhändiga anteckningar,* 1823, 204; Lindroth, "Legend och verklighet," 101–103. For Linnaeus' electro-medicine see Linnaeus, *Egenhändiga anteckningar,* 1823, 202; on his mystical medicine, see Linnaeus, *Clavis medicinae duplex* (1766), 1967.

5. Anders Jahan [sic] Retzius, "Tal hållit på Kongl," 10. See also Bäck, *Åminnelse-Tal,* 10; TMF I:32.

6. EA, 202.

7. FL, notes, 375.

8. See Schefferus, *Lapponia,* 1704, 4. *Lapponia* was translated into English (1674),

German (1675), French (1678), and Dutch (1678). The Lutheran missionary reports on which it is based are published in Roung and Fjellström, *Berättelser om samerna*, 1982.

9. Sandblad, "Clarke and Acerbi," 175.

10. Sydow, "Vet. Soc.," 153; Pålstig, *Myten om Lappland;* Broberg, "Lappkaravaner på villovägar," 61. Broberg, "Olof Rudbeck," 18. Rudbeck the Younger, *Iter Lapponicum*, 52, 59. Rudbeck also contended that the Finns were Turks, and that old North Germanic, or Gothic, was a form of Chinese (he worked out these linguistic kinships in *Lexicon harmonicum*, a never published manuscript in 10 vols.).

11. See also Schefferus, *Lapponia*, 3; Consett, *Tour through Sweden*, 62.

12. See Broberg, *Homo Sapiens*, 243-244, quotation on p. 244; the San once inhabited Southern Africa. A remnant, the !Kung, survive in the Kalahari desert, on the border of South Africa and Botswana.

13. Goldsmith, *The History of the Earth*, vol. 2.

14. Broberg, *Homo Sapiens*, 222. Linnaeus included the Sami in this catch-all category (tangential to his global racial taxonomy of [indigenous] Americans, Europeans, Asians and Africans) because he considered them unnaturally short. He also included in it humans supposedly deformed by custom, such as one-testicled San and corseted European women. See also Rudbeck the Elder, *Atlantica*, vol. 1, 201, also cited in Broberg, "Olof Rudbeck," 18; Schefferus, *Lapponia*, 24, for earlier arguments that the Sami were "Pigmies."

15. Broberg, "Lappkaravaner på villovägar," 62. The idea came most immediately from the German pathologist Rudolf Virchow (1821–1902).

16. Schefferus, *Lapponia*, 28; Kunze, "Lappen oder Finnen in der deutschen Flugschriften des 30-jährigen Krieges," *Ural-altaische Jahrbücher* 43 (1971), quoting the broadsheet "Beschreibung der Lapländer, die der König von Schweden mit auf Teutschen Boden gebracht."

17. Furley, "Three 'Lapland songs,'" 3–12.

18. FL, 302.

19. Sydow, "Vet. Soc.," 152–153.

20. Broberg, "Olof Rudbeck," 18; see Bäckman and Kranz, *Studies in Lapp Shamanism.* Today many Sami and homesteaders adhere to the hebraicizing revivalism of Lars Levi Laestadius, nineteenth-century charismatic preacher of part-Sami descent.

21. Rudbeck the Younger's handwritten instruction to Maupertius, undated, cited in Rudbeck the Younger, *Iter Lapponicum*, commentary, 27.

22. Broberg, *Homo Sapiens*, 266.

23. In later life, Rudbeck claimed that he had lost eleven manuscript volumes on Lapland in the 1702 Uppsala fire. Tomas Anfält, "Olof Rudbeck den Yngres lapska skissbok," in Rudbeck the Younger, *Iter Lapponicum*, commentary 7, in my view rightly questions this. FL (1737), 1905, preface, 1, suggests seven volumes were lost.

24. Sydow, "Vet. Soc.," 138, 142–144.

25. Linnaeus was commissioned by the Uppsala Royal Society of Arts and Sciences (*Societetas Regia Literaria et Scientiarum*), Sweden's first scientific society. Founded in 1710 on the initiative of Christopher Polhem as the Society of Curiosities (*Societetas Curiosorum*, also called *Societas Literaria* and *Bokwetts-Gillet*), it received its name and a royal patent in 1728. Later it was given its present name, *Kungliga Vetenskaps-Societeten i Uppsala*. In the text I refer to it as "the Uppsala Science Society" or "the Science Society." Quotations from IL, 5. On the ornithology manuscript, now kept in UUB, see BS I:1, 311, n1.

26. IL, 21, 5–12; 29, 36, 17, quotations from 103.

27. Rudbeck the Younger, *Iter Lapponicum*, 29.

28. See for example FL, 278; IL, 49; quotation from IL, 106–107. Also quoted in Selander, "Linné i Lule Lappmark," 9–20. See also Linnaeus, "Appendix N. I. A Brief Narrative . . ." (1811), 1971, 258.

29. IL, 263-264; Linnaeus, "Relation," 1732, 327–328.

30. EA, 210.

31. Bäck, *Åminnelse-Tal*, 16.

32. Quoted in TMF I:95.

33. Linnaeus, *Tour in Lapland* (1732), 1971, vol. 1, 249, 261, 270; vol. 2, 18, 197; Appendix, 245, 268.

34. Sandblad, "Clarke and Acerbi," 129. In 1732, Swedish Lapland comprised parts of present day Finland and Russia. It included areas east of Torne and Muonio rivers, and stretched north almost to the Arctic Sea at Varangerfjord. It was governed from Umeå, a city of then 900 inhabitants, in its southermost corner. Borders were redrawn in 1751 and 1809.

35. IL, 49. The Settlement Act was re-confirmed in 1695 and 1749, with additional privileges granted in 1741.

36. Wiklund, "Linné och lapparna," 59-93; FL, 287; Linnaeus, "Tankar," 1754, 183.

37. IL, 58-59; Linnaeus, "Relation," 319.

38. Wiklund, "Linné och lapparna," 68-79.

39. IL, 265; BS I:2, 312-314, Linnaeus to the Science Society, 14 April 1732, where Linnaeus estimates his journey at c. 3,400 miles. See also Selander, "Linné i Lule Lappmark," 18-20. Linnaeus began by miscalculating his mileage pay. (At one point, he reckoned 170 times 2 equals 240.) He then added on sufficient miles to make a profit of c. 20%.

40. Wiklund, "Linné och lapparna," 72-75, analyzes Linnaeus's false and actual travel routes.

41. Selander, "Linné i Lule Lappmark."

42. BS I:2, 340, n1.

43. Quoted in BS I:1, 329, added later in margin of Linneaus, "Relation," 1732.

44. IL, 5. See also BS I:5, 314. Linnaeus to Gustaf Cronhjelm, chancellor of Uppsala University. Undated and no place, but written in Stockholm, spring 1733. TMF I:119-120 unconvincingly argues Linnaeus's Sami guides misled him. In this 1733 petition to the Chancellor of Uppsala University, Linnaeus

claimed that he had traveled c. 4,000 miles, with c. 1,000 miles in the mountains, by foot.

45. DN, 216-222, bibliography; Wikman, *Lachesis and Nemesis*, 30.

46. DN, 133, 75; CD, 144.

47. Rudbeck the Younger, *Iter Lapponicum*, 28, 29, 32, 39.

48. Sydow, "Introduktion till dagboken," Rudbeck the Younger, *Iter Lapponicum*, commentary, 22, quoted from the royal letter announcing Rudbeck's grant for the voyage. Karl XI's 1694 Lapland voyage to view the midnight sun was in a similar representative mode, although the practical-minded king also inspected Bothnian forts.

49. Rudbeck the Younger, *Iter Lapponicum*, commentary, 71–104, esp. 75.

50. FL, 3. Linnaeus, "Relation," 323, for the four-week claim; see also Wiklund, "Linné och lapparna," 73. Linnaeus made this claim for his 16-day cross-over to Norway and back from Kvikkjokk, 6-21 July. FL, 107, more cautiously speaks of "en så lång tid."

51. COR I:269. Linnaeus to Haller. No date or place: probably written from Havercamp, summer 1737.

52. TMF I:104; quotation from *Vita* I, 66.

53. IL, 55.

54. Wiklund, "Linné och lapparna," 79.

55. COR I:269. Linnaeus to Haller. No date or place: probably written from Havercamp, summer 1737.

56. TMF I:109; IL, 43-44.

57. FL, 331.

58. Linnaeus, "Relation," 1732, 323; TMF II:243. Lapland's highest peak, Kebnekaise, at 2,111 meters, is not quite half the height of Mont Blanc, at 4,807 meters.

59. Now kept at LSL. Also reprinted in IL.

60. Linnaeus, "Relation," 316, 323, 322.

61. DN.

62. TMF I:234; Broberg, "Lappkaravaner på villovägar," 42; TMF I:193.

63. Quotation from Consett, *Tour Through Sweden*, 69.

64. On Sami clothes, see Fjellström, *Samernas*, 311-347.

65. Schefferus, *Lapponia*, 214.

66. Ibid., 217; Fjellström, *Samernas*, 341-346, 501–503.

67. Wiklund, "Linné och lapparna," 62.

68. Broberg, "Lappkaravaner på villovägar," 42.

69. He was helped by a "Lapp" dress and a Schefferian pot-boiler, Nicolaus Örn's *Kurze Beschreibung des Laplandes* (Bremen: 1707). See Broberg, "Järven-Filfrassen-frossaren," 20; "Lappkaravaner på villovägar," 42; and "Olof Rudbeck," 16.

70. On how Sami were displayed , see Gösta Berg, "Lappland och Europa," *Scandinavica et Fenno-urgrica* (Stockholm: 1954), 221–245.

71. Quote from BS I:3, 209. Pehr Bjerchén to Linnaeus. London, 14 June 1758.

For Linnaeus's offer to buy her, see BS I:3, 211. Linnaeus to Pehr Bjerchén. Uppsala, 1758 [n.d.].

72. Uddman, "Caroli Linnaei Arch. et Prof." 162–163.

73. Broberg, "Lappkaravaner på villovägar," 37; Broberg, "Olof Rudbeck," 16.

74. Wiklund, "Linné och lapparna," 64, quoting J. J. Björnståhl's travels in Holland. On Linnaeus' Dutch patron, Georg Clifford, see Fries, "Linné i Holland," 141–155. On the history of the portrait, see Wachenfelt, "Lapplandsporträtt," 21-35; Tullberg, *Linnéporträtt;* Soulsby, *Catalogue,* XIV A, "Portraits."

75. Stearn, *Botanical Gardens,* lx.

76. LN, 48.

77. Heller, *Studies,* 214; "Om Lappland och lapparne," in FL, appendix. Linnaeus also adorned most illustrations to *Flora Lapponica* with a motto from Ovid or Virgil. See FL, annotations, 380.

78. Quotation from FL, 229-230; see also FL, 145; Linnaeus, LN, 1907, 71; Linnaeus, CD, 130; DN, 82.

79. DN, 35, 85, 142, 61, 61, and 35.

80. TL I:200n, quoting from FL.

81. TL I:174, 179.

82. TL I:58, 169, 200, 348–353; II:157, 166, 169, 174, 179; FL, 353; quotation from Linnaeus, *De pane diaetetico* (1760), 1964, 17.

83. Linnaeus, *Ceres noverca arctorum* (1733), 1964, 7; IL, 128–130. As late as the 1930s, elderly northern Swedes remembered bread eked out with bark, straw, chaff, bone flour, heather, hazel buds, mash, and bran.

84. Ehrström, "Lefnadsförhållanden och sedvänjor beskrivna," describes eating *lunds* fish in Finland in the 1890s.

85. Linnaeus, "Om nödvändigheten af forskningsresor inom fäderneslandet" (1741), 1906, 77, also quoted in LD, 104. *Lunds* or *luns* fish meant a jumble of fermented fish.

86. FL, 283. See also Nelson and Svanberg, "Lichens as Food," 34. In 1867, famine again struck the Lapland settlers. They were advised to eat lichen, this time by the state-directed Economic Society and in a brochure entitled "Instructions for the Preparation of Food Stuffs from Pine Lichen, Iceland Moss, etc." Linnaeus, *Disputationer,* 219; TL II:166; quotation from TL I:77; II:166.

87. FL, 279, here contradicts *ibid,* 278; Linnaeus, *De varietate ciborum,* 15; TL I:239; quotation from IL, 252, from fragment "Om Lappland och lapparne," also quoted in LD, 82. See also FL, 236.

88. Linnaeus, "Om nödvändigheten af forskningsresor inom fäderneslandet" (1741), 1906, 78.

89. LD, 242, quoted from the 1754 dissertation *Cervus rheno.* See also Schefferus, *Lapponia,* 307 and 310, and Rudbeck, *Iter Lapponicum,* 41.

90. Quoted from TL I:28, 333; IL, 123; FL, 283; Linnaeus, LN, 1907, 26, 44, 46. See also TL I:86, 122, 192, 284; IL, 52; FL, 144, 236, and 237. See also Wiklund, "Linné och lapparna," 86-87, for summary of page references to Sami alcohol dependency in IL, FL, LN and CD. DN, 25, rule 48, warns against aquavit, and

rule 50 against tobacco. Linnaeus deplored Sami tobacco use: see FL, 37, 77, and 208.

91. Linnaeus, LN, 1907, 710; FD, 272; LD, 96, summary of Linnaeus' dissertation *Exanthemata viva* (1757); Heckscher, *Sveriges ekonomiska historia*, III:1, 39.

92. Quotation from IL, 149; see also TL I:127, 136–137; IL, 54-56, 149–151.

93. DN, 90; LD, 223, "*Colica Lapponum*" from Linnaeus' dissertation *Rariora Norwegia* (1768); FL, 1905, 73.

94. FL, 1905, 145; IL, 15; Linnaeus, *Nutrix noverca* (1752), 1947, introduction. Linnaeus's dietary lectures typically sang the praises of breast milk; see LN, 1907; CD, 1907.

95. DN, 36–38.

96. Quotation from DN, 41. See also LD, 131–133, summarizing Linnaeus's dissertation *Spiritus frumenti* (1764); DN, 98; and Consett, *Tour through Sweden*, 118, for the reaction of an amazed English visitor meeting aquavit-swigging Swedish women.

97. On how Linnaeus used the binary pair European-Sami, see DN, 65. Quotation from IL, 147.

98. Linnaeus, *Nutrix noverca* (1752), 1947, 12, 15. See however DN, 38.

99. Quotation from DN, 68. See also DN, 40; LD, 153; LN, 1907, 99.

100. CD; and IL, 143, respectively.

101. Quotations from FL, 263; also IL, 67; BS I:1, 337–341, reprint of Linnaeus, "Observat. 3:ia. De Lecto in desertis ex temporaneo." On Sami personal hygiene, see also FL, 31.

102. CD, 86; Uddman [Hansson], "Collegium Diateticum Habitum Diversis," 94; DN, 23, preface to Linnaeus' fragment manuscript *Oeconomia Lapponia*, quoted in Wiklund, "Linné och lapparna," 91; see also IL, 54, where Linnaeus improbably claims the Sami walked about naked, without shame.

103. Quotation from DN, 39. On Sami style of dress, see FL, 145, 219; IL, 120, 165–166; CD, 11; DN, 59, 61.

104. Cited from one such list, reproduced in full, in Hagberg, *Carl Linnaeus*, 212.

105. FL, 302.

106. Such as Linnaeus's DN; LN; CD; and Linnaeus, "Inledning till dieten," 1961.

107. Linnaeus, *Potus coffeae* (1761), 1966, 4. See also, "Påminnelse för den rasande ungdomen"; LD, 124, summarizing and quoting from Linnaeus' dissertation *Spiritus frumenti* (1764).

108. Quoted from a stray annotation in Linnaeus' hand, in Sydow, "Linné och de lyckliga lapparna," 78, "*docentibus lapponibus*."

109. On teeth and sugar, see CD, 1907, 147; *Vita* II, 65; V, 180; "Med. Doctorens Archiaterns . . ." 130.

110. Uddman, "Caroli Linnaei Arch. et Prof." 192, 195.

111. Linnaeus, *Nutrix noverca* (1752), 1947, 3.

112. Linnaeus, *De pane diaetetico* (1760), 1964, 16.

113. Linnaeus, *Circa fervidorum* (1765), 1968, 13.

114. Quotations from FL, 236; IL, 133. See also FL, 15, 254; DN, 89, 93.

115. DN, 82.

116. Schefferus, *Lapponia,* 228, 227, 182–183.

117. TL I:147, 89; IL, 59, 115, 38.

118. FL, 59 and 94; CD, 147; exchange between the Scottish physician James Lind and Linnaeus, 1754, reprinted in COR II:472–474.

119. IL, 121; CD, 198 and 144; and DN, 69 and 76.

120. IL, 66.

121. Such as Linnaeus *Plantae esculentae patriae* (1752), and *Fructus esculenti* (1763), 1965.

122. *Angelica archangelica* and *Epilobium angustifolium* were key. See for example Linnaeus, *Flora oeconomica,* 1749, 76. On traditional highland Sami foods, see Fjellström, *Samernas,* 193, 261, and 283-289. On Sami trading, see Schefferus, *Lapponia,* 182–183, 229.

123. Cited in TMF I:81. On Sami medicine see HU and Fjellström, *Samernas,* 348-358.

124. Quoted from *Posttidningen* in TMF I:85.

125. FL, 226; see also IL, 14.

126. IL, 55; see also Consett, *Tour through Sweden,* 63.

127. Fjellström, *Samernas,* 55-66 and 70; FL, 353; Schefferus, *Lapponia,* 31, 79, 173–174, 182–183, and 229; Wiklund, "Linné och lapparna," 86–88. Bergman-Sucksdorff and Vogel-Rödin, *Buores Buores,* 129, describes the greatest and oldest of these fairs, Alta, abandoned after the retreating German army burned the fairgrounds in 1945.

128. FL, 236, 167, 167; and Linnaeus, *Plantae esculentae patriae,* 1752, 29.

129. IL, 104, 439, 95. See also Schefferus, *Lapponia,* 229.

130. FD, 49; see also Rudbeck the Younger, *Iter Lapponicum,* commentary, 27; Linnaeus, "Relation," 1732, 329. On how the Sami dripped the stem sap of *Angelica archangelica* onto the tobacco, see TL II:108; see Fjellström, *Samernas,* 70; and FL, 71, on how the Sami made gin, or mixed juniper berries in their aquavit.

131. LD, 238–239, quoted from Linnaeus' self-review of *Cervus rheno* (1754) in *Lärda Tidningar* (1755).

132. FD, 273 and commentary, 513–514, 518, describes the insect attacks; Rudbeck the Younger, *Iter Lapponicum,* 42-43, describes the conditions of Sami laborers at the Kvikkjokk furnaces; see also commentary on p. 25.

133. IL, 17.

134. On the complementary relation between Swedish and Sami ecocultures see IL, 55.

135. Quote from Schefferus, *Lapponia,* 222; for Linnaeus' terms *fjällappar* and *skogslappar* see LD, 239.

136. South Samish (c. ten native speakers were still alive in 1980); West or North Samish, divided into the Lule and Norsk dialects; and East Samish, nearly extinct today and divided into Skolt, Enare, and Kola. Bergman-Sucksdorff and Vogel-Rödin, *Buores Buores,* 138, and Fjellström, *Samernas,* 170.

137. Broberg, *Homo Sapiens*, 273, quoted from DN.

138. DN, 11.

139. BS I:5, 317. Linnaeus to Gustaf Cronhjelm, chancellor of Uppsala University. Undated and no place, but written in Stockholm, spring 1733; the work is also mentioned in BS I:6, 240. Linnaeus to county governor Gabriel Gyllengrip. Uppsala, 1 October 1733. In this letter the work is founded "ex principiis Zoologicis."

140. Cited from Hagberg, *Carl Linnaeus*, 214.

141. Broberg, *Homo Sapiens*, 222; DN, 130. See also LD, 237, quoted from Linnaeus' self-review of *Cervus rheno* (1754) in *Lärda Tidningar* (1755).

142. Linnaeus, "Doctor," 418; and LD, 242, quoted from Linnaeus' self-review of *Cervus rheno* (1754) in *Lärda Tidningar* (1755).

143. Linnaeus's preface to the manuscript fragment *Oeconomica Lapponica*, quoted in Wiklund, "Linné och lapparna," 91.

144. IL, quoted in Broberg, *Homo Sapiens*, 262; compare Rudbeck the Younger, *Iter Lapponicum*, 46, "Elysios campos."

145. IL, 55.

146. DN, 133.

147. Quoted from a stray annotation in Linnaeus' hand, in Sydow, "Linné och de lyckliga lapparna," 78.

148. FL, 229.

149. Linnaeus, *Iter ad exteros* (1735–), 1888a, 386.

150. Wachenfelt, "Lapplandsporträtt," 34.

151. I use Linnaeus' term for tribal people here. See as one example Uddman, "Caroli Linnaei Arch. et Prof." 195. Note that Linnaeus distinguishes "wild nations" from "wild men." See Broberg, *Homo Sapiens*, which contains a Latin binomial from the 10th edition of *Systema naturae* (1758).

152. As Schefferus' English publisher summarizes the prevailing European view of the Sami in Schefferus, *Lapponia*, preface.

153. Broberg, "Olof Rudbeck," 15. Letter from the councillor of the realm Carl Bonde, to Gustav Adolf's chancellor Axel Oxenstierna.

154. Schefferus, *Lapponia*, 31; quotation on p. 36 discusses how Sami and the larger Scandinavian economy were linked.

155. TMF I:111, citing the Science Society's protocols; BS I:1, 325, n1.

156. BS I:1, 307. Linnaeus to the Science Society. Uppsala, 15 December 1731.

157. Linnaeus to the Directorate of the Lapland Ecclesiastical Bureau. 14 October 1747. Quoted in TMF II:243.

158. Ibid.

159. BS I:7, 119, n1, quote from edict of the Bureau of 24 February 1748. See also Linnaeus to the Directorate of the Lapland Ecclesiastical Bureau. Uppsala, 9 September 1748.

160. Linnaeus, "Tankar," 1754, 189, postscript. The saffron was to be planted on the southside of Gällivare's *fjäll*.

161. BS I:4, 336, n1. The Sparre prize medal was awarded for the first time on 22

February 1755. It was granted for Linnaeus' "Tankar," 1754, expanded into Linnaeus, *Flora alpina* (1756).

162. FL, 278.
163. LD, 218, quote from Linnaeus' dissertation *Usus muscorum* (1766), in turn quoted from FL.
164. Linnaeus, *Iter Dalekarlium* (1734), 1889, 283.
165. BS I:2, 27. Memorial from Linnaeus to the Swedish Academy of Sciences. N.d., but received in 1742.
166. IL, 172.
167. Ibid., 112. See also Schefferus, *Lapponia*, 32.
168. CD, 130.
169. DN, 128. See also the similar formulations in FL, 10; Schefferus, *Lapponia*, 32–34.
170. Linnaeus to the Directorate of the Lapland Ecclesiastical Bureau. Protocoled 14 October 1747. Quoted in TMF II: 243; I:111.
171. FD, 87, 88. On the voyage through Dalarna (1734), see TMF I:151.
172. FD, 69.
173. Linnaeus to the Directorate of the Lapland Ecclesiastical Bureau. Quoted in TMF II:243; Linnaeus, "Tankar," 1754, 187. Linnaeus to the Science Society, 15 December 1731. Quoted in TMF I:113.
174. TMF I:263.
175. IL, 49.
176. TMF I:139, and 139, n1. Lyme grass is *Elymus arenarius* or *strandråg*.
177. BS I:6, 242. Linnaeus to Gabriel Gyllengrip. Uppsala, October, 1733; and ibid., 5, 314. Linnaeus to Gustaf Cronhjelm, chancellor of Uppsala university. N.d. [written in Stockholm, spring 1733].
178. Ibid., 6, 243. Linnaeus to Gabriel Gyllengrip. Falun, 23 August 1734.
179. TMF I:139–140. Linnaeus finally understood this when he undertook his Skåne journey in 1749.
180. IL, 55. For references to tar and pitch, see IL, 62.
181. LD, 174, quoted from *Lärda Tidningar*'s summary of the 1756 *Flora alpina*.
182. Linnaeus, "Tankar," 1754, 185–186.
183. LD, 99, quoting from Linnaeus' dissertation *Exanthemata viva* (1757), an examination of skin diseases and the organisms that cause them, based on "Leuvenhök"'s microscope observations.
184. Linnaeus, "Tankar," 1754, 187, 184–185; and LD, 174, quoted from *Lärda Tidningar*'s summary of the 1756 *Flora alpina*.
185. TMF I:310. See also Linnaeus, "Tankar," 1754, 186.
186. "Register," *K. Vet. Akad. Handl.*, 1754, n.p., equivalent to 324–326. Saffron is also listed as a possible Öland cultivar in anon., "Förteckning på de örter, som kunna planteras på Öland" (ca 1749). 10 pp. Econ., LSL.
187. Linnaeus, "Tankar," 1754, 186.
188. DN, 128; IL, 55.
189. Linnaeus, "Tankar," 1754, 186; CD.

190. On Linnaeus' observations on Sami addiction to alcohol and nicotine, see TL I:86, 122, 192, 284. On the Christianization and pastoral care of the Sami in Swedish Lapland, see Haller, *Svenska kyrkans mission.* For Linnaeus' sparse comments on the Christianization process see TL I:158 and 364.

4 "God's Endless Larder"

1. Linnaeus, *Märkvärdigheter uti insekterna* (1739), 1939; Lindroth, "Legend och verklighet," 19; Sahlgren, "Linné som predikant," 40–55.
2. Linnaeus, "Naturaliesamlingars ändamål och nytta" (1754), 1939.
3. BS I:8, 28. Letter from Pehr Kalm to Linnaeus. London, 3 June 1748.
4. For one expression of this divine economy, see Linnaeus, "Upsala Academiae Rector Carl Linnaeus hälsar denna academiens och stadsens samteliga fäder och inbyggare, så af högre som lägre stånd" (11 December 1759), in "Två svenska akademiprogram," 1954–55, 105–108.
5. Linnaeus, "Skaparens avsikt" (1763), 1947, 85.
6. Ibid.
7. Linnaeus, *Wästgötaresa,* 1747, 225.
8. Quoted from a contemporary English translation of *Oeconomia naturæ, The Oeconomy of Nature,* in Linnaeus, *Miscellaneous Tracts* (1775), 1977, 119. See also Lindroth, "Legend och verklighet," 72–73.
9. Stauffer, "Ecology," and Lindroth, "Legend och verklighet," 73.
10. Linnaeus, "Collegium Oeconomicum," 1758.
11. Rydbeck, "'Tal, om planterings.'"
12. FD, 221.
13. Linnaeus, "Skaparens afsikt" (1763), 1947, 87.
14. TMF I:353.
15. Linnaeus, *Skånska resa* (1751), 1975, 197, 238. On coppicing see p. 189.
16. Quoted in Broberg, *Homo Sapiens,* 286.
17. Anders Löfman, untitled introductory poem, in Linnaeus, *Horticultura academica,* 1754.
18. Ibid.
19. Quoted in TMF II:96–97, from Linnaeus' 1750 congratulatory tract on the crown prince's birthday.
20. Bäck, *Åminnelse-Tal,* 39–40.
21. LD, 203, quote from Linnaeus' dissertation *Horticultura academica* (1754).
22. *Uppsala Nya Tidning,* 10 November 1933.
23. BS I:7, 17. Fredrik Hasselquist to Linnaeus. Smyrna, 17 April 1750.
24. Ibid., 215. Pehr Bjerchén to Linnaeus. Stockholm, 7 October 1759. See also p. 220, Bjerchén to Linnaeus.
25. Linnaeus, "Skaparens afsikt" (1763), 1947, 86.
26. See also Worster, *Nature's Economy,* 50.
27. Daniel Rolander to Linnaeus. "Paramaribo i Suriname," 11 July 1755, LSL.

28. Linnaeus to Carl Peter Thunberg and Pehr Löfling, partly reprinted in Sörlin, "Apostlarnas gärning," 82 and 83.
29. Löfling, *Iter Hispanicum*, preface by Linnaeus.
30. Barreiro, "Pehr Löflings kvarlåtenskap," 136, quoting from Linnaeus's biography over Löfling, in *Den Swenska Mercurius*, October 1757. See also Mörner, "Peter Löflings levnadshistoria," 93–94. The expedition left in 1754. By 1756, half of its members had died from fevers or starved to death.
31. Tärnström's travel diary, quoted in Grape, "Christopher Tärnström," 134, 139.
32. LSL, Daniel Rolander to Linnaeus. "Paramaribo i Suriname," 11 July 1755.
33. Linnaeus, "Tal" (1759), 1939, 94, and 150, notes. Linnaeus here mocked Rolander's supposed discovery.
34. LSL, Linnaeus to Johann Gottlieb Gmelin. 14 February 1747.
35. Linnaeus, "Markattan Diana," 1754, 210.
36. Linnaeus, *Menniskans cousiner* (1760), 1955, 4 ff.
37. Broberg, *Homo Sapiens*, 177, referring to Voltaire (François Marie Arouet), "Relation touchant un maure blanc," in *Oeuvres*, vol. 6 (Amsterdam: 1745), 238 ff.
38. Linnaeus, "Markattan Diana," 1754, 211.
39. Ibid., 215.
40. Quoted in Broberg, *Homo Sapiens*, 201–202. The student was Linnaeus' friend Johan Browallius. He was speaking with the governor of Dalarna, who financed Linnaeus' voyage through his province.
41. Quoted from Linnaeus, *Deliciae naturae*, 1773, in TMF II:22.
42. Quoted in Broberg, *Homo Sapiens*, 286.
43. Linnaeus, *De politia naturae*, 1760, paragraph 35. See also Lepenies, "Naturgeschichte," 31, and 37–39.
44. ND, 59.
45. Quoted in Broberg, *Homo Sapiens*, 146.
46. DN, 136 ff.
47. LN, 1907, 23; Linnaeus, "Skaparens afsikt" (1763), 1947, 49.
48. Lindroth, "Legend och verklighet," 70.
49. Linnaeus, "Företal" (1753), 1952, 1.
50. Bäck, *Åminnelse-Tal*, 64.
51. Quoted from an manuscript-fragment, LSL, transcribed in Broberg, *Homo Sapiens*, 144.
52. *Vita* II, 77; III, 126. On how Linnaeus wrote his *vitae* for his memorial services in scientific academies, see *Vita*, preface by Elis Malmeström and Arvid Hj. Uggla, 17.
53. *Vita* III, 127.
54. Letter by Samuel Linnaeus, reprinted in Virdestam, "Samuel Linnaeus," 125. See also LD, 96.
55. Linnaeus, *Metamorphosis humana* (1767), 1956, 7.
56. Linnaeus to Bäck, here cited from Hagberg, *Carl Linnaeus*, 132.
57. *Vita*, notes, 196.

58. Linnaeus, *Systema naturae,* 1758, preface, here quoted from Broberg, *Homo Sapiens,* 286.
59. Linnaeus, *Inledning till dieten,* n.d., 1961, 13. Also quoted in Broberg, *Homo Sapiens,* 284. See also ND, 101–102; and Linnaeus, *Metamorphosis humana* (1767), 1956, 9.
60. Linnaeus, *Plantae esculentae patriae,* 1752, 4.
61. Quotations from Linnaeus, *Flora Oeconomica,* 1749, 1, 76. See also Linnaeus, *Plantae esculentae patriae,* 1752, 40, for reference to greedy children.
62. Ehrström, "Minnesanteckningar," 70. Linnaeus, "Naturaliesamlingars ändamål och nytta" (1745), here quoted from partial reprint in TMF II:20. See also Linnaeus, "Skaparens afsikt" (1763), 1947, 74.
63. BS I:1, 3. Linnaeus to the Swedish king. N.d. Arrived 9 March 1757.
64. Ibid.
65. Ibid. See also Linnaeus, "Skaparens afsikt" (1763), 1947, 87.
66. Quote from Linnaeus, "Doctor," 416. The passage discusses famine foods. See also HU, 48.
67. Linnaeus, "Företal" (1753), 1952, 1.
68. "Extract af Prof. Linnaei bref til Secret. Elvius," first published in *Lärda Tidningar* (13 February 1746), and reprinted in BS I:2, 64–66. See also *Svenska Dagbladet,* 8 March 1954. Linnaeus' great-great-grandmother was burned as a witch in 1622, when his great-grandfather was ten years old. The boy studied at Copenhagen University and became a parson in Visseltofta, south Sweden.
69. Linnaeus, *Ölänska och Gotländska resa* (1745), 1962, here quoted from the English translation in Wikman, *Lachesis and Nemesis,* 68.
70. Linnaeus, *Skånska resa* (1751), 1975, 192. Tunbyholm, June 7.
71. BS I:2, 129. Linnaeus to the members of the Swedish Academy of Science. Uppsala, 29 August 1749.
72. HU, 47.
73. Linnaeus, "Tal" (1759), 1939, 92. The Renette is a cultivated apple (like Cox Orange and Blenheim).
74. Ibid. Cochineal is a costly crimson dyestuff derived from *Dactylopius coccus,* cactus-eating insects native to tropical and subtropical America. The *Tuber jalapae,* the secondary roots of *Exogonium purga,* are a laxative. Rhubarb, a native of Mongolia and China, was used medicinally.
75. Ibid. For a different translation, see Hildebrand, "The Economic Background," 28.
76. Ibid., 95.

5 "A New World—Pepper, Ginger, Cardamon"

1. Linnaeus, *Culina mutata* (1760), 1956, 13–14.
2. Linnaeus, *De potu chocolatae* (1756), 1934.
3. On the term "*ekonomi*" in physico-theological discourses, see Frängsmyr, "Den gudomliga ekonomien."

4. Heckscher, *Sveriges ekonomiska historia*, 853.
5. Linnaeus, *De potu chocolatae* (1756), 1934, editor's preface, quoted from an autograph manuscript in LSL.
6. Linnaeus, "Tankar om grunden til oeconomien" 1740, 407.
7. Heckscher, *Sveriges ekonomiska historia*, 731–811, summarizes the monetary systems. For the hyperinflation of the early 1760s, see 762. One barrel of gold meant c. 100,000 Swedish *daler* silver.
8. Christopher Tärnström to Linnaeus, quoted in Grape, "Christopher Tärnström," 126–144, 135.
9. EA, 61. See also TMF I:97.
10. Rydberg, *Svenska studieresor*, 100; Heckscher, "Linnés resor," 7.
11. See Small, *Cameralists*, 113, for Becher's argument on local monopolies.
12. BS I:1, 2. Linnaeus and Charles de Geer to the Swedish king. Permission to import foreign spirits was granted on 19 April 1757.
13. Quote from Heckscher, *Sveriges ekonomiska historia*, 854.
14. BS I:2, 58. Memorial from Linnaeus to the Swedish Academy of Science. 10 January 1746.
15. Linnaeus, *Oeconomia naturae* (1750), 1906, here from quote in Fredbärj, "Linné och vintern," 69–78, 77.
16. Linnaeus, "Doctor," 412.
17. Introduction to Linnaeus' *resa* through Dalarna, here cited from Hagberg, *Carl Linnaeus*, 142.
18. Cited in Heckscher, *Sveriges ekonomiska historia*, 842.
19. Quotation from Linnaeus, "Doctor," 418. On his attempt to refashion Uppsala University, see Linnaeus, "Linnés tankar" (1768), 1940, esp. 13–15.
20. Thunberg, "Inträdes-Tal," 30.
21. Iwao, "C. P. Thunbergs," 143.
22. Thunberg, "Inträdes-Tal," 4.
23. Svedelius, "Carl Peter Thunberg", 49.
24. Thunberg, "Inträdes-Tal," 9.
25. Ibid, 4.
26. Linnaeus, "Linnés tankar" (1768), 1940, 13–15, quote from 13.
27. Thunberg, "Inträdes-Tal," 5.
28. Rydbeck, "Tal."
29. Fernand Braudel, *Civilization and Capitalism, 15th–18th Centuries*, 3 vols., trans. and rev. Sian Reynolds (New York: Harper and Row, 1981), vol. 2, 134. Italics in the original.
30. FD, 5.
31. Linnaeus, *Wästgötaresa*, here quoted from a 1928 edition in Heckscher, "Linnés resor," 5. See also "Alingsås," *Svensk Uppslagsbok*, 2nd ed., 1955. The Alströmer family were paid some 413,000 *daler* silver between the parliament sessions of 1726/27 and 1765/66.
32. Linnaeus, "Upsala Academiae Rector Carl Linnaeus hälsar" (1759), in

Linnaeus, "Två svenska akademiprogram av Linné" (1750, 1759), 1954–55, 97–114, 106.

33. Linnaeus, "Doctor," 1740, 406.

34. Heckscher, "Linnés resor," 5. See also Linnaeus, "Linnés tankar" (1768), 1940, 8; and "Tankar om grunden til oeconomien," 1740, 406.

35. BS I:1, 129. Linnaeus to Uppsala University governing body, or "Consistorium academicum." 1746.

36. On Smith's relation to concepts of nature, see Alvey, New Adam Smith Problem. I owe this point to Mark Madison.

37. See Hildebrand, K. Vet. Akad., 185, quoting Anders Gabriel Duhre, Sweriges högsta wälstånd, bygdt uppå en oeconomisk grundwal, a pamphlet dedicated to the 1738 session of parliament.

38. Heckscher, Sveriges ekonomiska historia, 841. Berch argued that even if an eventual harmony is inherent to the economy, it is more efficient and just to regulate it (to avoid trials and errors).

39. See Berch, Sätt att igenom politisk arithmetica and Inledning.

40. Anders Falck to Johan Peter Falck. N.p., written in Uppsala. N.d., written in 1765.

41. Heckscher, Sveriges ekonomiska historia, 827–39, 877. Throughout the eighteenth century, the most important English economic writers in Sweden remained Josiah Child, Thomas Mun, and Charles Davenant. The most important German writer remained Johann Joachim Becher.

42. Both quoted in Heckscher, "Linnés resor," 4.

43. Linnaeus, "Tankar om grunden til oeconomien," 1740, 406.

44. Linnaeus, "Skaparens afsikt" (1763), 1947, 87.

45. Daniel Solander's certificate of competence, written by Linnaeus on 5 March 1759 and reprinted in Uggla, "Daniel Solander," 23–64, 52.

46. Quoted from Linnaeus' letter of recommendation for Lidbeck, reprinted in Stenström, "Pehr Osbeck och Lars Montin," 59–94, 72–73.

47. BS I:7, 27. Pehr Kalm to Linnaeus. London, 3 June 1748.

48. Linnaeus, "Två svenska akademiprogram," (1750, 1759), 1954–55, 106. Linnaeus here compared "our own" economy to the economy of nature. Compare the donation letter of the 1760 Uppsala chair from Eric Ericsson Borgström, probably written by Linnaeus, reprinted in Telemak Fredbärj's commentary to Linnaeus, "Två svenska akademiprogram," 111–112.

49. BS I:I, 26, annotations. Protocols from Riksens Högl. Ständers Cammar Oeconomie och Commercie-Deputations Förordnings-Utskott, on a memo by Linnaeus, read on 7 April 1741, and later published in K. Vet. Akad. Handl. of 1741 as "Uppsats på de Medicinalväxter . . . "

50. See Hildebrand, K. Vet. Akad., 185, quoting Duhre, Sweriges högsta wälstånd. . . .

51. Ibid., 174.

52. TMF I:258, 263.

53. Heckscher, "Linnés resor." Industrial subsidies were largely abolished when the Hats fell in 1765; see Hildebrand, "Economic Background," 19–21.

54. Linnaeus to Pehr Wargentin, on the founding of the Academy. N.d., written

1761. Reprinted in TMF I:17–18. (The prospective member was Jonas Meldercreutz, professor of mathematics at Uppsala university.) The Swedish Academy of Science was a brainchild of Mårten Triewald, a machine builder and popularizer of Newtonian physics who lived in England from 1716 to 1726. The other founders were Carl Linnaeus, Sten Carl Bielke, Johan Anders von Höpken, Carl Wilhelm Cederhielm, and Jonas Alströmer.

55. Svenska Vetenskapsakademien, *Protokoll*, 11 (9 June 1739).

56. Ibid., 4 (2 June 1739), Statement by Olof Celsius.

57. Ibid.

58. TMF I:17–18.

59. Svenska Vetenskapsakademien, *Protokoll*, 16 (16 June 1739).

60. Ibid., 4 (2 June 1739), "Oeconomisk Wetenskaps Societet." See Hildebrand, *K. Vet. Akad.*, 172–174, on how the terms "oeconomia" and "oeconomie" were used. Italics mine.

61. BS I:7, 59. Pehr Kalm to Linnaeus. Philadelphia, 5 December 1750.

62. This plea was repeated in Linnaeus, "Angående Oeconomiens och landthushåldningens uphielpande genom Historiae Naturalis flitiga läsande" (1746), reprinted in TMF II:195.

63. Linnaeus, quoted in Heckscher, "Linnés resor," 10.

64. BS I:1, 23–26. Memorial from Linnaeus to Riksens Ständers Utskott och Deputationer. Stockholm, 2 May 1744. Quotation on p. 26.

65. Linnaeus, "Tankar om grunden till oeconomien," 1740, 422. In Linnaeus' times, parsons were alloted landholdings as well as tithes.

66. Stavenow-Hidemark, *Anders Berch*, esp. 14–22. *Theatrum oeconomico-mechanicum.*

67. Quoted from donation letter of the 1760 Uppsala chair from Eric Ericsson Borgström, reprinted in Telemak Fredbärj's commentary to Linnaeus, "Två svenska akademiprogram," (1750, 1759), 1954–55, 111.

68. Linnaeus, "Upsala Academiae Rector Carl Linnaeus hälsar" (1759), in ibid., 107.

69. Small, *Cameralists*, 298, quoting J. H. G. von Justi.

70. Odhelius, *Åminnelse-Tal*, 15.

71. Linnaeus, "Två svenska akademiprogram," (1750, 1759), 1954–55, 107. See also Heckscher, "Linnés resor," 3.

72. IIU, 48.

73. BS I:4, 347. Linnaeus to Abraham Bäck. Uppsala, 18 July 1755. The comment was ironic, because Linnaeus had no faith in the tea-growing scheme he described there. But a productivity/territory swap suggested itself to him in relation to his attempts to abolish the Asia trade.

74. Lapland (1732), Dalarna (1734), Öland and Gotland (1741), Västergötland (1746), and Skåne (1749). Carl Linnaeus, "Merit-Lista: Att applicera Naturen till oeconomien och vice versa" (1775), reprinted in Strandell, "Patriotiska Sällskapet," 130–137, 132–133. Linnaeus' diploma from Patriotiska Sällskapet is now kept in Biographica, LSL.

75. EA, 202, 203.

76. Lindroth, "Legend och verklighet," 56–122, 88; Lepenies, "Naturgeschichte," 27; Linnaeus, "Företal," 1.
77. Linnaeus, *Skånska resa* (1751), 1975, 181.
78. Linnaeus, "Naturaliesamlingars ändamål och nytta," (1754), 1906, 47. See also TMF II:19.
79. Quoted from Johan Browallius' description of Linnaeus' student rooms in Uppsala, preface to Linnaeus, *Libellus amicorum* (1734-), 1958.
80. For descriptions of Linnaeus' rooms, see Tullberg, "Linnés Hammarby." esp. 19, 50, 52, 53; Varia, G3, LSL, *Calendae [sic] Botanicae Perpetuae,* ms in Linnaeus' hand.
81. Lepenies, *Autoren,* 195. Linnaeus' favorite motto was inscribed over his bedroom door, adorned the 12th edition of *Systema naturae* (1766–67), and concluded his "Skaparens afsikt" (1763), 1947, 96.
82. Tullberg, "Linnaeus Hammarby," 53, 66. Extant are six such statues of Africans, and three of Europeans.
83. Von Sydow, "Linné och Lappland," 22.
84. Linnaeus to Carl Gustaf Tessin. 11 April 1740. Quoted in TMF I:264–265.
85. Triewald, *Mårten Triewalds;* Hildebrand, *K. Vet. Akad.,* 136–171.
86. Linnaeus, "Tankar om grunden till oeconomien," 1740, 422–423.

6 "Should Coconuts Chance to Come into My Hands"

1. *Vita,* 71; quotation from *Vita* III, 147; see also *Vita* III, 122.
2. Other people could be considered Linnaeus' traveling disciples. Swedish aristocrats on their European grand tour called themselves Linnaeus' students and sent him collections of *naturalia.* Linnaeus, to the Swedish Academy of Science (1771), quoted in Sörlin, "Apostlarnas gärning," 79.
3. For the Linnaean travelers' joint historiography, see TMF II:1–86; Rob. E. Fries, "De Linneanska," 31–40; Strandell, "Linnés lärjungar"; and Sörlin, "Apostlarnas gärning."
4. TMF II:86.
5. Quotation from Sörlin, "Apostlarnas gärning," 77.
6. BS I:2, 60. Memorial from Linnaeus to the Academy of Science. Uppsala, 10 January 1746. See also BS I:2, 30. Letter from Linnaeus to Pehr Elvius. Uppsala, 28 August 1744.
7. BS I:7, 35. Linnaeus to Fredrik Hasselquist. Uppsala, December 1750.
8. LD, 42, quoting a letter from Linnaeus to Abraham Bäck.
9. Quoted in TMF I:82.
10. For example TMF II:42, quoting Linnaeus' preface to Löfling's *Iter Hispanicum.*
11. BS I:2, 59–60. Linnaeus, memento to the Academy of Science, recommending it to finance Kalm's voyage to North America. Uppsala, 10 January 1746.
12. On domestic journeys preparatory to longer ones, see El. Frondin, praes. 1745. *De Alandia, maris Baltici insula.* Resp. Christopher Tärnström. Uppsala, 1739: part I, 1745: part II, 1739–1745; TMF II: 27; Grape, "Christopher Tärnström,"

127; Stenström, "Pehr Osbeck och Lars Montin," 64–65; TMF II:29, and Hulth, "Kalm som student", 42; Svanberg, "Turkic ethnobotany," 53–118.

13. Linnaeus, "Oration concerning the necessity of traveling in one's own country" (1775), 1977.

14. Linnaeus, "Om nödvändigheten af forskningsresor inom fäderneslandet" (1741); BS I:2, 59. Linnaeus to the Academy of Science. Uppsala, 10 January 1746.

15. TMF I:141, 41.

16. Pehr Osbeck, "Nödvändigheten och Nyttan av Naturalhistoriens handhavande, särskilt vid Stranden," presented at the Gothenburg fraternity of Uppsala University, and reviewed in *Lärda Tidningar* (1749), here cited from Stenström, "Pehr Osbeck och Lars Montin," 64–65.

17. Hulth, "Kalm som student" 47–48, quoting letter from Pehr Kalm to Linnaeus. 18 April 1745.

18. See, for example, BS I:4, 210. Linnaeus to Bäck. Uppsala, 17 April 1753: BS I:2, 53–54. Linnaeus to Elvius. Uppsala, 22 November 1745. BS I:2, 58–62, "Memorial" from Linnaeus to the Academy of Science, to be passed on to Kalm. Uppsala, 10 January 1746.

19. Quoted from Manufactur Contoirets protocol of 7 May 1741, reprinted in TMF I:40, 356.

20. Quoted in ibid., 44.

21. Sörlin, "Apostlarnas gärning," 81; Mörner, "Peter Löflings levnadshistoria," 93; Linnaeus' letter to the Swedish Academy of Science in 1771, quoted in Sörlin, "Apostlarnas gärning," 79; Sandblad, "Bjerkander," 263, quoting a letter from Johan Peter Falck to Claes Bjerkander. Uppsala, 21 March 1763; BS I:2, 268. Linnaeus to Pehr Wargentin. No date, but written in late October or early November 1764.

22. Linnaeus to Osbeck, quoted in TMF II:38.

23. BS I:2, 53–54. Linnaeus to Elvius. Uppsala, 22 November 1745. BS I:2, 33. Linnaeus to Elvius. Uppsala, 10 September 1744.

24. Grape, "Christopher Tärnström," 130, citing Mårten Triewald's instruction list now kept at KB, and the Academy of Science's protocol.

25. Protocols of the Swedish Academy of Science, 20 July 1745, instructions for a proposed voyage of discovery to North-East America, reprinted in BS I:2, annotations, 56–57. Probably the Academy meant *Zizania aquatica*.

26. BS I:2, 60–61, "Memorial" from Linnaeus to the Academy of Science, to be passed on to Pehr Kalm. Uppsala, 10 January 1746.

27. Ibid., 61. Linnaeus to the members of the Swedish Academy of Science, Uppsala, 10 January 1746. See also Manufactur Contoirets protocol of 7 May 1741, reprinted in TMF I:40, on Linnaeus' 1741 Öland and Gotland journey.

28. Hulth, "Kalm som student," 43, 44, both quoting letters from Pehr Kalm to Sten Carl Bielke. 7 April 1742 and 22 October 1745.

29. Ibid., 44, quoting a letter from Pehr Kalm to Sten Carl Bielke. 24 January 1745.

30. Brief discussions on Linnaeus' views on climates can be found in Drake, "Linné och pärlodlingen," 109–123; Juel, "Om Kalms bemödande," 40–60; and Fredbärj, "Linné och vintern."

31. Roberts, *Swedish Imperial Experience*, 70–75.

32. Linnaeus, preface to Carl Renmarck, *De praestantia orbis sviogothici*, Uppsala PhD, 1747, praes., Petr. Ekerman, reprinted in Fredbärj, "Linné och vintern," 69–74, all quotes from 72.

33. Broberg, "Olof Rudbeck," 14; Fredbärj, "Linné och vintern"; Broberg, *Homo Sapiens*, 266–267; CD, 77. Linnaeus borrowed from Olof Rudbeck the Younger's *Nora Samolad* (1701) the age-old saying, traceable to Magnus (1555) and earlier, that no Sami died from cold, plague, or snakebites.

34. Linnaeus, praes., *De morbis ex hyeme* (1754), additional passage in *Amoenitates academicae*, 1756, translated and quoted in Fredbärj, "Linné och vintern," 78.

35. Linnaeus, *Calendarium florae*, 1757, from unpaginated calendral part and p. 7. On Linnaeus' belief that swallows wintered under water, see also LD, 156, quoting from Linnaeus' dissertation *Aer habitabilis* (1759).

36. EA, 66.

37. BS I:2, 58. Memorial by Linnaeus to the Swedish Academy of Science. 10 January 1746.

38. Linnaeus, "Doctor," 1740, 414.

39. BS I:2, 30. Linnaeus to Pehr Elvius. 28 August 1744.

40. Ibid., 8, 16–17. Linnaeus to Pehr Kalm. Uppsala, 7 December 1745.

41. Ibid., 37. Pehr Kalm to Linnaeus. "Philadelphia uti nya Swerige i America," 14 October 1758.

42. "Protocoller," 1760, "Australe," "Orientale," "Occidentale," "Mediterraneum," and "Boreale." This nonautograph manuscript, probably written by a student, also includes a sixth category, "Alpinum," discussed below.

43. LD, 190. Quoted from Linnaeus' review in *Lärda Tidningar* of Pehr Löfling's *Gemma arborum* (1749). As Nancy Slack kindly informed me in a letter of April 16, 1991, there are genetically different ecotypes of some trees that grow at different latitudes.

44. Linnaeus, *Deliciae naturae* (1773), 1939, 122.

45. Linnaeus, preface to Carl Renmarck, *De praestantia orbis sviogothici*, Uppsala Ph.D., 1747, praes., Petr. Ekerman, reprinted in Fredbärj, "Linné och vintern," 69–74, all quotes from 72. On Linnaeus' view on other benefits of a cold climate see his *De morbis ex hyeme* (1754), quoted in ibid., 78.

46. BS I:2, 58. Memorial from Linnaeus to the Swedish Academy of Science. 10 January 1746.

47. LD, 33, quoting the "Fägne-Rim till Herr Auctoren" in Linnaeus, *Plantae officinales* (1753). This dissertation elaborated Linnaeus, "Uppsats på de Medicinal Wäxter," 1741, listing 170 plants. Note that for clarity I have shifted the placement of the two parts of the quote. This does not alter its intended meaning.

48. Linnaeus, "Med. Profess. Ädel och wida berömde Herr Doct. C. Linnaei Anmärkningar . . . " (1746), reprinted in "Linnés almanacksuppsatser," 1928, 136.
49. BS I:7, 14. Fredrik Hasselquist to Linnaeus. Smyrna, 28 February 1750.
50. BS I: 8, 59. Pehr Kalm to Linnaeus. Philadelphia, 5 December 1750.
51. LD, 210, *Lärda Tidningar*'s review of Linnaeus' dissertation *Arboretum svecicum* (1759).
52. "Protocoller," 1760.
53. BS I:8, 24. Pehr Kalm to Linnaeus. London, 24 March 1748; LD, 33, quoting the "Fägne-Rim till Herr Auctoren" in Linnaeus's dissertation, *Plantae officinales* (1753); LD, 190, *Lärda Tidningar*'s review of Pehr Löfling's dissertation *Gemma arborum* (1749); Tessin, *En Gammal Mans Bref,* 97.
54. The annual records of the costs for the gardens are brief, and only extant for a few years. See Carl Linnaeus and Carl Linnaeus the Younger, "Räkning för Örtegårdsmedlen," (1754, 1766, 1778–1780), Lin. pater et fil, LSL. On Linnaeus' view of experiment vs. experience, see, for example, COR I:107. John Ellis to Linnaeus. London, 24 October 1758, 110; Linnaeus to John Ellis. Uppsala, 8 December 1758.
55. LD, 173. Quoted from *Lärda Tidningar*'s review of Linnaeus' dissertation *Flora alpina* (1756); ibid, 223–225, summary of Linnaeus' dissertation *Coloniae plantarum* (1768).
56. Triewald, "Anmärkningar wid utländska Frukt-."
57. BS I:8, 24. Pehr Kalm to Linnaeus. London, 24 March 1748.
58. LD, 29–34, quoting Linnaeus' dissertation *Plantae officinales* (1753); TMF I:346, quoting his *Skånska resa.*
59. Linnaeus to Abraham Bäck. Uppsala, 7 September 1750. Quoted in TMF II:97, n2. See also Linnaeus, *Hortus Upsaliensis,* 1748; TMF II:97, n2; BS I:2, 142, n3, quoting from an anonymous visitor's letter to the Uppsala botanic garden in 1750; LD, 190, quoting from Linnaeus' review in *Lärda Tidningar* of Pehr Löfling's *Gemma arborum* (1749); Pehr Osbeck to Linnaeus, 20 April 1759, LSL.
60. See for example BS I:4, 119. Linnaeus to Abraham Bäck. Uppsala, 13 March 1750.
61. Bäck, *Åminnelse-Tal,* 37, 38; Carl Renmarck, *De praestantia orbis sviogothici,* Uppsala PhD, 1747, with Petr. Ekeman as praes., cited in Fredbärj, "Linné och vintern," 70. Linnaeus wrote the preface (he knew the candidate's father from Lapland). Reiterating Rudbeck's *Atlantica,* a belated Gothicist work, the dissertation frequently quotes Linnaeus.
62. LD, Drake's introduction, 5–6.
63. LD, 190. Quoted from Linnaeus' review in *Lärda Tidningar* of Pehr Löfling's *Gemma arborum* (1749).
64. Ibid.
65. BS I:4, 105. Linnaeus to Abraham Bäck. Uppsala, 18 November 1749.
66. LD, 190.
67. BS I:4, 105. Linnaeus to Abraham Bäck. Uppsala, 18 November 1749.

68. TMF I:339.

69. Linnaeus, BS I:1, 55. Linnaeus to the Swedish king. Uppsala, 10 November 1749.

70. Lindroth, "Naturvetenskaperna," 185–187, on lieutenant-colonel Johan Bernhard Virgin, *Rön och försök om en underbar sädesarternes förwandling ifrån sämre til bättre slag* (Stockholm: 1757); Carl Linnaeus, praes., *De transmutatione frumentorum*, (Uppsala: 1758), and Pehr Forsskål's anti-transmutationist articles in *Stockholms Wekoblad* 1758: 32, 34, and 39, and 1759: 1 and 16. The transformation of oats to rye was a Swedish folk belief.

71. Linnaeus, "Tal, vid deras Kongl. Majesteters höga närvaro," (1759), 1939, 94.

72. LSL, Solander to Linnaeus. "Landscrona och Wästra Carleby," 28 February 1760.

73. Ibid.

74. FD, 5–6.

75. "Protocoller," 1760.

76. Linnaeus, *Oratio de Telluris*, (1744), 1906, 89–124. Linnaeus developed this argument also to explain what he and his contemporaries believed was a sinking sea-level (the Scandinavian peninsula is still rising after having been compressed by the last ice age's glaciers).

77. Linnaeus, "Tankar," 1754, 182.

78. LD, 173, quoted from *Lärda Tidningar*'s summary of the 1756 *Flora alpina*.

79. Ibid.

80. FD, 5.

81. LD, 177, quoting from *Lärda Tidningar*'s review of Linnaeus' dissertation *Flora Capensis* (1759).

82. Tessin, *En Gammal Mans Bref*, 96–97.

83. Odhelius, *Åminnelse-Tal*, 22, 25, 21.

84. *Dagens Nyheter*, 11 December 1951. Partial reprint of a 1755 report from Pehr Kalm to Åbo University's governing body (*akademiska konsistorium*).

85. See the following quotations from BS I:8, 14. Pehr Kalm to Linnaeus. Moscow, 3 May 1744; I:5, 309. Olof Collin to Linnaeus. Jena, 15 November 1750; I:3, 210. Pehr Bjerchén to Linnaeus. London, 14 June 1758.

86. TMF I:271. Quoted from the protocols of the Cammar-, Oeconomie-, och Commerce-deputationen.

87. BS I:4, 319. Linnaeus to Abraham Bäck. Uppsala, 25 November 1754; quotation from I:4, 109. Linnaeus to Abraham Bäck. Uppsala, 24 November 1759. On opium, see "Protocoller," 1760.

88. LD, 50 and summary of Linnaeus' dissertation *Meloe vesicatoria* (1762), and quoting Linnaeus' *Skånska resa* (1751).

89. BS I:6, 5. Augustin Ehrensvärd to Linnaeus. Stralsund, 6 May 1758.

90. LD, 26, discussing Linnaeus' dissertation *Rharbarbarum* (1752).

91. BS I:1, 34, one of Linnaeus' merit lists.

92. BS I:2, 240. Letter from Linnaeus to Pehr Wargentin. Uppsala, 15 January 1761.

93. BS I:8, 37–38. Pehr Kalm to Linnaeus. "Philadelphia uti nya Swerige i America," 14 October 1748; Juel, "Om Kalms bemödande," 44, 53, quote from 52. On almonds and olives, see BS I:8, 14. Fredrik Hasselquist to Linnaeus. Smyrna, 28 February 1750.

94. BS I:3, 11. Jonas Ahlelöv to Linnaeus. Gothenburg, 22 July 1752.

95. LD, 218. Quoted from preface of Linnaeus' dissertation *Hortus culinaris* (1764).

96. Sparrman, *Åminnelse-Tal*, 40.

97. BS I:6, 6. Augustin Ehrensvärd to Linnaeus. Stockholm, 26 April 1763.

98. Ibid. See also p. 7. Augustin Ehrensvärd to Linnaeus. Stockholm, 5 May 1763.

99. LD, 105, quoting from Linnaeus' self-review of *Plantae esculentae patriae* (1752), in *Lärda Tidningar*.

100. *Svenska Dagbladet*, 17 February 1915.

101. LD, 50–51, quoting from a 1757 letter from Linnaeus to Peter Jonas Bergius.

102. LD, 51; summarizing Linnaeus' dissertation *Purgantia indigena* (1766).

103. Linnaeus, *Medicamenta graveolentia* (1758), 1968, 17.

104. LD, 29. This list encompassed c. 560 plants, arranged alphabetically and in three categories: plants Linnaeus wished to make official; official plants which he accepted; and official plants which he wished to remove.

105. Linnaeus, *Potus coffeae* (1761) 1966, 14, relates a frightful case of a man who drank eight cups of coffee in one day. Coffee-drinking, Linnaeus warned, causes sudden death, premature aging, impotence, blindness, miscarriage, hemorrhoids, and stunted growth. In *De varietate ciborum* (1767), 1966, 14, he added that coffee drinkers are emaciated and constipated, too.

106. Linnaeus, *Potus coffeae* (1761) 1966, 6.

107. Ibid., 11.

108. *Svenska Dagbladet*, 29 January 1924.

109. Linnaeus, *Potus coffeae* (1761), 1966, 10–12. Such substitutes were used during the Second World War. See also TMF II:182.

110. Anders Falck to Johan Peter Falck. Upsala and Hummelsta, 4 August 1766, published in Bergquist, "Anders Falcks brev," 144.

111. LD, 105, quoting from Linnaeus' self-review of *Plantae esculentae patriae* (1752), in *Lärda Tidningar*.

112. Linnaeus to Carl Peter Thunberg. 29 October 1773. Quoted in Sörlin, "Apostlarnas gärning," 88.

113. On how to survive famines, see Linnaeus, *Flora oeconomica*, 1749; *Plantae esculentae patriae*, 1752; and *Berättelse Om The Imhemska wäxter*, 1757; "Upsats på några örter, som norruht . . . wäxa och i nödfall kunna brukas endels såsom kåhl, och endels såsom bröd," n.d., folio, 5 leaves, Econ. LSL; "På följande sätt giöra Bönderna bröd," n.d., 2 leaves, Econ. LSL; and Bergius, *Rön om Spannemåls-Bristens Ärsättjande*.

114. BS I:1, 3. Linnaeus to the Swedish king. N.d. Arrived 9 March 1757.

115. Bäck, *Åminnelse-Tal*, 78.

116. BS I:1, 3. Linnaeus to the Swedish king. N.d. Arrived 9 March 1757.

117. Linnaeus, "Skaparens afsikt" (1763), 1947, 87.

118. Linnaeus, *Plantae esculentae patriae*, 1752, 21, 19, 22, 28, 26, 30. See also Linnaeus, *Flora oeconomica*, 1749, 32.

119. Ibid., 37, 36, 37, 40, 40, 35, 40, 39, 38.

120. Svenska vetenskapsakademien, "Wetenskapsakademiens berättelse," given in response to a royal order of 10 October 1775.

121. HU, 41.

122. For these recommendations, see comments in Juel, "Om Kalms bemödande"; Osvald, "Linnés höfrö," 119–129; "Protocoller," 1760; IL, 133; Kalm, "Beskrifning Huru Socker'; Svanberg, "Turkic ethnobotany," 86; and BS I:2, 195.

123. FD, 128.

124. LD, 247, quoting from Linnaeus' dissertation *Mus Indicus* (1754), probably written by respondent [J. J. Nauman?], "Berättelse om Marsuin," n.d., LSL.

125. Linnaeus, *Märkvärdigheter uti insecterna* (1739), 1942, 33.

126. Undated letter from 1778 from Linnaeus the Younger to Abraham Bäck, reprinted in Linnaeus the Younger, "Linné den Yngres brev," 162.

127. BS I:2, 268. Linnaeus to Pehr Wargentin. N.d., but written in late October or early November 1764. On soy beans, see Sparrman, *Åminnelse-Tal*, 41; BS I:2, 269, n4, quoting from letter from Peter Jonas Bergius to the Academy of Science, 1764; and Ekeberg, "Om Chineska Soyan."

128. Heckscher, *Sveriges ekonomiska historia*, 594.

129. Nicander, *Åminnelse-Tal*, 3. On Alingsås dye manufacture, see Kruse, "Beskrivning om åtskillige." For sanguine hopes of a Swedish textile industry, see Christopher Polhem, et al., "Handlingar rörande intressentskapet 'Floors linnemanufakturi'," 1731, (including "anmärkningar vid nya manufakturers anläggande"), KB Teknol. V, Rål. Ms saml., no. 117 in fol.

130. On the textile industry, see Heckscher, *Sveriges ekonomiska historia*, 600–601, 605, 615, 652.

131. LD, 247. Quoted from *Lärda Tidningar*'s review of Linnaeus' dissertation *Phalaena bombyx* (1756). See also Triewald, "Tankar," and Kalm, "Beskrifning på nord-Americanska."

132. Ibid., 248.

133. BS I:8, 37–38. Pehr Kalm to Linnaeus. "Philadelphia uti nya Swerige i America," 14 October 1748. On Kalm's work on mulberries, see Kalm, "Beskrifning på nord-Americanska."

134. LD, 247. Quoted from *Lärda Tidningar*'s review of Linnaeus' dissertation *Phalaena bombyx* (1756).

135. On Lidbeck's silk project, see Törje, *I den oförlikneliga nyttans tjänst*, and *Svenskt Biografiskt Lexikon*, XXII, 769 ff. On silk in Sweden, see Stavenow-Hidemark, *Anders Berch*, 29, 170.

136. Stavenow-Hidemark, *Anders Berch*, 170; quotation from p. 174.

137. BS I:4, 65. Linnaeus to Abraham Bäck. Uppsala, 29 September 1746; BS I:4, 71. Linnaeus to Abraham Bäck. Uppsala, 25 November 1746. See also TMF II:98, n3.

138. Svanberg, "Turkic ethnobotany," 87; BS I:4, 1304.

139. FL, 271; IL, 62 and 143; HU, 258; Consett, *Tour through Sweden,* 92.

140. See FD, n70, on a "Peruvian goat"; BS I:2, 61, "Memorial" from Linnaeus to the Academy of Science, to be passed on to Pehr Kalm. Uppsala, 10 January 1746; FD, 71.

141. "Färgbok," n.d. (mid-1700s), Teknol. Rål. Ms saml., no. 78 in qvo., KB.

142. LD, 208, quoting *Lärda Tidningar*'s 1760 review of Linnaeus' dissertation *Plantae tinctoriae* (1759).

143. BS I:3, 303–312. Four letters from Erik Brander to Linnaeus. Algers, 23 August 1756; 8 November 1756; 26 February 1757; 12 September 1758. Quoted from the letter of 23 August 1756, 305.

144. TMF II:101.

145. Wolf, *Europe,* 254, 255.

146. Glamann, "European Trade," 448.

147. Linnaeus, *Potus coffeae* (1761), 1966, 4.

148. Linnaeus, *Västgöta resa* (1747), 1978, 166.

149. Linnaeus, "Professor Linnaei upsats på de Medicinal-Wäxter," 1741, 96, also quoted in Drake, "Linnés försök," 70.

150. Linnaeus, "Med. Profess. Ädel och wida berömde Herr Doct. C. Linnaei Anmärkningar" (1746), 1928, 137.

151. COR I:107. John Ellis to Linnaeus. London, 24 October 1758. See also 110, Linnaeus to John Ellis. Uppsala, 8 December 1758; and "Rules for Collecting and Preserving Specimens of Plants," n.d., 3 pages, 1773. Misc. LSL copied from John Ellis, *Some additional observations on the method of preserving seeds,* London: 1773.

152. Linnaeus, *Skånska resa* (1751), 1959, 20.

153. BS I:4, 347. Linnaeus to Abraham Bäck. Uppsala, 18 July 1755.

154. Linnaeus, "Med. Profess. Ädel och wida berömde Herr Doct. C. Linnaei Anmärkningar" (1746), 1928, 136. Linnaeus quotes Hippocrates as proof.

155. Ibid.

156. Ibid.

157. Drake, "Linnés försök"; Torén, "Olof Torén," 17–56; TMF II:38.

158. Quoted from *Samling af Rön och Uptäkter* (Göteborg: 1781), vol. 1, 253, in Drake, "Linnés försök," 74. Note the tea bush was purchased, not discovered.

159. Linnaeus had earlier appealed to Ellis for a tea plant. On Ellis see Smith, *Correspondence of Linnaeus,* vol. 1, 110. Letter from Linnaeus to John Ellis. Uppsala, 8 December 1758. Originally in Latin. Translated by Smith. On Solander see letter from Daniel Solander to Linnaeus, 19 December 1760. London. LSL.

160. Daniel Solander to Linnaeus, London, 19 December 1760, LSL.

161. Quoted from *Samling af Rön och Uptäkter* (Göteborg: 1781), vol. I, 253, in Drake, "Linnés försök," 74.

162. BS I:6, 10. Linnaeus to Carl Gustav Ekeberg. Uppsala, 18 August 1763.

163. Ibid., 4, 12. Linnaeus to Carl Gustav Ekeberg. Uppsala, 17 September 1763.

164. Sparrman, *Åminnelse-Tal,* 23.

165. BS I:6, 11. Linnaeus to Carl Gustav Ekeberg. Uppsala, 17 September 1763.

166. Sparrman, *Åminnelse-Tal*, 25.

167. BS I:7, 166. Anders Johan von Höpken to Linnaeus. Stockholm, 24 November 1763.

168. Linnaeus, *Västgöta resa* (1747), 1978, 166, also quoted from earlier edition in TMF II:99; BS I:4, 348. Linnaeus to Abraham Bäck. Uppsala, 18 July 1755; "Protocoller," 1760.

169. Linnaeus, "Linné's botaniske 'Praelectiones privatissimae'" (1770), 1911, 45.

170. TMF II:182.

171. Juel, "Om Kalms bemödande," 48.

172. Svanberg, "Turkic ethnobotany," 85.

173. Quoted from the academy protocols of 15 June 1754, in BS I:2, 196, n2. See also I:4, 349, n1.

174. BS I:2, 196. Linnaeus to Pehr Wargentin.

175. Ibid., 347, 348. Linnaeus to Abraham Bäck. Uppsala, 18 July 1755.

7 "The Lord of All of Sweden's Clams"

1. BS I:7, 137–138. Carl Hårleman to Linnaeus. Stockholm, 21 July 1748, answering a now lost letter from Linnaeus to Hårleman.

2. Linnaeus, quoted in Drake, "Linné och pärlodlingen," 114.

3. BS I:1, 27. Linnaeus to Baron Colonel Carl Funck, chair of *Riksens Höglofl. Ständers Oeconomie och Commercie Deputations Cammar Oeconomie Utskott*. Uppsala, 6 February 1761, answering a letter from Funck to Linnaeus of 29 January 1761, asking Linnaeus to comment on a proposal to parliament made by the Åbo professor in chemistry, Pehr Adrian Gadd, on how to improve Swedish pearl-fishing.

4. Ibid., 28. Linnaeus means the wild river pearl mussel, *Margaritana margaritifera L.*, sometimes referred to as *Mya margaritifera*.

5. Fyris ån, then called Sala ån. Forsstrand, "Uppsala," 13–31, 19.

6. BS I:1, 21, annotations.

7. Ibid., 22, annotations.

8. Linnaeus wrote a technical description of his pearl-making techniques in two copies. One was lodged in cabinet archives, the other in the Bagge family archives. Both are lost. He also described his techniques to *Riksens Höglofliga Ständers Oeconomie och Commercie Deputations Cammar Oeconomie Utskott* on 27 July 1761; there are extant minutes of this meeting. See annotations in BS I:1, 32.

9. BS I:1, 29. Linnaeus to Carl Funck. Uppsala, 6 February 1761.

10. Ibid., 18. Linnaeus to Cammar-Collegium. Uppsala, 25 January 1751.

11. Drake, "Linné och pärlodlingen," 109. *Anodonta plicata*.

12. BS I:1, 19. Linnaeus to Cammar-Collegium. Uppsala, 25 January 1751. See also Drake, "Linné och pärlodlingen," 114.

13. BS I:1, 29. Linnaeus to Carl Funck. Uppsala, 6 February 1761.

14. BS I:7, 137–138. Hårleman to Linnaeus. Stockholm, 21 July 1748.
15. Ibid.
16. Malmer, "Om Pärlemusslor och Pärlefiskerier"; Fischerström, "Om Perle-Musslors Fortplantning"; and Grill, "Berättelse om Chinesernes."
17. BS I:1, 19, quoted in annotations.
18. *Secreta utskottets Cammar Oeconomie deputation.*
19. BS I:1, 19. Linnaeus to Cammar-Collegium. Uppsala, 25 January 1751.
20. Ibid., 29–30. Linnaeus to Carl Funck, Uppsala, 6 February, 1761.
21. Ibid., 19. Linnaeus to Cammar-Collegium. Uppsala, 25 January 1751.
22. Ibid., 18–21.
23. Ibid., 22. Linnaeus to the secretary of the Cammar-Collegium. Uppsala, 29 January 1751. This letter must be dated incorrectly, however, since Linnaeus only posed the questions that the pearl-inspector should be asked on 25 January 1751, and here he gives thanks for the answers.
24. BS I:1, 33. Undated *"pro memoria"* in Linnaeus' hand.
25. Ibid., 29. Linnaeus to Carl Funck. Uppsala, 6 February 1761.
26. BS I:5, 57. Linnaeus to Abraham Bäck. Uppsala, 22 December 1758. Linnaeus borrowed from *consistorium academicum,* Uppsala University's governing body.
27. Ibid.
28. Drake, "Linné och pärlodlingen," 118–121.
29. From the deputation minutes of 27 July 1761, quoted in the annotations in BS I:2, 31.
30. TMF II:382, n3.
31. The cabinet minutes of 28 January 1762, quoted in the annotations in BS I:2, 31.
32. The Gothenburg merchant Pehr Samuel Bagge, acting as director of the Swedish Greenland Company, invited Linnaeus' student Anton Rolandsson Martin to travel on the Company's whaling ship to Spitsbergen. Bagge sat on the committee that evaluated Linnaeus' pearl-producing techniques.
33. TMF II:351–352.
34. Drake, "Linné och pärlodlingen," 109–123; TMF II:380–384.
35. In order of death dates, the students who perished as a result of their travels were: Tärnström, age 43, fever, outside Cambodia, 1746; Hasselquist, age 30, lung troubles, Smyrna, 1752; Torén, age 34, lung troubles, Gothenburg, on returning from China, 1752; Löfling, age 27, fever, present-day Venezuela, 1756; Adler, age 41, fever, Java, 1761; Forsskål, age 31, fever, Yemen, 1763; Berlin, age 27, fever, East Africa, 1773; Falck, age 42, suicide, Kazan, 1774. Perhaps we should add Rothman, crippled after African voyage (1773–1776) (d. 1778, age 39). On mortality rates, see Philip D. Curtin, *Death by Migration: Europe's Encounter with the Tropical World in the Nineteenth Century* (Cambridge: Cambridge University Press, 1989).
36. Linnaeus to Thunberg, quoted in Sörlin, "Apostlarnas gärning," 82. On how Thunberg survived, see Thunberg, *Resa uti Europa;* Arne, "Svenska läkare,"

132; Svedelius, "Carl Peter Thunberg," 3. On Linnaeus' omens about Tärnström see TMF II:28, and Grape, "Christopher Tärnström," 132, quoting a letter from Linnaeus to Sten Carl Bielke. Uppsala, 13 February 1746.

37. ND, 92.
38. BS I:4, 211. Linnaeus to Bäck. Uppsala, 18 May 1753.
39. Arne, "Svenska läkare," 132.
40. Linnaeus, quoted in TMF II:35.
41. With his children, Olof Rudbeck the Elder spent decades working on *Campus Elysii* (1669-), an illustrated edition of Bauhin's *Pinax* (1623). The 1702 Uppsala fire consumed many thousands of his completed but never printed woodcuts.
42. Juel, "North American Flora," 77. Lovisa Ulrika bought it from Hasselquist's Smyrna debtors.
43. Grape, "Christopher Tärnström," 142, quoting a letter from Magnus Lagerström to Linnaeus, Gothenburg, 6 July 1748.
44. LSL, Osbeck to Linnaeus, n.p., n.d. Later hand added, "written day after arrival [in Gothenburg] 26 June 1752." See also Osbeck to Linnaeus. Gothenburg, 26 July 1752. LSL. Osbeck sent Linnaeus some 150 seeds from China.
45. Stenström, "Pehr Osbeck och Lars Montin," 78. Osbeck diagnosed himself as suffering from "Malo Hypochondriaco."
46. Martin, *Dagbok*, 104, 110. See also Svanberg, "Turkic ethnobotany," *passim*. During his travels in the Russian empire (1768–1774), Falck made elaborate ethnobotanical annotations, yet none led to Swedish usage.
47. Pehr Osbeck to Linnaeus. 5 September 1765. LSL.
48. "Protocoller," 1760.
49. TMF I:147. Cited from a 1728 letter from Commerce Collegium to the Swedish king.
50. Linnaeus, *Skånska resa* (1751), 1975, 187. June 7, on Tunbyholm, the estate of Baron Esbjörn Reuterholm.
51. Ibid.
52. *Svenska Vetenskapsakademiens protokoll*, vol. 1, 44 (18 August 1739).
53. SBL, "Eva Ekeblad," quoting the protocols of the Academy of Science, 19 November 1748. See also "Försök at tilverka bröd, brännvin, stärkelse och puder af potatos gjorde af Eva De la Gardie," *K. Vet. Akad. Handl.* (1748); 277–278, and "Beskrifning på tvål, som är tjenlig til bom-ulls-garns blekning. Af Eva De la Gardie," *K. Vet. Akad. Handl.* (1752): 57–59. Born into the highest nobility, she married a chancellor of the realm (*riksråd*).
54. LD, 250, annotations; Törje, *Eric Gustaf Lidbeck*, 57.
55. SBL XXII: 769 ff.
56. LD, 203. Quoting from *Lärda Tidningar*'s review of Linnaeus' dissertation *Stationes plantarum* (1754), which taxonomizes the domestic plants in *Flora Svecica* into 24 different plant habitats.
57. BS I:2, 206. Linnaeus to Pehr Wargentin. Uppsala, 28 September 1756.
58. EA, 57; BS I:3, 206.

59. BS I:3, 303–312. Four letters from Erik Brander to Linnaeus. Algers, 23 August 1756; 8 November 1756; 26 February 1757; 12 September 1758.

60. *Vita* III, 88.

61. Drake, "Linnés försök," 78.

62. Pehr Kalm to Sten Bielke. 31 October 1746. Quoted in Hulth, "Kalm som student i Uppsala," 43.

63. Sparrman, *Åminnelse-Tal,* 21.

64. BS I:3, 228. Jacob Jonas Björnståhl to Linnaeus. Paris, 16 January 1768.

65. BS I:6, 10. Linnaeus to Carl Gustav Ekeberg. Uppsala, 18 August 1763.

66. Linnaeus, *Deliciae naturae* (1773), 1939, 122.

67. Sparrman, *Åminnelse-Tal,* 21–22.

68. Linnaeus, "Linné's botaniske 'Praelectiones privatissimae' (1770)," 1911, 45.

69. Palmstruck, *Svensk botanik,* first page. See also Kihlman, *Chinesiska théet,* for a belated effort to find native teas.

70. TMF I:263.

71. Linnaeus, 1733 outline of *Ceres Lapponica,* also letters to the county governor of Västerbotten, quoted in TMF I:139. See also BS I:5, 314. Linnaeus to Gustaf Chronhjelm, chancellor of Uppsala University. N.d. [Stockholm, spring 1753].

72. LD, 201, citing *Lärda Tidningar*'s summary of Linnaeus' dissertation *Demonstrationes plantarum in horto Upsaliensi* (1753).

73. BS I:5, 206. Linnaeus to Abraham Bäck. Hammarby, 3 August 1773. See also EA, 16; LD, 173, quoted from *Lärda Tidningar*'s summary of the 1756 *Flora alpina;* LD, 195; and BS I:4, 129, on Linnaeus' dissertation *Plantae rariores Camschatcenses* (1750), discussing Georg Wilhelm Steller's Kamchatka collections during the 1741 Bering expedition; LD, 286, citing Linnaeus' *Necessitas promovendae historia naturalis in Rossia* (1766), nominally written by one of Linnaeus' few ethnically Russian students, Alexander Karmyschev.

74. BS I:1, 129. Linnaeus to Uppsala University's governing body, or Consistorium academicum, sometime in 1746.

75. BS I:6, 8. Augustin Ehrensvärd to Linnaeus. Saris, 28 October 1768.

76. Linnaeus, BS I:1, 103–104, 103. Memorial from Linnaeus to Consistorium academicum. Undated but written 1769.

77. TMF II:78, quoting a letter from Linnaeus to Lidbeck. Uppsala, 30 March 1756.

78. TMF II:118.

79. Rob. E. Fries, "De Linneanska," 32; *Vita,* introduction, 24. Linnaeus introduced this expression in 1750.

80. BS I:4, 149. Linnaeus to Bäck. Uppsala, 28 May 1751.

81. Bergquist, "Anders Falcks brev," 123–158, 126. Quoted from a "Bref, om Prof. Joh. Pet. Falck," in *Upfostrings-Sälskapets Allmänna Tidningar* (1787), I, 108 ff.

82. Hulth, "Kalm som student i Uppsala," 40, quoting a letter from Pehr Kalm to his patron Sten Bielke. Uppsala, 10 February 1742.

83. Hedin, *Minne,* 90.

84. Quoted from a 1757 letter written by J. O. Hagström, a close friend to Löfling, in TMF II:41.

85. Linnaeus often assigned such jobs to his students. Kalm, for example, wrote the fair copy of *Flora Svecica* in 1745, and prepared the index for *Västgöta resa*. See Hulth, "Kalm som student," 41.

86. From Linnaeus' preface to Löfling's *Iter Hispanicum*, quoted in TMF II:43.

87. *Vita* III, 103.

88. *Vita*, 194, quoted from Swedish translation of reprint of letter from Linnaeus to Albrecht von Haller. 12 September 1739.

89. On Johan Stensson Rothman, doctor and teacher at Växsjö grammar school, see *Vita* I, 46, and *Vita* II, 60; on Kilian Stobaeus, professor of history at Lund university and a physician, see *Vita* I, 48 and *Vita* II, 62; on Olof Celsius the Elder, see *Vita* IV, 157; on Albrecht von Haller, see *Vita* appendix, Swedish translation of reprint of letter from Linnaeus to Haller. 12 September 1739; on George Clifford, see *Vita* II, 70. See also *Vita* II, 69, on Johan Heinrich von Spreckelsen (died 1764), secretary of the Hamburg city council and collector of natural objects.

90. *Vita* III, 110. In Linnaeus' mind, this deathbed scene replaced the one which he did not attend, namely his father's death in 1748, when his younger brother Samuel, then curate of Stenbrohult, stood by the bed.

91. Linnaeus to Kilian Stobaeus, written sometime during Linnaeus' first Uppsala year, quoted in TMF I:51.

92. *Vita* II, 60. See also *Vita* III, 92.

93. *Vita* IV, 156.

94. Solander, Falck, and Löfling all taught Linnaeus the Younger. See also TMF II:78, n1. In letters to colleagues, Linnaeus broke off his botanic discourse to insert verbatim the babble of his toddlers.

95. *Vita* IV, 155.

96. *Vita* III, 100; quotation from I, 56.

97. ND, 94; See also 99.

98. *Vita*, introduction, 33, quoting from the memorial Linnaeus wrote for the Swedish Academy of Science. See also III, 104.

99. *Vita* III, 121.

100. *Vita* III, 97; Linnaeus, "Linnés almanacksanteckningar" (1735), 1935, 148, 27/28 September.

101. *Vita* III, 134; BS I:4. Linnaeus to Abraham Bäck. [14] September 1753. See also BS I:270–271. Linnaeus to Bäck. Uppsala, 9 April 1754. Linnaeus here asked Bäck to bring up his son Johannes, if Linnaeus were to die.

102. COR II:23. Peter Collinson to Linnaeus. London, 30 October 1748. See also 435. Letter from Albrecht von Haller to Linnaeus. Göttingen, 23 October 1749; *Vita* IV, 174; *Vita* V, 183; and Svedelius, "Carl Peter Thunberg," 53, quoting from a letter from Linnaeus the Younger to Thunberg.

103. Barreiro, "Pehr Löflings kvarlåtenskap," 136, quoting Linnaeus' biography over Pehr Löfling, published in *Den Swenska Mercurius*, October 1752. Italics mine.

104. Bäck, *Åminnelse-Tal*, 79, 45.

105. Linnaeus to Pehr Löfling and Carl Peter Thunberg, quoted in Sörlin, "Apostlarnas gärning," 83.

106. TMF II:29, n3.

107. BS I:4, 149. Linnaeus to Bäck. Uppsala, 28 May 1751. Linnaeus' use of a go-between suggests that he doubted his own powers over Kalm, or that he was too proud to ask Kalm directly.

108. BS I:4, 151–154. Four letters from Linnaeus to Bäck. Uppsala, 18 June 1751; Uppsala, 20 June 1751; Uppsala, 21 June 1751; Uppsala, 28 June 1751.

109. BS I:4, 155. Linnaeus to Bäck. Uppsala, n.d, but probably written in July 1751. This episode is misread in TMF II:31 to mean that Kalm was not able to visit Linnaeus because he was ill.

110. EA, 65.

111. LSL. Daniel Solander to Carl Linnaeus. 2 November 1759. Wästra Carleby [Skåne]. See also Uggla, "Daniel Solander," 24, 37.

112. Uggla, "Daniel Solander," 46. Solander also stopped writing to his other Swedish friends and his family. See Sandblad, "Bjerkander," 1979–80, 263, quoting a letter from Johan Petter Falck to Claes Bjerkander. 21 March 1763.

113. Uggla, "Daniel Solander," 62, quoting another letter by Solander.

114. Ibid., 47. Linnaeus tried to build a relation with the Forsters instead: the letter he wrote to them was published by C. M. Wieland, in *Der Neue Teutsche Merkur,* Bd 2 (1805): 261–268.

115. Hunter A. Dupree, "Sir Joseph Banks and the Origins of Science Policy," The James Ford Bell Lectures, no. 22, Minneapolis: University of Minnesota, n.d., 11.

116. COR I:270. Linnaeus to John Ellis. Uppsala, 22 October 1771.

117. LSL. Murray to Linnaeus. Göttingen, 24 October 1771.

118. Ibid., 3 October 1774.

119. Linnaeus the Younger, "Linnés den Yngres brev," 138–165, 157.

120. COR II:575. Joseph Banks to Linnaeus the Younger. Soho Square, 5 December 1778.

121. Ibid., I:264. Linnaeus to John Ellis. Uppsala, 8 August 1771.

122. Anders Falck to Johan Petter Falck, Uppsala, 6 September 1770, reprinted in Bergquist, "Anders Falcks brev," 132.

123. Ibid.

124. Pehr Wargentin, the secretary of the Swedish Academy of Science, to Albrecht von Haller. 13 November 1764, quoted in Lindroth, "Legend och verklighet," 117.

125. ND, preface. On the manuscript's relation to Linnaeus' theodicy, see Lepenies, *Autoren,* 27. On *Nemesis divina*'s dating, see Uggla, "Nemesis Divina."

126. Quoted from Hedin, *Minne,* vol. 2, 64, in Uggla, "Nemesis Divina," 14.

127. Linnaeus often cites F. Chr. Friis, *Theologisk och Historisk Afhandling, Om Jus Talionis Divinum . . .* Trans. Glof Rönigk, Stockholm: 1756. On this, see ND, preface, 26.

128. ND, 94. See also 75, 76, 99, 100.

129. ND, 198.
130. See for example ND, 60; also 147, 160, 161.
131. ND, 190, 124, 153.
132. ND, 194, 142.
133. TMF I:257–258.
134. Ibid., 258, 263.
135. ND, 116.
136. ND, 162. See also 193.
137. ND, 89, 110, 86. Christology meant little to Linnaeus. He cites the Old Testament 150 times in ND, and the New Testament 15 times.
138. *Vita* III, 135.
139. Bäck, *Åminnelse-Tal,* 75; EA, 231–234.
140. Rob. E. Fries, "150-årsminnet", 162–172, esp. 168; *Vita* III, 128; ND, 35. On Linnaeus the Younger see also Krook, *Angår oss Linné?,* 94–104.
141. *Vita* III, 129.
142. TMF II:406; BS I:1, 105. Linnaeus to Uppsala university chancellor. Hammarby, 23 August 1775.
143. TMF II:406. Fries's source are letters from Ad. Murray, a Linnaeus student and an Uppsala professor in medicine, to his brother J. A. Murray, also a Linnaeus student and a Göttingen professor in natural history. But as Arvid Hj. Uggla points out in "Linnés den Yngres brev," Ad. Murray had quarreled with Linnaeus the Younger. See also Svedelius, "Carl Peter Thunberg," 55. On Thunberg's forty-four-year-long career at Uppsala see Annerstedt, *Upsala universitet historia,* III:1, 600–602.
144. Letters from Anders Falck to Johan Petter Falck. 135, 29 December 1763; 135, 6 September 1770; 135, 29 December 1763; 150, 17 December 1766, reprinted in Bergquist, "Anders Falcks brev."
145. Ibid., Anders Falck to Johan Petter Falck, 29 December 1763.
146. P. N. Christiernin, according to the de la Gardie family archives, quoted in Forsstrand, "Uppsala," 23. See also Annerstedt, *Upsala universitets historia* III:1, 496–497.
147. Bergquist, "Anders Falcks brev," 155, reprint of letter from Anders Falck to Johan Petter Falck. Stockholm, 6 September, 1770. See also Annerstedt, *Upsala universitets historia* III:2, chap. 5, 338–483.
148. Quoted from a contemporary's description in Forsstrand, "Uppsala," 22. See also TMF I:37.
149. Linnaeus to his wife, on the occasion of the birth of Linnaeus *fils.* in 1741, reprinted in TMF I:22.
150. Linnaeus, "Rop ur grafwen" (1776), 1910. See also BS I:4, 107. Linnaeus to Abraham Bäck. Uppsala, 18 November 1749.
151. Anders Falck to Johan Peter Falck. Stockholm, 6 September 1770. In Bergquist, "Anders Falcks brev," 155.
152. Linnaeus the Younger, "Linné den Yngres brev," 155. Linnaeus the Younger to

Abraham Bäck. Undated, but from 1778, and before July, and also 157, Uppsala, 21 July 1778.

153. Samuel Linnaeus to S. Ch. Duse, *akademieombudsman* at Uppsala University. 1783. Reprinted in Virdestam, "Samuel Linnaeus," 115–125, 125.

154. Quoted from the reprint of Samuel Linnaeus' letter in Virdestam, "Samuel Linnaeus," 125. Banking on kinship obligation, Samuel Linnaeus introduced the five-year-old as a "client" of the university-employed husband of one of Linnaeus' daughters.

155. Linnaeus had four daughters who survived infancy, Elisabet Christina (1743–1782), Lovisa (1749–1839), Sara Christina (1751–1835), and Sophia (1757–1830). About his fifth daughter, Sara Lena (1744; d. in the same year), see *Vita* II, 77. Samuel Linnaeus' daughters are mentioned in his 1783 letter to his niece's husband, reprinted in Virdestam, "Samuel Linnaeus," 125: Christina Helena, married to a pastor in the nearby town of Växsjö; Emerentia Juliana, married to an iron-master near Stenbrohult; and Elisabeth and Ulrica Scharlotta, unmarried at the time of the writing.

8 "His Farmers Dressed in Mourning"

1. *Vita* III, 135.

2. J. S. [Johan Salberg], "Öfver Herr Archiatern och Riddaren C. v. Linné's återvundna hälsa, sedan han af en kall-feber varit angripen vid sit 69 år." *Tidningar utgifne i Uppsala.* 7 September 1776.

3. Rob. E. Fries, "150-årsminnet," 162–172, esp. 168, quoting from Adam Afzelius' memory of the funeral.

4. Brovallius, *Den första biografien;* Beckmann, *Schwedische Reise;* Linnaeus, *Linnés Minnesbok,* 1919; Agardh, "Äreminne öfver Arkiatern," 47–108; Hedin, *Minne;* Afzelius' commentary in EA; and A. Ehrström, "Minnesanteckningar," 69–72.

5. Condorcet, "Eloge de M. de Linné"; Smith, "Introductory Discourse"; Millin, *Discours sur l'origine.* Aubin-Louis Millin (1759–1818) was one of the founders of the short-lived 1788 Société Linnéenne de Paris, which reappeared under the name Société d'Histoire Naturelle in August 1790.

6. Bergquist, "Anders Falcks brev," 135; quote from Rydberg, *Svenska studieresor,* 263.

7. Heckscher, *Sveriges ekonomiska historia,* 593, 604, 605, 701.

8. Törje, *Eric Gustaf Lidbeck,* 57.

9. *Stockholmsposten* (1781): 2, "Lärda nyheter ifrån Nådendal, 1:0: Beskrifning öfver en resa til Guadalaxa år 1736, af Henric Durr, med 1500 Tabeller," quoted in Ehnmark, "Linné traditionen," 57.

10. Ibid.

11. Quoted from Anders Berch, article in *Posten,* 20 May 1769, by Uggla, "När Linné fick en Venus," 96, and also in his preface to Linnaeus, *Herbationes Upsaliensis* (1747), 1950–51, 103. *Posten* was published 1768–69 by Anders

Berch the Younger, son of Linnaeus' old enemy, the Uppsala professor of cameralism, Anders Berch.

12. Cederborgh, "Strödde Reflexioner."
13. Ibid.
14. LSL, Pehr Osbeck to Linnaeus. 18 August 1752. Göteborg.
15. Wallenberg, *Min son på galejan,* 58.
16. Ehnmark, "Linnétraditionen," 57. The professor, Claes Blechert Trozelius (1719–1794), wrote over 100 dissertations, mainly in Swedish, and in the Linnaeo-cameralist tradition.
17. [Baron Ignaz von Born], *Joannis Physiophili Specimen Monachologiae methodo Linnaeana* . . . (Vienna: 1783). See Soulsby, *Catalogue,* 237.
18. Arndt, *Reise durch Schweden,* vol. 1, 202, 197, 205.
19. Ibid., vol. 2, 20, 26–34.
20. Ibid., vol. 1, 205.
21. Ibid., 86–87.
22. Sandblad, "Bjerkander," 194.
23. Arndt, *Reise durch Schweden,* vol. 1, 83.
24. Annerstedt, *Upsala universitets historia* III:1, 601–602. Thunberg donated his collections, including his Asian library, to Uppsala university in 1785. A 1786 letter from Gustav III stresses that the university was not obliged to display the collections or provide storage space. In 1787, however, Gustav III donated land for a new botanical garden, by the old royal castle.
25. Arndt, *Reise durch Schweden,* vol. 1, 83.
26. Edward D. Clarke, quoted in Sandblad, "Clarke and Acerbi," 189.
27. Agardh's diary, quoted in Schoultz, "C. A. Agardhs resa," 171. Agardh's unpublished diaries of 1807–1810, in two folios, are now kept in Värmlands musei arkiv, Karlstad.
28. Thomas Thomson, *Travels in Sweden During the Autumn of 1812* (1813), quoted in Förberg, "Hortus Upsaliensis", 120–135, 127.
29. Stenström, "Pehr Osbeck och Lars Montin," 61; Osbeck, *Utkast till Beskrifning öfver Laholms Prosteri.*
30. See Schoultz, "C. A. Agardhs resa."
31. The first volume appeared in 1820; Agardh was preparing the second volume for print in 1826. However, only the first half of this appeared, in 1828.
32. Carl Adolph Agardh, "Afhandling om foderbetans användande till socker," Ph.D. Lund University, 1812; *idem,* "Beskrifning på de tång-arter, som finnas vid stränderne af Göteborgs och Bohus län samt om sättet och värdet af deras användning i landthushållningen." In *Göteborgs och Bohus läns Kongl. Hushålln.-Sällsk. Handl.* (1816): 60–99. *Idem, "Conspectus specierum Nicotianae. Några ord om tobaksodlingens förbättring,"* Lund, 1819.
33. Carleson, *Åminnelse-Tal,* 15. Parsons typically used translators for their sermons. Odhelius, *Åminnelse-Tal,* 28.
34. Ibid., 25.
35. It is the last known letter to Bäck. We cannot know with absolute certainty if it

was Linnaeus' last letter, since he did not keep copies of his correspondence. See BS I:4 and I:5 for Linnaeus' letters to Bäck.

36. Afzelius, in his commentary in EA, 232, incorrectly states that the letter was written on 9 December 1776.

37. See BS I:5, 238–239, annotations; BS I:5, 216. Linnaeus to Bäck. Hammarby, 20 May 1774, "lille Bror"; BS I:5, 214. Linnaeus to Bäck. "Hamarby," 3 April 1774.

38. Agardh, *Antiquitates Linnaeanae*. Compare BS I:5, 239, BS I:5, 238, for how Linnaeus copies the sentence beginning "Gud hafwor" from Bäck's letter to Linnaeus of 2 December 1776, where Bäck tells of his son's death.

39. "Protocoll."

40. "Öfversigt af de Natur-Alster, som egentligast böra fästa Handlandes, Hushållares, Slöjd-idkares, Vext-odlares och Natur-Älskares uppmärksamhet." Read in Samfundet 6 October and 3 November 1832. Bound with "Protocoll."

41. Ibid.

42. Ibid.

43. *Dagens Nyheter*, 15 May 1921. Felix Bryk, "Ett bortglömt Skansen: Då Humlegården var Linnépark." References to the 1837 ms. *Vägledning genom Linnés park. Första delen*, authored by Gustaf Johan Billberg (1772–1844); *Dagens Nyheter*, 22 May 1921. Bryk notes that in KB there is one page with an "Explication de la Carte du Parc de la Société Linnéenne. Förklaring till Kartan öfver Linnéska Samfundets Park." "Tryckt hos Carl Deleen 1837." The map itself is lost.

44. "Protokoll, Hållet i Linnéska samfundet den 12 Januari 1842." Bound with "Protocoll."

45. Sandblad, "Clarke and Acerbi," 179–180.

46. Ibid., 183, 165.

47. Arndt, *Reise durch Schweden*, vol. 2, 12–14; 251, enthusiastically describes meeting a mother with six blond children. Quotation from vol. 1, 244.

48. Sandblad, "Bjerkander," 70–72.

49. Sandblad, "Clarke and Acerbi," 181.

50. Gage and Stearn, *Bicentenary History*, 5; Lyte, *Sir Joseph Banks*, 241. On Joseph Banks's advice, Smith bought the collection from Linnaeus' widow after Linnaeus the Younger died in 1783. Banks had offered to buy it in 1778, after Linnaeus the Elder's death. But Linnaeus the Younger, who inherited his father's chair, said no. The collection consisted of c. 19,000 plant specimens; 3,000 insects; 1,500 shells; 800 coral pieces, 2,500 minerals; and 2,500 books.

51. Sandblad, "Clarke and Acerbi," 185, 182.

52. Broberg, "Lappkaravaner på villovägar," 55.

53. Ibid., 40, 36, 72. The great Sami artist Nils Nilsson Skum was exhibited there.

54. Ibid., 40.

55. Soulsby, *Catalogue*, 168. It remained the sole official commemoration of Linnaeus larger than a medal.

56. *Upsala Nya Tidning*, "Vårt Linné-monument," May 1930, from Birger

Strandell's newspaper cutting collection, now in HBL. Unless stated otherwise, newspaper articles cited below are from that collection.

57. Soulsby, *Catalogue*, 170. A small plaque of Linnaeus was set up at Lund University in 1811.

58. Letter written by *landshövdingen* Hans Järta, 5 October 1829, quoted in Uggla, "Linnéstatyn," 153–164, 161. The statue was designed and sculpted by Johan Niclas Byström; letter by Hjerta of 19 October. Linnaeus' three surviving daughters were among the honorary guests.

59. Agardh, *Antiquitates Linnaeanae*.

60. *Dagens Nyheter*, 4 December 1955.

61. On the names see Floderus and Forsstrand, "Linnés ättlingar," 115–116.

62. *Stockholms-Tidningen*, 23 May 1907. See also *Helsingborgs Dagblad*, 22 May 1907.

63. *Helsingborgs Dagblad*, 24 May 1907. This was a political gesture by a member of the ethnic Swedish minority in Finland, then under Russian rule.

64. *Stockholms-Tidningen*, 23 May 1907, referring to Lundasångarna.

65. *Helsingborgs Dagblad*, 23 May 1907. See also *Stockholms-Tidningen*, 23 May 1918.

66. *Helsingborgs Dagblad*, 23 May 1907. See also 24 May 1907.

67. Ibid., 24 May 1907.

68. *Smålandsposten*, dated 17 May 1919 by HBL, but probably 23 May 1919.

69. See Sernander, "I Linnés fotspår," for a meager and contrived collection of "traditions" about Linnaeus and his family. Quotation from *Malmö-Tidningen*, 23 May 1907, according to note on cutting in HBL, and *Stockholms-Tidningen* 23 May 1907, according to text above cutting.

70. *Aftonbladet*, 22 May 1920.

71. Lindman, "Ett besök vid Råshult."

72. *Fyns Venstreblad*, 20 August 1931, citing poem in Råshult "Linné-Stugan."

73. *Barnens Tidning*. vol. 20. nrs. 9, 10. Sept.-Oct.: 105–107, 115–120, 1877; E. J. Lindberg, *En allmänn svensk natur-studie och planteringsdag på blomsterkonungens födelsedag*, Borlänge: 1905; J. G. Laurell, kyrkoherde, *Biblens liljor i ord och bild. Bidrag till blomsterkonungens Carl von Linnés 200-åriga födelseminne*, Strängnäs: 1907; Amalia Björck, *Blomsterkonungen på Hammarby. Liten svensk komedia i tre akter*, Stockholm, 1922.

74. See for example, Bergstedt, "Minnesfesten öfver Carl von Linné."

75. This genre was initiated by Esaias Tegnér in 1836. For later versions, see the poem in the broadsheet "Till Minnet af Carl von Linné med anledning af hans bildstods aftäckning i Stockholm den 13 Maj 1885"; and *Smålandsposten*, 18 May 1907, poem by "C. G. S."

76. *Smålandsposten*, 19 May 1919. See also Müller, "Svenska Linnésällskapets," 173–176, on the Goethe Society as model; *Svenska Dagbladet*, 25 November 1917; *Dagens Nyheter*, 30 May 1937, Arvid Hj. Uggla, speech at the inauguration of Linnaeus' Uppsala home as a museum; same occasion, speech by Prince Eugen cited in *Svenska Dagbladet*, 31 May 1937.

77. *Dagens Nyheter,* 15 May 1885.

78. Stavenow, speech, "Kväde och tal," 4.

79. *Nya Växsjöbladet,* 19 April 1918. The article's program "till vägledning vid eventuella minnesfester" was used at Falun Linnéfest, 23 May 1917.

80. *Stockholms-Tidningen,* 23 May 1907.

81. *Vita* II, 86.

82. *Borås Tidning,* article by Rob. E. Fries, 4; *Svenska Dagbladet,* 19 May 1907; *Nya Växsjöbladet,* 19 April 1918; Stavenow, "Kväde och tal," 5.

83. Hultgren, "Francois-Auguste Biard," 65–88.

84. *Svenska Familje-Journalen,* 1874, 331–332.

85. *Läsebok för folkskolan* (Stockholm: Norstedt, 1901), 651.

86. Ellen Fries, *Den svenska odlingens stormän. Lefnadsteckningar för skola och hem,* vol. 3, *Karl von Linné* (Stockholm, 1899); Isak Fehr, *Karl von Linné. En lefnadsteckning* (Stockholm, 1907).

87. *Stockholms-Tidningen,* 25 May 1926.

88. Ibid.

89. *Göteborgs Handels och Sjöfarts Tidning,* 28 May 1926.

90. For example, *Svenska Dagbladet,* 1 June 1919.

91. *Metallarbetaren,* 27 February 1932; *Social-Demokraten,* 16 November 1934.

92. *Smålandsposten,* 4 June 1935; *Nya Växsjöbladet,* 3 June 1935; *Stockholms-Tidningen,* 3 June 1935; *Stockholms Dagblad,* 3 June 1935; *Nya Dagligt Allehanda,* 3 June 1935; *Dagens Nyheter,* 3 June 1935.

93. *Stockholms-Tidningen,* 14 December 1948. See also *Dagens Nyheter,* c. 1928, undated cutting in HBL. Article by Dr Gunnar Beskow; *Aftonbladet,* 23 October 1932.

94. *Stockholms-Tidningen,* 24 February 1957.

95. *Arbetet,* 2 February 1957. See also *Kvällsposten,* 2 February 1957.

96. This point I owe to Märit Rausing.

97. *Smålandsposten,* 18 May 1957.

98. *Vestmanlands Läns Tidning,* 14 May 1957; *Morgon-Tidningen,* 14 May 1957; *Gotlands Allehanda,* 29 April 1957.

99. *Stockholms-Tidningen,* 24 February 1957.

100. *Svenska Bokvännen,* Nr 7, 1963. The Carl-Otto von Sydow edition, illustrated by Gunnar Brusewitz.

101. *Svenska Dagbladet,* 24 July 1950; 25 July 1950; *Upsala Nya Tidning,* 25 July 1950; 21 July 1950.

102. *Stockholms-Tidningen,* 24 February 1957.

103. *Nya Dagligt Allehanda,* 20 September 1938; *Sydsvenska Dagbladet,* 13 July 1946; *Dagens Nyheter,* 12 June 1947. See also *Smålandsposten,* 29 June 1948; and *Arbetet,* 12 July 1948.

104. *Stockholms-Tidningen,* 18 May 1945.

105. *Allehanda* [as marked in HBL; possibly *Nya Dagligt Allehanda*], 17 May 1957; *Göteborgs-Tidningen,* 22 May 1957.

106. *Upsal* [as marked in HBL; possibly *Upsala Nya Tidning*], 18 May 1957; *Göteborgs Handels- och Sjöfarts-Tidning* [not dated in HBL, but from 1969]; *Värnamo-Tidningen,* 18 April 1969; *Gefle Dagblad,* 27 June 1969.

107. Eric Hobsbawn and Terence Ranger, *Invention of Tradition* (Cambridge: Cambridge University Press, 1983).

Conclusion

1. Heckscher, *Mercantilism,* 263. Spiegel, *Growth of Economic Thought,* 714, argues the tradition ends with Sonnenfels.

2. Family resemblances, as well as intellectual genealogies, link early cameralist tracts such as Johann Joachim Becher, *Politische Discurs von den eigentlichen Ursachen des Auf- und Abnehmens der Städt, Länder und Republicken* . . . (Frankfurt: 1668; reprint, Klassiker der Nationalökonomie, Düsseldorf: Verlag Wirtschaft und Finanzen, 1990); and Linnaeus, "Tankar om grunden til oeconomien genom naturkunnogheten ock physiquen," 1740; with Enlightenment economists such as Johann Heinrich Gottlob von Justi, *Staatswirthschaft, oder systematische Abhandlung aller Oekonomischen und Cameral-Wissenschaften* . . . , 2nd ed. (Vienna, 1758); with Romantic and post-Romantic cameralists such as Johann Gottlieb Fichte, *Der geschlossne Handelsstaat* . . . (Tübingen: J. G. Cotta, 1800); Friedrich List, *Das nationale System der politischen Oekonomie* (Stuttgart and Tübingen: J. G. Cotta, 1841, 1844); and with later Marxist theorists, especially neo-Marxist dependency theorists such as Arghiri Emmanuel, *Unequal Exchange: A Study of the Imperialism of Trade* (New York: 1972); and Immanuel Wallerstein, *The Modern World System,* 2 vols. (New York: Academic Press, 1980).

3. Quoted in Whittaker, *History of Economic Ideas,* 305. The debate on whether the early cameralists were also cultural reactionaries hinges on Johann Joachim Becher's rare Utopian tracts, *Moral Discurs von den eigentlichen Ursachen des Glücks und Unglücks* (1669) and *Psychosophia oder Seelen-Weisheit* (1678). Becher was an avid anti-Semite who turned against alchemy because he worried gold production might disrupt the money supply. In later life he turned against money entirely.

4. See, for example, Robinson, *The Economics of Imperfect Competition.*

5. See, for example, E. Hagen, "An Economic Justification of Protectionism," *Quarterly Journal of Economics,* November 1958; A. Kafka, "A New Argument for Protectionism," *Quarterly Journal of Economics,* February 1962. On how the Stolper-Samuelson Theorem has been used to motivate "declinist" and "hyperindustrial" neo-protectionism, see, for example, Johnson, Tyson, and Zysman, *Politics and Productivity.*

6. H. Myint, "Infant Industry Arguments for Assistance to Industries in the Setting of Dynamic Trade Theory," in Roy Harrod, ed., *Trade in a Developing World* (London: Macmillan, 1963); W. W. Rostow, "The Take-Off into Self-Sustained Growth," *Economic Journal* 66, 1956; Alexander Gerschenkron, *Economic Back-*

wardness in Historical Perspective (Cambridge: Harvard University Press, 1962).

7. This point I have borrowed from Ernest Gellner, "What Do We Need Now? Social Anthropology and Its New Global Context." *Times Literary Supplement,* 16 July 1993:3.

8. Marxist or not, cameralists are not *a priori* committed to an authoritarian model of statehood. Robinson, in *Economics of Imperfect Competition,* arguably the groundwork for mid-twentieth-century neo-protectionism in the West, even argues that though it is a natural fact that industrial monopolies utilize production factors more efficiently, and develop more innovative new production technologies, for political reasons the enfranchised citizenries of the West may still opt for continued free economies. She adds that professional economists should not intervene in this political choice, however irrational they know it to be. As a matter of historical fact, however, cameralist economic policies and authoritarian political structures tend to be correlated, since rent-seeking state elites typically seek to fashion the rule of politics so as to protect their monopoly of rents.

9. See Hildebrand, "Economic Background," 23, for a more positive view on the ideological links between Linnaeus' economic thinking and present-day cameralism.

10. The classic theory of comparative advantage is, of course, the Ricardian theorem of comparative factor productivity. Its most important modern refinement is the Heckscher-Ohlin theorem of relative factor price. For an analysis of comparative advantage as a function of factor abundance relative to other domestic factors, see W. Leontief, "Domestic Production and Foreign Trade: The American Capital Position Reexamined," *Proceedings of the American Philosophical Society,* 28 September 1953. For arguments that free trade means higher technical efficiency and greater social utility than restricted trade or no trade, see M. C. Kemp, "The Gains from International Trade," *Economic Journal* 72 (December 1962); P. A. Samuelson, "The Gains from International Trade Once More," and Jagdish Bhagwati, *Trade, Tariffs and Growth* (Cambridge, Mass.: MIT Press, 1969). Bhagwati argues that trade is superior to autarky *even* if, as he puts it on p. 171, "the country has no monopoly power in trade but has a non-economic objective which consists of requiring production to be maintained at a certain level in a specific activity."

11. For a typical popular defense of neo-cameralism, drawing on a moral argument about trading profits, see Leif Drambo, "'Utbudsekonomerna' saknar en egen teori," *Dagens Nyheter,* 27 October 1981.

12. Kreuger, "The Political Economy of the Rent-Seeking Society," and Bhagwati, "Directly Unproductive Profit-Seeking Activities" (see p. 216 n26 above). They have also turned to neo-Malthusianist sustainable-growth theory, which has its Luddite family resemblances, but which nonetheless, in its radical environmentalist stance, is a genuinely new moral intuition.

13. See for example Emmanuel, *Unequal Exchange;* and Wallerstein, *The Modern World System,* both cited in note 2 above and for a more recent example, Mary

Louise Pratt, *Imperial Eyes: Travel Writing and Transculturation* (London: Routledge, 1992). For critiques of this position see Wolfgang Mommsen, *Theories of Imperialism* (London: Weidenfeld and Nicolson, 1981), esp. 29–65 and 113–141 and Philip D. Curtin, *Cross-Cultural Trade in World History* (Cambridge: Cambridge University Press, 1984). Curtin espouses the economic location theory of August Lösch, translated as *The Economics of Location* (New Haven: Yale University Press, 1954).

14. To use the terminology of Philip D. Curtin.

Works Cited

Afzelius, Adam. *Sierra Leone journal 1795–1796.* Ed. Alexander Peter Kuj. Uppsala: Institutet för allmänn och jämförande etnografi, 1967.

Agardh, Carl Adolph. *Antiquitates Linnaeanae.* . . Lund: litteris Berlingianis, 1826.

Almer, Tage. *Linné om skaparen.* Lund: Gleerups, 1968.

Almkvist, Johan. "Studier över Carl von Linnés verksamhet som läkare, ett bidrag till syfilidologiens historia." *SLÅ* VI (1923): 61–119.

Alvey, James. "A New Adam Smith Problem: The Teleological Basis of Commercial Society." Ph.D. diss., University of Toronto, 1995.

Ambjörnsson, Ronny. "'Guds Republique'. En utopi från 1789." *Lychnos* (1975–76): 1–57.

Annerstedt, Claes. *Upsala universitets historia.* 10 vols. Uppsala and Stockholm: various publishers, 1877–1914.

Arndt, Ernst Moritz. *Reise durch Schweden im Jahr 1804. Erster Teil.* Berlin: G. A. Lange, 1806.

Arne, Ture J. "Svenska läkare och fältskärare i Holländska Ostindiska Kompaniets tjänst." *Lychnos* (1956): 132–146.

Ashworth, William B., Jr. "The Persistent Beast: Recurring Images in Early Zoological Illustration," in *The Natural Sciences and the Arts,* ed. Allan Ellenius. Stockholm: Almquist & Wiksell, 1985.

Atran, Scott. *Cognitive Foundations of Natural History. Towards an Anthropology of Science.* Cambridge: Cambridge University Press, 1990.

Bäck, Abraham. *Åminnelse-Tal öfver . . . Carl von Linné, . . . För Kongl. Vetenskaps Academien, Den 5 December, 1778* . . . Stockholm: Tryckt hos Joh. George Lange, 1779.

Bäckman, Louise, and Åke Hult Kranz. *Studies in Lapp Shamanism.* Stockholm: Almqvist & Wiksell, 1978.

Barreiro, Augustin Jesus. "Pehr Löflings kvarlåtenskap och eftermäle." *SLÅ* X (1927): 131–136.

Beckmann, Johann. *Schwedische Reise in den Jahren 1765–1766: Tagebuch.* Ed. Th. M. Fries. Uppsala: Almqvist & Wiksell, 1911.

Benedikt, E. "Goethe und Linné." *SLÅ* XXVIII (1945): 49–54.

Benjamin, Marina, ed. *Science and Sensibility: Gender and Scientific Enquiry 1780–1945.* Oxford: Basil Blackwell, 1991.

Berch, Anders. *Inledning til allmänna hushåldningen innefattande grunden til politie, oeconomie och cameral wetenskaperne.* Stockholm: Trykt hos L. Salvius, 1747.

——*Sätt att igenom politisk arithmetica utröna länders och rikens hushåldning.* Stockholm: Trykt hos L. Salvius, 1746.

Beretta, Marco. "The Sociéte Linnéenne de Paris (1787–1827)." *SLÅ* (1990–91): 151–171.

Bergius, Peter Jonas. *Rön om Spannemåls-Bristens Ärsättjande medelst Qwikrot.* Stockholm: no publ. noted, 1757.

Bergman Sucksdorff, Astrid, and Vogel-Rödin, Gösta. *Buores Buores. Medmänniskor i norr.* Höganäs: Bra Böcker, 1975.

Bergquist, Olle, ed. "Anders Falcks brev till Petersburg." *Lychnos* (1965–66): 123–158.

Bergstedt, Jacob A. "Minnesfesten öfver Carl von Linné den 25 maj 1907." *Kungliga Vetenskapsakademiens Årsbok.* Stockholm: Almqvist & Wiksell, 1910.

Blunt, Wilfried. Ass. W. T. Stearn, *The Compleat Naturalist. A Life of Linnaeus.* London: Collins, 1971.

Bobrov, E. G. "On the Works by and on Linnaeus Published in Russia and the Soviet Union." *SLÅ* (1978): 265–275.

Bodenheimer, F. S. *The History of Biology: An Introduction.* London: Dawson and Sons, 1958.

Boerman, Albert Johan. *Carolus Linnaeus als middelaar tussen Nederland en Zweden.* 1953 Ph.D. Medical Faculty, University of Utrecht. Utrecht: Pressa Trajectina, 1953.

——"Linnaeus and the Scientific Relations between Holland and Sweden." *SLÅ* (1978): 42–56.

Bondeståndets riksdagsprotokoll. På Riksgäldskontorets uppdrag utgivna av Sten Landahl. 1751–1756. Vol. 7. Stockholm: Bok och Reklamtryck, 1963.

Brading, D. A. and Cross, Harry E. "Colonial Silver Mining: Mexico and Peru." *Hispanic American Historical Review* 52 (1972): 545–579.

Bremekamp, C. E. B. "Linnés' Views on the Hierarchy of the Taxonomic System." *Acta Botanica Neerlandica,* 2 (1953): 242–253.

Briggs, Robin. "The Académie Royale des Sciences and the Pursuit of Utility." *Past and Present* 13 (May 1991): 34–88.

British Museum. *Memorials of Linneaus.* Bicentenary exhibition catalogue. London: 1907.

Broberg, Gunnar. *Homo Sapiens L. Studier i Carl von Linnés naturuppfattning och människolära.* Motala: Borgströms, 1975.

——"Lappkaravaner på villovägar. Antropologin och synen på samerna fram emot sekelskiftet 1900." *Lychnos* (1981–82): 27–86.

——ed. *Linnaeus: Progress and Prospects in Linnaean Research.* Pittsburgh: Hunt Botanical Library, 1980.

————"Olof Rudbecks föregångare." In Olof Rudbeck the Younger. *Iter Lapponicum. Skissboken från resan till Lappland 1695* II Kommentardel, 11–21. Stockholm: René Coeckelbergh, 1987.

Browallius, Johannes and Linnaeus, Carl. *Den första biografien öfver Linné. Curriculum vitae af Johannes Brovallius.* Ed. and pub. Johan Bergman. Stockholm: 1920.

Browne, Janet. "Botany for Gentlemen. Erasmus Darwin and *The Loves of the Plants.*" *Isis* 80 (1989): 593–621.

Bryk, Felix. "Die Stellung der Polnischen Botanik zu Linnés System." *SLÅ* VI (1923): 35–51.

————"Var Linné svensk?" *Götesborgs Handels- och Sjöfartstidning*, 8 May 1942.

Carleson, Carl. *Åminnelse-Tal öfver . . . Johannes Browallius, efter Kongl. Vetenskaps Acad:ns Befallning, Hållit uti Stora Riddarehus-Salen, den 18. Martii 1756.* Stockholm: Lars Salvius, 1756.

Cederborgh, Fredrik. "Strödde Reflexioner ur T-gs Resa uti Europa, Africa och Asia från år 1770 til 1779," in *Uno von Trasenberg. . .* 4th ed., Strängnäs: C. E. Ekmarck, 1814.

Colclough, Christopher and James Manor, eds. *States and Markets: Neo-liberalism and the Development-Policy Debate.* Oxford: Oxford University Press, 1992.

Condorcet, Marquis de. "Eloge de M. de Linné." *Histoire de l'Académie royale des sciences.* Paris: 1781.

Consett, Matthew. *A Tour through Sweden, Swedish Lapland, Finland and Denmark.* London: printed for J. Johnson . . . , 1789.

Corsi, Pietro. *The Age of Lamarck. Evolutionary Theories in France 1790–1830.* Rev. and trans. ed. Berkeley: University of California Press, 1989.

Dahlgren, Ossian K. V. "Philosophia Botanica, ett 200-årsminne." *SLÅ* XXXIII-XXXIV (1950–51): 1–30.

Daudin, Henri. *Les méthodes de classification et l'idée de série en botanique et en zoologie de Linné à Lamarck (1740–1790).* Paris: Félix Alcan, 1926.

Delaporte, François. *Nature's Second Kingdom: Explorations of Vegetality in the Eighteenth Century.* Trans. Arthur Goldhammer. Cambridge, Mass.: MIT Press, 1979.

Drake, Gustaf. "Linné och pärlodlingen." *SLÅ* XIII (1930): 109–123.

————"Linnés försök till inhemsk teodling." *SLÅ* XIII (1927): 68–83.

Dupree, Hunter A. "Sir Joseph Banks and the Origins of Science Policy." The James Ford Bell Lectures, no. 22. Minneapolis: University of Minnesota, n.d.

Ehnmark, Elof. "Linnétraditionen och naturskildringen." *SLÅ* XIV (1931): 41–57.

Ehnmark, Erland. "Linnés Nemesis-tankar och svensk folktro." *SLÅ* XXIV (1941): 29–63.

Ehrström, Anders. "Lefnadsförhållanden och sedvänjor beskrivna." *SLÅ* XXX (1947): 36–47.

————"Minnesanteckningar om Linné av kapellanen i Kronoby (Österbotten, Finland), meddelade av Robert Ehrström." *SLÅ* XXIX (1946): 69–72.

Ekeberg, Carl Gustaf. "Om Chineska Soyan." *K. Vet. Akad. Handl.* 1764: 38–40.

Ekeblad, Eva (de la Gardie). "Försök at tilverka Bröd, Brännvin, Stärkelse och Puder af Potatos," *K. Vet. Akad. Handl.* 1748: 277–278.

Ellis, Howard S. and Lloyd A. Metzler, eds. *Readings in the Theory of International Trade.* Philadelphia: Blakiston, 1949.

Eriksson, Gunnar. "The Botanical Success of Linnaeus. The Aspect of Organization and Publicity." *SLÅ* (1978): 57–66.

———*Botanikens historia i Sverige intill år 1800.* Stockholm: Almqvist & Wiksell, 1969.

———"Olof Rudbeck d.ä." *Lychnos* (1984): 77–119.

Fabricius, Johann Christian. "Einige nähere Umstände aus dem Leben des Ritters von Linné." In *Deutsches Museum.* Leipzig: in der Weygandschen Buchhandlung, 1780. I Band: 431–441, II Band: 39–49.

Faggot, Jakob. "Fäderneslandets känning och beskrifwande." *K. Vet. Akad. Handl.,* 1741: 1–29.

Fischerström, Joh. "Om Perle-Musslors Fortplantning, Natur och Lefnads-sätt." *K. Vet. Akad. Handl.* 1759: 139–146.

Fjellström, Phebe. *Samernas samhälle i tradition och nutid.* Stockholm: Norstedt, 1985.

Floderus, Matts. "Några traditioner från det Linnéska huset." *SLÅ* II (1919): 111–114.

Floderus, Matts and Forsstrand, Carl. "Linnés ättlingar. Några slututredningar." *SLÅ* II (1919): 115–125.

Florin, Magnus. *Trädgården.* Stockholm: Bonniers, 1995.

Förberg, Elof. "Hortus Upsaliensis. Några anteckningar om dess öden under 180 år." *SLÅ* VI (1923): 120–135.

Forsskål, Pehr. *Resa till lycklige Arabien; Petrus Forsskåls dagbok, 1761–1763.* Uppsala: Almqvist & Wiksell, 1950.

Forsstrand, Carl. "Linnés ekonomi och hans kvarlåtenskaps öden." *SLÅ* V (1922): 70–89.

———"Uppsala på Linnés tid." *SLÅ* VII (1924): 13–31.

Foucault, Michel. *The Order of Things: An Archeology of the Human Sciences.* New York: Vintage Books, Random House, 1973.

Frängsmyr, Tore. "Den gudomliga ekonomien: Religion och hushållning i 1700-talets Sverige." *Lychnos* (1971–72): 217–244.

———ed. *Linnaeus, The Man and His Work.* Berkeley: University of California Press, 1983.

———"Swedish Polar Exploration." In Tore Frängsmyr, ed., *Science in Sweden. The Royal Swedish Academy of Sciences 1739–1989.* Canton, Mass.: Science History Publications, 1989, 177–189.

Fredbärj, Telemak. "Linné och vintern." *SLÅ* XXXXII (1959): 69–78.

Friedenthal, Richard. *Goethe. Sein Leben und seine Zeit.* Munich: Piper, 1963.

Fries, Elias. "Carl von Linnés Anteckningar öfver *Nemesis Divina.*" Uppsala: Kungliga Universitet, 1848.

Fries, Rob. E. "150-årsminnet av Linnés död." *SLÅ* XI (1928): 162–172.

———"De Linneanska 'apostlarnas' resor. Kommentarer till en karta." *SLÅ* XXXIII-XXXIV (1950–51): 31–40.

———"Linné i Holland." *SLÅ* II (1919): 141–155.

Fries, Th. M. *Linné. Lefnadsteckning.* 2 vols. Stockholm: Fahlcrantz, 1903.

Frondin, El. praes. *De Alandia, maris Baltici insula.* Resp. Christopher Tärnström. Uppsala: 1739, part I; 1745, part II.

Furley, Frank Edgar. "Three 'Lapland songs'." *Publication of the Modern Language Association of America* XXI (1906): 1–39.

Gage, A. T. and Stearn, W.T. *A Bicentenary History of the Linnean Society of London.* London: Academic Press for the Linnean Society of London, 1988.

Geertz, Clifford. "Thick Description: Toward an Interpretive Theory of Culture." In Clifford Geertz, *The Interpretation of Cultures.* New York: Basic Books, 1973, 3–30.

———"The Way We Think Now: Toward an Ethnography of Modern Thought." In Clifford Geertz, *Local Knowledge. Further Essays in Interpretive Anthropology* (New York: Basic Books, 1983), 146–163.

Geoffroy, Saint-Hilare É., and Serres, A.-E.-R.-A. "Sur quelques changements observés dans les animaux domestiques transportés de l'ancien monde dans le nouveau continent." *Mémoires du Musée National d'Histoire Naturelle* 17 (Paris: 1828): 201–208.

Gertz, Otto. "Olof Celsius D. Ä. och Flora Uplandica." *SLÅ* III (1920): 36–56.

Gillespie, Charles Coulston. *Science and Polity in France at the End of the Old Regime.* Princeton: Princeton University Press, 1980.

Giseke, Paul. *Caroli a Linné praelectiones in ordines naturales plantarum.* Hamburg: bei Benj. Gottl. Hoffman, 1792.

Glacken, Clarence J. *Traces on the Rhodian Shore. Nature and Culture in Western Thought from Ancient Times to the End of the Eighteenth Century.* Berkeley: University of California Press, 1967.

Glamann, Kristof. "European Trade 1500–1750." In Carlo M. Cipolla, ed., *Fontana Economic History of Europe. The Sixteenth and Seventeenth Centuries.* Glasgow: Collins/Fontana Books, 1974.

Gledhill, D. *The Names of Plants.* Sec. ed. Cambridge: Cambridge University Press, 1989.

Goerke, Heinz. *Carl von Linné: Arzt, Naturforscher, Systematiker, 1770–1778.* Stuttgart: Wissenschaftliche Verlagsgesellschaft, 1966.

Goerke, Heinz. "Linnaeus' German Pupils and Their Significance." *SLÅ* (1978): 223–239.

Goethe, Johann Wolfgang von. *Schriften zur Naturwissenschaft.* Der Deutschen Akademie der Naturforscher Leopoldina. Weimar: Hermann Böhlaus Nachfolger, 1947-.

Gombocz, E. "Linné und die Ungarische Botanik." *SLÅ* IV (1921): 24–32.

Granit, Ragnar, ed. *Utur stubbotan rot—essäer till 200årsminnet av Carl von Linnés död.* Svenska Vetenskapsakademien. Stockholm: Nordstedt, 1978.

Grape, Anders. "Om Christopher Tärnströms resejournaler." *SLÅ* I (1918): 126–144.

Grape, Erik. *Nya Handlingar för år 1803 och 1804.* Stockholm: Svenska Vetenskapsakademien, 1804.

Grill, Claes. "Berättelse om Chinesernes sätt at eftergöra Äkta Perlor." *K. Vet. Akad. Handl.* 1772: 93–95.

Hagberg, Knut. *Carl Linnaeus.* Stockholm: Natur och Kultur, 1939.

Hallenberg, Jonas. "Stam tafla öfver Linnaeiska Slägten." In C. C. Gjörwell, *Collectio Gjörwelliana*. Stockholm: 1777.

Haller, R. *Svenska kyrkans mission i lappmarken under frihetstiden*. Stockholm: Carlson, 1896.

Hasselquist, Frederik. *Iter palaestinum, eller Resa till Heliga landet förättad ifrån år 1749 til 1752 . . . utgifven av Carl Linnaeus*. Stockholm: L. Salvii, 1757.

Heckscher, Eli F. "Linnés resor—den ekonomiska bakgrunden." *SLÅ* XXV (1942): 1–11.

———*Mercantilism*. Trans. Mendel Shapiro, ed. E. F. Söderlund, 2 vols, rev. ed. New York: Macmillan, 1955.

———*Sveriges ekonomiska historia från Gustav Vasa. Det moderna Sveriges grundläggning.* Vol. 2, part 2. Stockholm: Bonniers, 1949.

Hedin, Sven Anders. *Minne af von Linné fader och son*. 2 vols. Stockholm: Nordström, 1808.

Heilbroner, Robert L. *The Worldly Philosophers. The Lives, Times and Ideas of the Great Economic Thinkers.* 4th ed. New York: Simon and Schuster, 1972.

Heller, John Lewis. *Studies in Linnaean Method and Nomenclature.* Marburger Schriften zur Medizingeschichte. Band 7. Frankfurt Am Main: Peter Lang, 1983.

Hildebrand, Bengt. *Kungl. Svenska Vetenskapsakademien. Förhistoria, grundläggning, och första organisation.* Stockholm: Svenska Vetenskapsakademien, 1939.

Hildebrand, Karl-Gustav. "The Economic Background of Linnaeus. Sweden in the Eighteenth Century." *SLÅ* (1978): 18–29.

Hirschman, Albert O. *The Passions and the Interests: Political Arguments for Capitalism before Its Triumph.* Princeton: Princeton University Press, 1977.

Hobhouse, Henry. *Seeds of Change. Five Plants That Transformed Mankind.* New York: Harper & Row, 1985.

Hofsten, Nils von. "Linnaeus' Conception of Nature." *Kungl. Vetenskaps-Soc. Årsbok* (1957): 65–105.

———"Linné och Goethe." *SLÅ* XXXXVI (1963): 1–4.

———"Linnés dubbla bokföring." *SLÅ* L (1967): 1–12.

———"Systema Naturae, ett 200-årsminne." *SLÅ* XVIII (1935): 1–15.

Hort, Sir Arthur. "Linnaeus and the Naming of Plants." *Blackwood's Magazine* 230 (1931): 682–700.

Hultgren, E. O. "Francois-Auguste Biard's tavla 'La Jeunesse de Linnè'." *SLÅ* IV (1921): 65–68.

Hulth, J. M. "Kalm som student i Uppsala och lärjunge till Linné åren 1741–1747." *SLÅ* VII (1924): 39 ff.

———"Linnés första utkast till Species Plantarum." *Svensk Botanisk Tidskrift.* 6 (1912): 627–631.

Iwao, S. "C. P. Thunbergs ställning i japansk kulturhistoria." *SLÅ* XXXVI (1953): 135–147.

Jackson, Benjamin Dayton. *Linnaeus.* London: Witherby, 1923.

———"The Visit of Carl Linnaeus to England in 1736." *SLÅ* IX (1926): 1–11.

Johnson, Chalmers, Laura D'Andrea Tyson, and John Zysman. *Politics and Productiv-*

ity: How Japan's Development Strategy Works. New York: HarperBusiness, 1989.

Johnson, Duncan S. "The Evolution of a Botanical Problem: The History of the Discovery of Sexuality in Plants." *Science* N.S. 39 (1914): 299–319.

Jonsell, Bengt. "Linnaeus and Spain—Before Mutis." *SLÅ* (1990–91): 145–150.

Jörberg, Lennart. "The Nordic Countries 1850–1914." In *The Fontana Economic History of Europe*, gen. ed. Carlo M. Cipolla, vol. 4, part 2 (Glasgow: Fontana/Collins, 1973): 375–485.

Juel, H. O. "Early Investigations of North American Flora, with Special Reference to Linnaeus and Kalm." *SLÅ* III (1920): 61–79.

——*Hortus Linnaeanus. An Enumeration of Plants Cultivated in the Botanical Garden of Uppsala during the Linnean Period*. Skrifter utgivna av Svenska Linné-Sällskapet. Nr 1. Uppsala: Almqvist & Wiksell, 1919.

——"Om Kalms bemödande att i vårt land införa nordamerikanska växter." *SLÅ* XIII (1930): 40–60.

Jussieu, Antoine-Laurent de. *Genera plantarum secundum ordines naturales disposita*. Paris: Herissant, 1789.

Kalm, Pehr. "Beskrifning Huru Socker göres uti Norra America af åtskilliga slags trän." *K. Vet. Akad. Handl.* 1751: 143–159.

——"Beskrifning på nord-Americanska Mullbärsträdet, Morus rubra kalladt." *K. Vet. Akad. Handl.* 1776: 143–163.

——*En Kårt Berättelse, Om Naturliga stället, nyttan, samt skötseln af några wäxter, utaf hwilka frön nyligen blifwit hembragte från Norra America, til deras tjenst, som hafwa nöje, at i wårt Climat göra försök med de sammas cultiverande*. På Kongl. Wetenskaps Academiens befallning upsatt. Stockholm: 1751.

——*En Resa Til Norra AMERICA På Kongl. Swenska Wetenskaps Academiens befallning, Och Publici kostnad*. 3 vols. Stockholm: Salvii, 1753, 1756, and 1761. Modern critical edition: Elfving, F.; Schauman, G. *Pehr Kalms resa till Norra America*. Helsingfors: Skrifter utgifna af Svenska Litteratursällskapet i Finland, 1904.

——*Norra americanska färgeörter*. Åbo: Trykt hos J. C. Frenckel, 1763.

——*Resejournal över resan till Norra America*. 3 vols. Published by Martti Kerkkonen. Helsingfors: Skrifter utgivna av Svenska Litteratursällskapet i Finland, 1966.

——*Västgöta och bohuslänska resa; förrättad år 1742*. Ed. Claes Krantz. Stockholm: Wahlström & Widstrand, 1960.

Kihlman, J. H. *Chinesiska théet til dess skada och nytta samt några inländska wäxter hwaraf det kan beredas*. Stockholm: B. M. Bredberg, 1830.

Klaus, Joachim. "Johann Joachim Bechers Universalsystem der Staats- und Wirtschaftspolitik." Commentary to Johann Joachim Becher, *Politische Discurs von den eigentlichen Ursachen des Auff- und Abnehmens der Städt, Länder und Republicken*.(1668), reprint, Klassiker der Nationalökonomie. Düsseldorf: Verlag Wirtschaft und Finanzen, 1990.

Knight, David. *Ordering the World. A History of Classifying Man*. London: Burnett Books, 1981.

Koenig, Johan Gerard. "Utdrag af et bref ifrån Hr Konig Zvanquebar 15 february 1777." LSL.

Kotz, Annemarie and Oehling, Helmut, eds. *Carl von Linné und die deutschen Botaniker seiner Zeit*. Exhib. Cat. Tübingen. 3–10 June 1977. Tübingen: 1977.

Krook, Hans. *Angår oss Linné?* Stockholm: Rabén & Sjögren, 1971.

Kruse, Jonas G. "Beskrivning om åtskillige slags främmande färgmaterialers planterande uti Sverige, författad under des utländska resor, giorde på Alingsåhs manufactoriets omkostnad ifrån åhr 1728 till 1731." KB X. 335.

Kungliga Svenska Vetenskapsakademien. *Carl von Linnés betydelse såsom naturforskare och läkare*. Uppsala: Almqvist & Wiksell, 1907.

Labat, Jean Baptiste. *Nouvelle relation de l'Afrique occidentale*. Paris: Chez G. Cavelier, 1728.

Lamarck, J.-B. A. P. M. de. *Flore françoise*, vol. 1. Paris: Imprimerie Royale, 1778.

Lanham, Url. *Origins of Modern Biology*. New York: Columbia University Press, 1968.

Larson, James L. "Linné's French Critics." *SLÅ* (1978): 67–79.

———"Linnaeus and the Natural Method." *Isis* 58 (1967): 304–320.

———*Reason and Experience. The Representation of Natural Order in the Work of Carl von Linné*. Berkeley: University of California Press, 1971.

Lepenies, Wolf. *Autoren und Wissenschaftler im 18. Jahrhundert. Buffon, Linné, Winckelmann, Georg Forster, Erasmus Darwin*. Munich: Carl Hanser Verlag, 1988.

———*Das Ende der Naturgeschichte. Verzeitlichungen und Enthistorisierung in der Wissenschaftsgeschichte des 18. und 19. Jahrhunderts*. Munich: Carl Hanser Verlag, 1976.

———"Naturgeschichte und Anthropologie im 18. Jahrhundert." *Historische Zeitschrift*, Band 231, Heft 1 (August 1980): 21–41.

Lery, Jean de. *Histoire d'un voyage fait en la terre du Brésil, autrement dite Amérique*. La Rochelle: pour Antoine Chuppin, 1578.

Letwin, William. *The Origins of Scientific Economics*. Garden City, N.Y.: Doubleday, 1964.

Levertin, Oscar. "Carl von Linné." In *Samlade skrifter*. Stockholm: Bonnier, 1908.

Lindman, C. A. M. "Ett besök vid Råshult." *Nya Växsjöbladet Julläsning*. 19 December 1924.

Lindroth, Sten. "Adam Afzelius. En linnean i England och Sierra Leone." *Lychnos* (1944–45).

———"Linnaeus in His European Context." *SLÅ* (1978): 9–17.

———"Linné—legend och verklighet." *Lychnos* (1965–66): 56–122.

———*Kungl. Svenska Vetenskapsakademiens historia 1739–1818*. 2 parts. 3 vols. Stockholm: Svenska Vetenskapsakademien, 1967.

———"Naturvetenskaperna och kulturkampen under frihetstiden." *Lychnos* (1957–58): 180–193.

———*Svensk lärdomshistoria: Frihetstiden*. Stockholm: Norstedt, 1978.

Linnaeus, Carl. *Adonis Uplandicus . . . 13 May 1731*. In *Ungdomsskrifter*.

———*Amoenitates academicae*. Ed. Carl Linnaeus, vols. 1–7. Stockholm, 1749–1769; and *idem*, vols. 8–10, ed. Joh. Chr. Dan. Schreber. Erlangen, 1785–1789.

———"Angående Oeconomiens och landthushåldningens uphielpande genom Historiae Naturalis flitiga läsande . . ." 1746. TMF 2: 195.

————*Berättelse Om The Imhemska wäxter, som i brist af Säd kunna anwändas til Bröd- och Matredning Efter Hans Kongl. Maj:ts Allernådigste befallning Til Trycket befordrad Af Thess Collegio Medico.* Stockholm: Tryckt uti Kongl. Tryckeriet, 1757.

————*Botaniska exkursioner i trakten av Uppsala (Herbationes Upsalienses).* 1753. Resp. Anders Niclas Fornander. Trans. C. A. Brolén. In *Valda*, 1921.

————*Bref och skrifvelser af och till Carl von Linné.* Pub. T. M. Fries, series I: vols. 1–8, and J. M. Hulth, series II: vols. 1–2. Printed in Stockholm, Uppsala, and Berlin: by Aktiebolaget Ljus, Lundequistska Bokhandeln, Akademiska Bokhandeln, and R. Friedländer & Sohn, 1907–1943.

————"A Brief Narrative of a Journey to Lapland, Undertaken with a View to Natural History, in the Year 1732," Appendix n. I. In Linnaeus, *A Tour in Lapland.*

————*Calendarium Florae, Eller Blomster-Almanach.* Stockholm: Tryckt uti Kongl. Tryckeriet, 1757.

————*Carl von Linnés ungdomsskrifter,* ed. Ewald Ährling. Svenska Vetenskapsakademien. Stockholm: P. A. Norstedt, 1888. (Entries abbrev. as *Ungdomsskrifter.*)

————*Catalogus plantarum rariorum Scaniae item catalogus plantarum rariorum Smolandiae.* December 1728. In *Ungdomsskrifter.*

————*Ceres noverca arctorum.* 1733. Ed. Telemak Fredbärj. In *Valda*, 1964.

————*Circa fervidorum.* 1765. Resp. Carl Ribe (ennobled Ribben). Trans. Ejnar Haglund. Ed. Telemak Fredbärj. In *Valda*, 1968.

————*Classes plantarum.* Leyden: Apud Conradum Wishoff, 1738.

————*Clavis medicinae duplex.* 1766. Trans. Albert Boerman and Telemak Fredbärj. In *Valda*, 1967.

————*Collegium diaeteticum.* See Linnaeus, *Linnés dietetik.*

————*Critica botanica.* Leyden: Apud Condradum Wishoff, 1737.

————*Cui bono?* 1753. Trans. Christopher El. Gedner. In *Skrifter* II.

————*Culina mutata.* 1760. Resp. Magnus Gabriel Österman. Trans. Ejnar Haglund. Ed. Telemak Fredbärj. In *Valda*, 1956.

————*Curiositas naturalis.* 1748. Resp. Olof Söderberg. Ed. and trans. Th. M. Fries. In *Skrifter* II.

————*Cynographia.* Resp. and trans. Erich M. Lindekrantz. Västerås: Joh. Laur. Horrn, 1756.

————*Deliciae naturae. Tal, hållit uti Upsala Dom-kyrka år 1772 den 14 decemb. vid rektoratets nedläggande.* Stockholm: Joh. Georg Lange, 1773. Ed., trans., Th. M. Fries. In *Skrifter* II.

————*De morbis ex hyeme.* 1752. Trans. Arvid Hj. Uggla. In *Valda*, 1950.

————*Den lärda världens omdöme om Med Dr C Linnaei skrifter.* 1740. Ed. Telemak Fredbärj. In *Valda*, 1952.

————*De pane diaetetico.* 1760. Resp. Isaac Svensson. Trans. Arvid Hj. Uggla and Telemak Fredbärj. In *Valda*, 1964.

————*De potu chocolatae.* 1756. Resp. Anton Hoffman. Trans. Arvid Hj. Uggla and Telemak Fredbärj. In *Valda*, 1934.

————*De varietate ciborum.* 1767. Resp. Adolph Fredric Wedenberg. Trans. Ejnar Haglund. Ed. Telemak Fredbärj. In *Valda,* 1966.

————*Diaeta naturalis.* 1733. Ed. Arvid Hj. Uggla. Svenska Vetenskapsakademien. Uppsala: Almqvist & Wiksell, 1958.

————*Disputationer.* See Linnaeus, Carl. *Linnés disputationer.*

————"Doctor Linnaei Tankar om Grunden til Oeconomien genom Naturkunnogheten ock Physiquen." *K. Vet. Akad. Handl.* 1740: 405–423.

————*Egenhändiga anteckningar af Carl Linnaeus om sig sielf . . .* Ed. Adam Afzelius. Uppsala: Palmblad, 1823.

————*L'equilibre de la nature.* Ed. Camille Limoges. Paris: Vrin, 1972.

————*Flora Lapponica.* 1737. Ed. and trans. Th. M. Fries. In *Skrifter* I.

————*Flora oeconomica, Eller Hushålls-Nyttan Af de i Swerige Wildt wäxande Örter . . .* 1748. Resp. and trans. Elias Aspelin. Stockholm: Tryckt på Lars Salvii egen kostnad, 1749.

————*Föreläsningar öfver djurriket.* 1741–1770s. Ed. Einar Lönnberg. Uppsala: Uppsala Universitet and Akademiska Bokhandeln; Berlin: Friedländer, 1913.

————*Fructus eculenti.* 1763. Resp. Johan Salberg. Trans. Ejnar Haglund and Telemak Fredbärj. In *Valda,* 1965.

————*Fundamenta botanica.* Amsterdam: Apud Salomonem Schouten, 1736.

————*Herbationes Upsalienses. Protokoll över Linnés exkursioner i Uppsalatrakten. I. Herbationerna 1747.* Ed. Åke Berg. Introd. Arvid Hj. Uggla. Pub. Svenska Linné-Sällskapet. Uppsala: Almqvist & Wiksell, 1952.

————*Horticultura academica.* Resp. Johan. Gust. Wollrath. Uppsala: L. M. Höjer, Reg. Acad. Typogr. XVIII Decemb. 1754.

————*Hortus Cliffortianus.* Amsterdam: n.p., 1737.

————*Hortus Uplandicus.* Uppsala: 1730. Ed. Ewald Ährling. In *Ungdomsskrifter.*

————*Hortus Uplandicus. Methodo Tournefortiana.* N.d., c. 1730. Ed. Ewald Ährling. In *Ungdomsskrifter.*

————*Hortus Uplandicus . . . methodum propriam . . .* Marked Stockholmiae 1731, written 13 May 1731 in Uppsala. In *Ungdomsskrifter.*

————*Hortus Upsaliensis . . . ab anno 1742 in annum 1748,* vol. 1. Stockholm: Lars Salvius, 1748.

————*Inledning till dieten. Fyra föreläsningskoncept.* N.d. Ed., partly trans., Arvid Hj. Uggla and Telemak Fredbärj. In *Valda,* 1961.

————*Instruktion för resande naturforskare.* 1759. Resp. Erik Andreas Nordblad. In *Skrifter* II.

————*Iter ad exteros.* 1735-. In *Ungdomsskrifter.*

————*Iter Dalekarlium.* 1734. In *Ungdomsskrifter.*

————*Iter Lapponicum.* 1732. Ed. Th. M. Fries. In *Skrifter* V.

————*Lachesis naturalis.* 1907. See Linnaeus, *Linnés dietetik.*

————*Libellus amicorum.* 1734–. Trans. and ed. Telemak Fredbärj and Arvid Hj. Uggla. In *Valda,* 1958.

————"Linnés almanacksanteckningar för år 1735." Ed. Arvid Hj. Uggla. *SLÅ* XVIII (1935): 134–148.

———"Linnés almanacksuppsatser." 1741–1749. *SLÅ* XI (1928): 116–146.

———"Linné's botaniske 'Praelectiones privatissimae' paa Hammarby 1770." Ed. Jens Holmboe. *Bergens Museums Aarbog* 1910. Bergen: John Griegs, 1911.

———*Linnés dietetik . . . Collegium diaeteticum* (1741–1770s). Ed. A. O. Lindfors. Uppsala: mediciniska fakulteten and Akademiska Boktryckeriet, 1907.

———*Linnés disputationer. En översikt.* Ed. and comp. Gustaf Drake af Hagelsrum. Nässjö: Nässjö-Tryckeriet, 1939.

———"Linnés företal till Species Plantarum 1753." *SLÅ* XXXV (1952): 1–4.

———*Linnés Minnesbok.* Pub. and ed. Felix Bryk. Stockholm: 1919.

———"Linnés tankar om den akademiska ungdomens uppfostran. Ett okänt betänkande från 1768." 1768. Ed. Arvid Hj. Uggla. *SLÅ* XXIII (1940): 1–16.

———"Markattan Diana." *K. Vet. Akad. Handl.* 1754: 210–217.

———*Märkvärdigheter uti insekterna.* 1739. Reprint in Carl Linnaeus, *Carl von Linné: Fyra skrifter,* ed. Arvid Hj. Uggla, ill. Harald Sallberg, pub. Nordiska Bibliofilsällskapet. Stockholm: Esselte, 1939.

———*Medicamenta graveolentia.* 1758. Resp. Jonas Theodor Fagraeus. Trans. Ejnar Haglund. Ed. Telemak Fredbärj. In *Valda,* 1968.

———"Med. Profess. Ädel och wida berömde Herr Doct. C. Linnaei Anmärkningar, om Thée och Thée-drickandet." 1746. Reprint Carl Linnaeus, "Linnés almanacksuppsatser" (1741–1749). *SLÅ* XI (1928): 116–146.

———*Menniskans cousiner.* Trans. Carl Linnaeus. Ed. Telemak Fredbärj. In *Valda,* 1955.

———"Merit-Lista: Att applicera Naturen till oeconomien och vice versa" (1775). Reprint, Birger Strandell, "Patriotiska Sällskapet och Linné." *SLÅ* XXXIX-XL (1756–1957): 130–137.

———*Metamorphosis humana.* 1767. Resp. Johan Adolph Wadström. Trans. Nils Dahllöf. In *Valda,* 1956.

———*Miscellaneous Tracts Relating to Natural History.* Rev. ed. Trans. Benj. Stillingfleet. London: J. Dodsley, Baker and Leigh, and T. Payne, 1775. Reprint History of Ecology Series, New York: Arno, 1977.

———"Naturaliesamlingars ändamål och nytta." 1754. In *Skrifter* II.

———*Nemesis divina.* Late 1750s–1765. Ed. Elis Malmeström and Telemak Fredbärj. Stockholm: Bonniers, 1968.

———*Nutrix noverca.* 1752. Resp. Fredrik Lindberg. Trans. Sven-Olof Thulin. Ed. Telemak Fredbärj. In *Valda,* 1947.

———*Nya bevis för sexualitet hos växterna.* 1759. In *Skrifter* IV.

———*Obstacula medicinae.* 1752. Resp. Johan Gerg Beyersten. Trans. Sven-Olof Thulin. Ed. Telemak Fredbärj. In *Valda,* 1948.

———*Oeconomia naturae.* Trans. Isaac J. Biberg. Stockholm: Kiesewetters boklådor, 1750. In *Skrifter* II.

———*Öländska och gotländska resa år 1741,* (1745) ed. Carl-Otto von Sydow. Stockholm: Wahlström & Widstrand, 1962.

———"Om nödvändigheten af forskningsresor inom fäderneslandet." 1741. In *Skrifter* II.

————"Om nyttan af wäxternas olika kiön wid åkerbruk och trägårdar." 1744 and 1745. In *Skrifter* IV.

————*Oratio de telluris habitabilis incremento.* Leyden: Cornelium Haak, 1744. Trans. as Carl Linnaeus, "Om den beboeliga jordens tillväxt." Ed. and trans. Th. M. Fries. In *Skrifter* II.

————"Oration concerning the necessity of travelling in one's own countrey, made by Dr. Linnaeus at Upsala, Oct. 17, anno 1741 . . ." (1741), in Linnaeus, *Miscellaneous Tracts.*

————*Philosophia botanica.* Stockholm: Apud Godofr. Kiesewetter, 1751, and Amsterdam: Z. Chatelain, 1751.

————*Plantae esculentae patriae.* Stockholm: Lars Salvii kostnad, 1752.

————*Politia naturae.* 1760. Resp. Henric Christian Daniel Wilcke. Ed. and trans. Th. M. Fries. In *Skrifter* II.

————*Potus coffeae.* 1761. Resp. Henrik Sparschuch. Trans. Ejnar Haglund. Ed. Telemak Fredbärj. In *Valda,* 1966.

————*Praeludia sponsaliorum plantarum.* 1730. Ed. and trans. Th. M. Fries. In *Skrifter* IV.

————"Professor Linnaei upsats på de Medicinal-Wäxter som i Apothequen bewaras, och hos oss i Fäderneslandet wäxa." *K. Vet. Akad. Handl.* 1741: 81–82.

————"Relation angående den af Kongl. Wettenskaps Societeten anstälda resa uti Historia Naturali till Lappland." 1732. In BS I:1, 314–329.

————"Rop ur grafwen till min i tiden kiära hustru." Autograph manuscript dated 2 March 1776. Reprint *Svenska Dagbladet,* 1 April 1910.

————*Skånska resa år 1749* (1751). Ed. Carl-Otto von Sydow. Stockholm: Wahlström & Widstrand, 1975.

————"Skaparens afsikt med naturens verk. En promotionsföreläsning af Linné 1763." Ed. and trans. Arvid Hj. Uggla. *SLÅ* XXX (1947): 71–96.

————*Skrifter af Carl von Linné.* Pub. Kungl. Svenska Vetenskapsakademien. 5 vol. Upsala: Almqvist & Wiksell, 1905–1913.

————*Spolia botanica.* 1729. In *Ungdomsskrifter.*

————*Sponsalia plantarum.* C. 1744–45. Ed. and trans. Th. M. Fries. In *Skrifter* IV.

————*Sponsalia plantarum.* 1746. Resp. and trans. Johan Gustav Wahlbom. Ed. Th. M. Fries. In *Skrifter* IV.

————"Swenskt Höfrö." *K. Vet. Akad. Handl.* 1742: 191–198.

————*Systema naturae.* Leyden: Apud Theodorum Haak, 1735.

————"Tal, vid deras Kongl. Majesteters höga närvaro, hållit uti Upsala, på Stora Carolinska Lärosalen den 25 septemb. 1759." Reprint, in Carl Linnaeus, *Carl von Linné: Fyra skrifter,* ed. Arvid Hj. Uggla, ill. Harald Sallberg, pub. Nordiska Bibliofilsällskapet. Stockholm: Esselte, 1939.

————"Tankar om nyttiga växters planterande på de Lappska fjällen." *K. Vet. Akad. Handl.* 1754: 182–189.

————*A Tour in Lapland* . . . . Ed. James Edward Smith. 2 vols. London: Printed for White and Cochrane . . . by Richard Taylor, 1811. Facs. reprint, New York: Arno Press & The New York Times, 1971.

————"Två svenska akademiprogram av Linné." 1750, 1759. Ed. Telemak Fredbärj. *SLÅ* XXXVII-XXXVIII (1954–1955): 97–114.

————*Valda avhandlingar av Carl von Linné: Översättning av Svenska Linnesällskapet.* Ed. most commonly by Telemak Fredbärj; various dates, cities, publishers.

————*Viridarium Cliffortianum.* Amsterdam: n.p., 1737.

————*Vita Caroli Linnaei.* 1720s–1770s. Eds. Elis Malmeström and Arvid Hj. Uggla. Stockholm: Almqvist & Wiksell, 1957.

————*Vorlesungen über die Cultur des Pflanzen.* Ed., trans., and pub. M. B. Swederus. Uppsala: Uppsala University, 1907.

————*Wästgötaresa, på Riksens Högloflige Ständers Befallning Förrättad år 1746.* Stockholm: L. Salvii, 1747. Modern scholarly edition: Carl Linnaeus. *Västgöta resa 1746* (1747), ed. Sigurd Fries and Lars-Erik Edlund. Stockholm: Wahlström & Widstrand, 1978.

Linnaeus the Younger, Carl. "Linné den Yngres brev till Abraham Bäck 1778." Ed. Arvid Hj. Uggla. *SLÅ* XXXIX-XL (1956–1957): 138–165.

————*Supplementum plantarum.* Lichfield: John Jackson, for Leigh and Sotheby; London: 1783.

Löwegren, Yngwe. *Naturaliekabinett i Sverige under 1700-talet.* Stockholm: Almqvist & Wiksell, 1952.

Lyon, John, and Phillip R. Sloan. *From Natural History to the History of Nature: Readings from Buffon and His Critics.* Notre Dame: University of Notre Dame Press, 1981.

Lyte, Charles. *Sir Joseph Banks: 18th Century Explorer, Botanist and Entrepreneur.* London: David & Charles, 1980.

Magnus, Olaus. *Historia de gentibus septentrionalibus.* 1555. Trans., ed. and comm., John Granlund, with fasc. ill. Stockholm: Institutet för folklivsforskning vid Nordiska Muséet, Stockholms universitet, and Gidlunds, 1949–1951.

Malmer, Olof. "Om Pärlemusslor och Pärlefiskerier," *K. Vet. Akad. Handl.* 1742: 214–225.

Martin, Anton Rolandsson. "Dagbok hållen vid en resa till Norrpolen" (1758). Ed. Simon Nordström. In *Ymer* (1881): 102–141.

May, Robert M. "How Many Species Inhabit the Earth?" *Scientific American* (October 1992): 42–48.

Mayr, Ernst. *The Growth of Biological Thought: Diversity, Evolution, and Inheritance.* Cambridge, Mass.: Harvard University Press, 1982.

————*Towards a New Philosophy of Biology.* Cambridge, Mass.: Harvard University Press, 1981.

"Med. Doctorens Archiaterns Caroli Linnaei föreläsningar öfwer Diaeten hållne i Upsala Åhr 1747." KB X.55.

Millan, Aubin-Louis. *Discours sur l'origine et le progrès de l'histoire naturelle en France, servant d'introduction aux "Memoires de la Société d'histoire naturelle."* Paris: Creuze, 1792.

Möller, Josef. "Die Rassenforschung in den nordischen Ländern." *Kölnische Volkszeitung,* 15 April 1937.

Mörner, Magnus. "Ett bidrag till Peter Löflings levnadshistoria." *SLÅ* XXXI (1948): 92–94.

Müller, Erik. "Linné och Abraham Bäck." *SLÅ* IV (1921): 80–118.

———"Om Svenska Linnésällskapets utvecklingslinjer." *SLÅ* I (1918): 173–176.

Mun, Thomas. *England's Treasure by Forraign Trade or The Ballance of Our Forraign Trade Is the Rule of Our Treasure.* London: J. G. for T. Clark. 1664.

Nannfeldt, J. A. "Species Plantarum. Ett 200-årsminne." *SLÅ* XXXVI (1953): 1–9.

Nauman, J. J. "Berättelse om Marsuin i anledning af thet jag enfaldigt kunnat förfara om theras art och egenskaper the tu åren som jag af them warit ägare." N.d. Zool., LSL.

Nelson, Marie Clark and Svanberg, Ingvar. "Lichens As Food. Historical Perspectives on Food Propaganda." *SLÅ* (1986–87): 1–51.

Nicander, Henric. *Åminnelse-Tal, öfver . . . Patr. Alströmer, Hållet: Kongl. Vetenskaps-Akademien den 15 December 1810.* Strengnäs: Segerstedt, 1811.

Nordhagen, R. "Linnés förbindelse med norske naturforskere. Foredrag holdt på Hammarby 22 maj 1949." *SLÅ* XXXII (1949): 5–18.

Nordström, Simon. "Anton Rolandsson Martin. Biografiska anteckningar." *Ymer* (1881): 8–101.

Odhelius, Johan Lorentz *Åminnelse-Tal, Öfver Kongl. Vetensk. Acad. Ledamot Herr Pehr Kalm, . . . Hållet för Kongl. Vetenskaps Academien, den 15 Novemb. 1780.* Stockholm: Tryckt hos Joh. Georg Lange, 1780.

Olsson, Torsten. "En Linnean om Linné." *SLÅ* XXXII (1949): 68–70.

Osbeck, Pehr. *Anledningar til nyttig upmärksamhet under chinesiska resor, upgifne i Kongl. Vet. academien, uti et inträdes-tal, den 25 februari 1758.* Stockholm: Salvius, 1758.

———*Dagbok öfwer en Ostindisk resa åren 1750, 1751, 1752.* Stockholm: Trykt hos L. L. Grefing, 1757.

———*Utkast till Beskrifning öfver Laholms prosteri.* 1796. Lund: Gleerup, 1992.

Osvald, Hugo. "Linnés höfrö." *SLÅ* XXXIX-XL (1956–57): 119–129.

Palmstruck, J. W. *Svensk botanik,* 2nd ed. Vol. 1, part 1. Stockholm: Carl Delén, 1803.

Pålstig, Johan. *Myten om Lappland.* Stockholm: Natur och kultur, 1963.

"Påminnelse för den rasande ungdomen." N.d. Autograph margin annotations, but not autograph. Originally found in *Lachesis* ms. Misc., LSL.

Petander, K. *De nationalekonomiska åskådningarna i Sverige såsom de framträda i litteraturen.* Vol. 1: 1718–1765. Stockholm: Kungl. Boktryckeriet, P. A. Norstedt, 1912.

"Protocoller vid Linnaei exiursioner." 1760. KB X.651.

"Protocoll, hållet i Linnéska Samfundet å dess Stiftelsedag den 23 Maj 1832," KB Nat. vet. Allm. Sällsk. Ex A.

Pulteney, R. *A General View of the Writings of Linnaeus.* London: W. G. Maton, 1805.

R.H.L.S.A. "Då Archiatern Herr Doct. Carl Linnaeus blef dubbad til Riddare af Konglige Nordstjärne-Orden den 27 April 1753. Yttrade deröfver sin fägnad några Natural Wettenskapens idkare." Uppsala: L. M. Höjer, [1753].

Ramsbottom, J. "Presidential Address: Linnaeus and the Species Concept." *Proceedings of the Linnean Society of London,* 150 (1938): 192–209.

Ray, John. *Historia plantarum.* London: Smith and Walford, 1686.

———*The Wisdom of God Manifested in the Works of Creation.* London: Smith and Walford, 1686.

Regnard, Jean Francois. *Voyage de Laponie.* 1681, pub. 1731. Trans. as *Resa i Lappland* by P. E. Öhman. Tammerfors: 1946.

Retzius, Anders Jahan [sic]. "Tal hållit på Kongl. Carolinska Academiens Naturalkammare d. 11 Junii 1811, då Framlidne Archiaterns, Professorns och Riddarens af Kongl. Nordstjerne Orden D. Carl von Linné's Bröstbild därstädes upsattes." Lund: Berlinska Boktryckeriet, 1811.

Ritterbush, Philip C. *Overtures to Biology: The Speculations of Eighteenth-Century Naturalists.* New Haven: Yale University Press, 1964.

Roberts, Michael. *The Swedish Imperial Experience 1560–1718.* Cambridge: Cambridge University Press, 1979.

Robinson, Joan. *The Economics of Imperfect Competition* (1933), 2d ed. London: St. Martin's Press, 1969.

Roung, Israel and Fjellström, Phebe, eds. *Berättelser om samerna i 1600-talets Sverige.* Umeå: Umeås universitet, 1982.

Rudbeck the Elder, Olof. *Atland el. Manheim* . . . . Vol. 1, 1679; vol. 2, 1689; vol. 3, 1698; vol. 4 at the printer's during the 1702 Uppsala fire. Reprint of first ed. Ed. Axel Nelson. Uppsala: 1937–1950.

Rudbeck the Younger, Olof. "Rudbeck d. y:s resedagbok." In *Iter Lapponicum. Skissboken från resan till Lappland 1695.* II Kommentardel, 28–59. Stockholm: René Coeckelbergh, 1987.

Rydbeck, Eric O. "'Tal, om planterings nytta och nödvändighet i synnerhet på slättbygden i Öster-Göthland,' hållet vid de studerandes af Öst-Götha Nationen allmänna sammankomst i Upsala den 31 Martii 1762." Stockholm: Lars Salvius, 1762.

Rydberg, Sven. *Svenska studieresor till England under frihetstiden,* Uppsala: Almqvist & Wiksell, 1951.

Rydén, Sven. "José Celestino Muti och hans förbindelser med Linné och hans krets. Med anledning av en brevpublikation." *SLÅ* XXXV (1952): 31–38.

Sachs, Julius. *Geschichte der Botanik vom 16. Jahrhundert bis 1860.* Munich: Oldenbourg, 1875.

Sahlgren, Jöran. "Ett Linnébibliotek." *SLÅ* VI (1923): 152–165.

———"Linné som predikant." *SLÅ* V (1922): 40–55.

———"Linnés talspråk." *SLÅ* III (1920): 25–35.

Sandblad, Henrik. "Bjerkander, J. P. Falck och Tessin. Några brev till och om Kinnekulle." *Lychnos* (1979–1980): 260–68.

———"Daniel E. Naezén, kulturpionjär i 1700-talets Västerbotten." In *Svenska turistföreningens årskrift.* Ed. Jan Jonason. Nacka: Esselte Herzogs, 1980.

———"Edward D. Clarke and Giuseppe Acerbi, upptäcktsresande i Norden 1798–1800," *Lychnos* (1979–1980): 155–205.

Schauman, Georg. "Studier i Frihetstidens nationalekonomiska litteratur." Ph.D.. Helsingfors University. Helsingfors: 1910.

Schefferus, Johannes. *Lapponia.* 1673. Trans. as *The History of Lapland* . . . . London: Printed for Tho. Newborough and R. Parker etc., 1704.

Scheller, Johann Gerhard. *Reise-Beschreibung von Lappland und Bothnien.* Jena: n.publ. noted, 1727.

Schlesch, H. and Venmans, L. A. W. C. "Carl von Linné og Nederlandene. Nogle Bidrag." *SLÅ* XXVII (1944): 106–112.

Schmoller, Gustav Friedrich von. *The Mercantile System and Its Historical Significance Illustrated Chiefly from Prussian History.* Reprint in trans. New York: Macmillan, 1896.

Schoultz, Gösta von. "C. A. Agardhs resa 1809 genom Uppland, Gästrikland och Dalarna." *Lychnos* (1989): 163–193.

Schulzenheim, David von. *Åminnelse-Tal, Öfver . . . Nils Rosen von Rosenstein . . . Den 17 November 1773* . . . Stockholm: Lars Salvii Tryckeri, 1773.

Schuster, Julius. *Linné und Fabricius. Zu ihrem Leben und Werk. Drei Faksimiles zu Linnés 150 Todestag.* Munich: Verlag der Münchner Drucke, 1928.

Selander, Sten. "Linné i Lule Lappmark." *SLÅ* XXX (1947): 9–20.

————*Linnélärjungar i främmande länder: essayer.* Stockholm: Bonniers, 1960.

Sernander, Rutger. "Hårleman och Linnaei Herbationes Upsalienses," *SLÅ* IX (1926): 78–86.

————"I Linnés fotspår." *SLÅ* IV (1921): 33–64.

————"Linnaeus och Rudbeckarnes Hortus Botanicus." *SLÅ* XIV (1931): 126–157.

Sheets-Johnstone, Masine. "Why Lamarck Did Not Discover the Principle of Natural Selection." *Journal of the History of Biology* 15 (3): 1982.

Sloan, P. "The Buffon-Linnaeus Controversy," *Isis* 67 (1976): 356–375.

Small, Albion W. *The Cameralists. The Pioneers of German Social Policy.* Chicago: University of Chicago Press, 1909.

Smith, Adam. *An Inquiry into the Nature and Causes of the Wealth of Nations.* 1776. Reprint. Ed. Edwin Cannan. 2 vols. London: Methuen, 1924.

Smith, James E. "Introductory Discourse on the Rise and Progress of Natural History. Delivered . . . April 8, 1788," Linnean Society of London, *Transactions*, vol. 1 (1791): 1–55.

Smith, Sir James Edward. *A Selection of the Correspondence of Linnaeus, and Other Naturalists.* London: Longman, Hurst, Rees, Orme and Brown, 1821.

Sörlin, Sverker. "Apostlarnas gärning. Vetenskap och offervilja i Linné-tidevarvet." *SLÅ* (1990–91): 75–89.

Soulsby, B. H. *A Catalogue of the Works of Linnaeus.* 2nd ed. London: The British Museum, 1933.

Sparrman, Anders. *Åminnelse-Tal Öfver . . . Carl Gust. Ekeberg, Hållet För Kongl. Vetenskaps Academien Den 1 December 1790.* Stockholm: Johan A. Carlbohm, 1791.

————*Resa till Goda Hopps-udens, Södra, pol-kretsen och omkring jordklotet, samt till Hottentott- och Caffer-landen, åren 1772–76.* 2 vols. Stockholm: Tryckt hos. A. J. Nordström, 1783–1818; also vol. II: 2, *Resa omkring jordklotet i sällskap med kapit. J. Cook och her Forster.* Stockholm: Tryckt hos Carl Delen, 1802–1818.

Spiegel, Henry William. *The Growth of Economic Thought.* 3rd ed. Durham: Duke University Press, 1991.

Sprague, T. A. "Linnaeus as a Nomenclaturist." *Taxon* 2 (1953): 40–46.

Stafleu, Frans Antonie. *Linnaeus and the Linnaeans. The Spreading of Their Ideas in Systematic Botany, 1735–1789.* Utrecht: Oosthoek, 1971.

Starbatty, Joachim. "Johann Joachim Becher—ein merkantilistischer Klassiker." Commentary to Johann Joachim Becher, *Politische Discurs . . .* (1668). Reprint, Klassiker der Nationalökonomie. Düsseldorf: Verlag Wirtschaft und Finanzen, 1990.

Stauffer, R. C. "Ecology in the Long Manuscript Version of Darwin's *Origin of Species* and Linnaeus' *Oeconomy of Nature.*" *Proc. of the Am. Phil. Society* (1960) 104: 2.

Stavenow, Ludvig. "Kväde och tal." Göteborg: Wettergren och Kerber, 1907.

Stavenow-Hidemark, Elisabet, ed. *1700-tals textil. Anders Berchs samling i Nordiska Museet.* Stockholm: Nordiska Museets Förlag, 1990.

Stearn, W. T. "The Background of Linnaeus' Contributions to the Nomenclature and Methods of Systematic Biology." *Systematic Zoology* 8 (1959): 4–22.

———"Botanical Gardens and Botanical Literature in the Eighteenth Century," in A. Stevenson, comp., *Catalogue of Botanical Books in the Collection of Rachel McMasters Miller Hunt.* Pittsburgh: Hunt Botanical Library. Vol. 2 (1), xli-cxl.

———"An Introduction to the *Species plantarum* and Cognate Botanical Works of Carl Linnaeus." Preface to vol. 1 of Carl Linnaeus, *Species plantarum,* facs. London: Ray Society, 1957, 1–176.

Stenström, Fritz. "Pehr Osbeck och Lars Montin. Två halländska Linnélärjungar." *SLÅ* XVIII (1935): 59–94.

Stevens, P. F. "Metaphors and Typology in the Development of Botanical Systematics 1690–1960, or the Art of Putting New Wine in Old Bottles." *Taxon* (May 1984), 33(2): 169–211.

Stevens, P. F. and Cullen, S. P. "Linnaeus, the Cortex-Medulla Theory, and the Key to His Understanding of Plant Form and Natural Relationships." *Journal of the Arnold Arboretum* 71 (April 1990): 179–220.

Stöver, Dietrich Heinrich. *Leben des Ritters Carl von Linné. . . .* 2 vols. Hamburg: Benj. Gottl. Hoffmann, 1792.

Strandell, Birger. "Linnés lärjungar. Varifrån kom de och vart tog de vägen?" *SLÅ* (1979–1981): 105–142.

———"Patriotiska Sällskapet och Linné." *SLÅ* XXXIX-XL (1956–57): 130–137.

Svanberg, Ingvar. "Turkic Ethnobotany and Ethnozoology as Recorded by Johan Peter Falck." *SLÅ* (1986–87): 53–118.

Svedäng, Bruno. "'Plinii näktergal'-anteckningar om stilen i Linnés Öländska och Gotländska resa." *SLÅ* (1990–91): 7–34.

Svedelius, Nils. "Carl Peter Thunberg 1743–1828. Ett tvåhundraårsminne." *SLÅ* XXVII (1944): 29–64.

Svenska Vetenskapsakademien. "Wetenskapsakademiens berättelse till Kungl. Maj:t ang. brännvinsämnen utom spannmål, af d. 6 december 1775 jämte särskilda yttranden av [P.J.] Bergius, T. Bergman, P. D. Gadd, P. Kalm." (1775). KB X.378.

Svenska Vetenskapsakademien. *Svenska Vetenskapsakademiens protokoll för åren 1739, 1740 och 1741.* Ed. E. W. Dahlgren. Vol. 1: *Protokoll och grundregler.* Vol. 2: *Anmärkningar och register.* Uppsala: Almqvist & Wiksell, 1918.

Svenskt Biografiskt Lexikon. Stockholm: Bonniers, 1918–.

Svenson, H. K. "On the Descriptive Method of Linnaeus." *Rhodora* 47 (1945): 273–302.

Swederus, Magnus Bernhard. *Botaniska trädgården i Upsala 1655–1807.* Falun: n.p., 1877.

Sydow, Carl-Otto von. "Linnaeus and Gmelin." *SLÅ* (1978): 212–222.

———"Linné och de lyckliga lapparna. Primitivistiska drag i Linnés Lapplandsuppfattning." In Ragnar Granit, ed., *Utur Stubbotan Rot. Essäer till 200-årsminnet av Carl von Linnés död,* 72–78. Stockholm: Norstedt, 1978.

———"Linné och Lappland. Hans uppfattning av landet och dess invånare." *SLÅ* (1972–74): 22–71.

———"Vetenskapssocieteten och Henric Benzelius' Lapplandsresa 1711." *Lychnos* (1962): 138–161.

Tessin, Carl Gustaf. *En Gammal Mans Bref, til En Ung Prints. Förra Delen.* Stockholm: Lars Salvius, 1756.

Thunberg, Carl Peter. *Inträdes-Tal, om de mynt-sorter, som i äldre och sednare tider blifvit slagne och varit gångbare uti kejsaredömet Japan, hållet för Kongl. Vetenskaps-Academien, den 25. aug. 1779.* Stockholm: Johan Georg Lange, 1779.

———"Om utlänske träd, buskar och blomster-växter, som kunna tåla svenska klimatet." Ph.D. diss. Uppsala: 1820.

———*Resa uti Europa, Africa, Asia, förättad åren 1770–1779.* 4 vols. Uppsala: J. Edman, 1788–1793.

———*Tal, om japanska nationen, hållet för Kongl. Vetensk. academien, vid praesidii nedläggande, den 3 novemb. 1784.* Stockholm: Johan Georg Lange, 1784.

Torén, Carl-Axel. "Carl von Linné som rusthållare vid livregementet till häst." *SLÅ* XXXIX-XL (1956–57): 48–56.

———"Om Olof Torén. Hans färder till Kina och Indien som skeppspredikant samt om växtsläktet Torenia L." *SLÅ* XXXVI (1953): 17–56.

Torén, Olof. *En ostindisk resa.* 1750–1752. Reprint. Stockholm: Tiden, 1961.

Törje, Axel. *I den oförlikneliga nyttans tjänst.* Lund: Skånska Centraltryckeriet, 1973.

Tournefort, Joseph Pitton de. *Eléments de botanique.* Paris: Imprimerie Royale, 1694. See also Tournefort, 1719, for a Latin 3rd ed.

———*Institutiones rei herbariae.* Paris: Imprimerie Royale, 1719. See also Tournefort, 1694, for the French 1st ed.

Triewald, Mårten. "Anmärkingar wid utländska Frukt- och andra Träds planterande: Swerige, af egna rön och försök." *K. Vet. Akad. Handl.* 1740: 204–206.

———*Mårten Triewalds år 1728 and 1729 håldne Föreläsningar, på Riddarehuset i Stockholm, öfwer nya Naturkunnigheten.* 2 vols. Stockholm: Tryckt uti P. J. Nyströms, 1736–1758.

———"Rön och Försök, angående möjligheten at Svea Rike kunde äga egit rådt Silke anstälte." *K. Vet. Akad. Handl.* 1745: 22–29.

Tullberg, Tycho. "Familjetraditioner om Linné." *SLÅ* II (1919): 1–20.

———*Linnéporträtt. Vid Uppsala universitets minnesfest på tvåhundraårsdagen af Carl von Linnés födelse.* Stockholm: Aktiebolaget Ljus, 1907.

———"Linnés Hammarby." *SLÅ* I (1918): 1–79.

Uddman [Hansson], Isaacus. "Caroli Linnaei Arch. et Prof. . . . Collegium Diateticum." N.d. KB X.53.

Uggla, Arvid Hj. "Daniel Solander och Linné." *SLÅ* XXXVII-XXXVIII (1954–55): 23–64.

———"Linnéstatyn i Uppsala Botanicum. Ett hundraårsminne. Föredrag vid Svenska Linné-Sällskapets sammankomst 6 december 1929." *SLÅ* XIII (1930): 153–164.

———"När Linné fick en Venus-staty till akademiens trädgård." *SLÅ* XXIX (1946): 91–98.

———"Om förhistorien till *Species Plantarum*." *SLÅ* XXXVI (1953): 10–16.

———"Om Linnés *Nemesis Divina*. I synnerhet de ursprungliga Nemesis-anteckningarna i London." *SLÅ* L (1967): 13–19.

———"Tvenne nya dokument till kännedom om Linnés engelska förbindelser." *SLÅ* XXI (1938): 85–94.

Uppsala universitet. *Skrifter med anledning af Linnéfesten den 23dje och 24de maj. Uppsala Universitets Årsskrift*. Uppsala: Akademiska Boktryckeriet, 1907.

Vaillant, Sébastien. *Sermo de structura florum*. Lugduni Batavorum (Leyden): Apud Petrum Vander Aa, 1718; rpt. 1727.

Veendorp, H., L. Becking, and L. G. M. Baas. *1587–1937: Hortus Academicus Lugdano Batavus. The Development of the Gardens of Leyden University*. Harleem: Typographia Enschedaiana, 1938.

Virdestam, Gotthard. "Kring några brev från Samuel Linnaeus." *SLÅ* XIV (1931): 115–125.

"Vita Caroli a Linné." *Nova Acta Regiae Societatis Scientiarum Upsaliensis*. Vol. 5 (1792): 335–344.

Wachenfelt, Miles von. "Ett återfunnet Lapplandsporträtt av Carl von Linné." *SLÅ* XXX (1947): 21–35.

Wallenberg, Jacob. *Min son på galejan*. (1769), 1781. Reprint. Höganäs: Bra Böcker, 1979.

Weinstock, John, ed. *Contemporary Perspectives on Linnaeus*. Lanham, Md: University Press of America, 1985.

Whittaker, Edmund. *A History of Economic Ideas*. New York: Longmans, Green, 1940.

Wiklund, K. B. "Linné och lapparna." *SLÅ* VIII (1925): 59–93.

Wikman, Karl Robert. *Lachesis and Nemesis*. Stockholm: Almqvist & Wiksell, 1970.

Wilmott, A. J. "Systematic Botany from Linnaeus to Darwin." *Lectures on the Development of Taxonomy*. London: The Linnean Society of London, 1948–49, 33–45.

Wiselgren, O. "'Yppighets nytta.' Ett bedrag till de ekonomiska åskådningarnas historia i Sverige under det adertonde århundradet." Skrifter utg. af K. Humanistiska Vetenskaps-samfundet i Uppsala 14:3. Uppsala: 1912.

Wittrock, Veit Brecher. "Några ord om Linné och hans betydelse för den botaniska vetenskapen." *Acta Horti Bergiani*. Band 4. N:o 1. (1907): 1–32. Stockholm: Isaac Marcus, 1907.

Worster, Donald. *Nature's Economy: A History of Ecological Ideas*. Cambridge: Cambridge University Press, 1977.

Acknowledgments

This book came out of my 1993 Harvard dissertation on Carl Linnaeus. As I was working on it I thought it would have been much jollier to have authored a dissertation *under* Linnaeus than *about* him. For it was Linnaeus' practice to write the theses he ostensibly directed. Indeed, most of the 186 dissertations he advised at Uppsala University between 1741 and 1776 are by him, not by his students.

Linnaeus dictated the text in Swedish to the doctoral candidate, whose task was to turn them into passable Latin (and, of course, to pay the professor for his troubles). Most dissertations were between fifteen and forty pages long. For scientific terms, Linnaeus already used Latin, so the student was not overly burdened. From start to finish, "writing" a dissertation under Linnaeus could take as little as two to three hours.

Ordinarily Linnaeus chose the topic. If the student wished to address some interest, however, the professor welcomed a discussion before he began his dictation. It happened, too, that the candidate copied a passage of someone else's published work. Or he could pay a poorer student to ghostwrite it. (Linnaeus himself was such a ghost-writer in his youth.) Ambitious pupils might venture to submit a few handcrafted pages: for example, one of Linnaeus' students penned a few lines on dogs. He noted that dogs are better companions than valets, and that they enjoy home-cooked meals but dislike violin music. Another student described how he raised guinea pigs. Even this seems to have been a sloppy, half-hearted endeavor. "If starving, she eats paper and clothes."

In writing this book, I am profoundly grateful to the many scholars who have inspired me: Gunnar Eriksson, Tore Frängsmyr, John Lewis Heller, James L. Larson, Wolf Lepenies, W. T. Stearn, Peter F. Stevens, Carl-Otto von

Sydow, Sverker Sörlin, and others. Gunnar Broberg has been greatly important to me. I am also indebted to the librarians at the Hunt Botanical Library in Pittsburgh, Kungliga Biblioteket in Stockholm, Lund University Library, Uppsala University Library, the Uppsala Linnaeus Museum, the Natural History Museum in London, the Linnean Society of London, and the Houghton Library and the Arnold Arboretum at Harvard. I owe particular thanks to Kristina Eriksson at Kungliga Biblioteket and to Gina Douglas, director of the library of the Linnean Society.

Nor could I have written this work without my teachers, colleagues, friends, and family. I first wish to thank Simon Schama, the director of the doctoral thesis from which this book originated. Friends and colleagues who have helped me include Warwick Anderson, Thomas H. Baker, Annie Berglöf, Ann M. Blair, Kenneth J. Boss, Michael Brava-Dobrée, Pia Bungarten, Janet Browne, Caroline Castaglioni, I. Bernard Cohen, Natalie Zemon Davis, Michael Dettelbach, Moti Feingold, Daniel L. Gordon, Hunter A. Dupree, Edward Eigen, Mark A. Kishlansky, Mark Madison, Ernst Mayr, Bruce Mazlish, Patricia Princehouse, Peter Reill, Margaret Schabas, Nancy Slack, Stefan Schrader, Sverker Sörlin, Mary Terrall, Jay Tribby, Corinna Treitel, Honor Woodward, Telford H. Work, Urban Wråkberg, and Richard Zipser.

I am grateful to Donald Fleming and to Peter F. Stevens; to two anonymous readers for their insightful commentary; and to Anita Safran and Michael Fisher for their skilled editing. Thanks also to Simon Schaffer and John Brewer, Peter F. Stevens and Kenneth J. Boss, Pnina Abir-Am, Michael Fischer, Sungook Hong, Stephen Greenblatt and Carla Hesse, Margaret Schabas, and Nicholas Jardine, James Secord, and Emma Spary, who kindly invited me to talks.

I also warmly thank my colleagues in Harvard's Department of the History of Science. I feel honored to be associated with this wonderful group of scholars. Most of all, thanks to Everett Mendelsohn—mentor, gentleman, and friend.

Helen Vendler and Neva Goodwin sustained my spirit with poetry, while Gillian Bushell, Brid L'Esperance, Fran Mavko, Phillippa Mayers, Sharon Richards, Steven Sarcione, Victoria Wardle, and Kathryn Young relieved me of many duties.

My mother-in-law, Joan Koerner, kindly and selflessly helped. My parents, Hans and Märit Rausing, my sister and brother-in-law, Sigrid and Dennis Hotz, and my brother and sister-in-law, Hans K. and Eva Rausing, have inspired me by their thoughtfulness and friendship.

My grandfather Ruben Rausing's love of learning, and his respect, reverence even, for women, exhilarated me as a girl and still sustain me. At the same time, the learning, profundity, and imagination—as well as the formidable street-smarts and drive—of both my father and my husband, in all tasks, provide my constant inspiration. My greatest thanks by far are to Benjamin and Sigrid's father, my dearest and most beloved husband, Joseph Leo Koerner.

Index

Koerner, Lisbet.

Linnaeus.

DATE			